Hybrid Soft Computing for Image Segmentation

Siddhartha Bhattacharyya
Paramartha Dutta · Sourav De
Goran Klepac
Editors

Hybrid Soft Computing for Image Segmentation

Springer

Editors
Siddhartha Bhattacharyya
Department of Information Technology
RCC Institute of Information Technology
Kolkata, West Bengal
India

Sourav De
Department of Computer Science and Engineering
Cooch Behar Govt. Engineering College
Cooch Behar, West Bengal
India

Paramartha Dutta
Department of Computer and System Sciences
Visva-Bharati University
Santiniketan, West Bengal
India

Goran Klepac
Department of Strategic Development
Raiffeisenbank Austria
Zagreb
Croatia

ISBN 978-3-319-83684-3 ISBN 978-3-319-47223-2 (eBook)
DOI 10.1007/978-3-319-47223-2

Softcover reprint of the hardcover 1st edition 2016

Printed on acid-free paper

This Springer imprint is published by Springer Nature
The registered company is Springer International Publishing AG
The registered company address is: Gewerbestrasse 11, 6330 Cham, Switzerland

To my parents, the late Ajit Kumar Bhattacharyya and the late Hashi Bhattacharyya, my beloved wife, Rashni, my maternal uncle, Mr. Arabinda Kumar Banerjee, and my maternal aunt, Mrs. Maya Banerjee

Siddhartha Bhattacharyya

To my parents, the late Arun Kanti Dutta and Mrs. Bandana Dutta

Paramartha Dutta

To my parents, Mr. Satya Narayan De and Mrs. Tapasi De, my wife, Mrs. Debolina Ghosh, my son, Mr. Aishik De, and my sister, Mrs. Soumi De

Sourav De

To my wife, Antonija Klepac, and my children, Laura Klepac, Viktor Klepac, Oliver Klepac and Gabrijel Klepac

Goran Klepac

Foreword

Imparting intelligence has become the focus of various computational paradigms. Thanks to the evolving soft computing and artificial intelligent methodologies, scientists have been able to explain and understand real-life processes and practices, which formerly remained unexplored by dint of their underlying imprecision, uncertainties and redundancies, as well as the unavailability of appropriate methods for describing the inexactness, incompleteness, and vagueness of information representation.

Image segmentation is no exception in this regard, when it comes to the manifestation of uncertainty and imprecision in varied forms. The situation becomes more severe when images become corrupt with noise artifacts. Adding to it, the time complexity involved in processing images exhibiting the different color gamuts is also a challenging thoroughfare. A plethora of literature exists involving the classical domains for coping with the complex task of image segmentation, particularly for handling uncertainty and imprecision prevalent in image processing. Soft computing has been applied to deal with these unanswered uncertainties to a great and appreciable extent. However, as we know that the notable soft computing methodologies often fall short of deriving at a robust solution to the problem of image segmentation.

The present treatise is targeted at introducing to the readers the evolving hybrid soft computing paradigm, which is a confluence of the existing soft computing tools and techniques aimed at achieving more efficient and robust solutions to uncertain real-life problems with a special regard to image segmentation. An in-depth analysis of the different facets of the hybrid soft computing models has been discussed. Illustrative examples (with case studies) of the applications of the hybrid soft computing techniques for the purpose of handling uncertainty in image segmentation has been presented for the sake of understanding of the subject under consideration.

The edited volume comprises a total of 12 well-versed chapters focused on the varied aspects of image segmentation using the hybrid soft computing paradigms.

Chapter "Hybrid Swarms Optimization Based Image Segmentation" explores the application of an admixture of Firefly Algorithm (FA) and Social Spider Optimization (SSO) for handling multilevel thresholding in image segmentation. The authors have shown the superiority of their proposed technique in terms of lower time complexity.

Chapter "Grayscale Image Segmentation Using Multilevel Thresholding and Nature-Inspired Algorithms" deals with another hybridization approach named GSA-GA, which is a fusion of Gravitational Search Algorithm (GSA) and Genetic Algorithm (GA). The authors demonstrate the usage of the hybrid algorithm on multilevel image thresholding.

The objective of Chapter "A Novel Hybrid CS-BFO Algorithm for Optimal Multilevel Image Thresholding Using Edge Magnitude Information" is to use the optimal edge magnitude information of an image to obtain multilevel threshold values on the basis of the Gray-Level Co-occurrence Matrix (GLCM) of the image. Here, the authors have used a novel hybrid algorithm comprising cuckoo search and bacterial foraging optimization in the form of CS-BFO algorithm for obtaining the optimal threshold values.

In Chapter "REFII Model and Fuzzy Logic as a Tool for Image Classification Based on Image Example", the author demonstrates the effectiveness of REFII model coupled with fuzzy logic on image classification tasks.

Chapter "Microscopic Image Segmentation Using Hybrid Technique for Dengue Prediction" offers a hybrid methodology capable of providing an automated platelet counting system for the efficient, easy, and fast detection of dengue infection. This is achieved by segmentation of platelets from microscopic images of a blood smear.

Chapter "Extraction of Knowledge Rules for the Retrieval of Mesoscale Oceanic Structures in Ocean Satellite Images" focuses on the extraction of the rules for the oceanic structures of mesoscale dimension, where the imaging modality used are ocean satellite images.

Chapter "Hybrid Uncertainty Based Techniques for Segmentation of Satellite Imagery and Applications" presents an approach to establish the effectiveness of hybridization of soft computing techniques for achieving segmentation of satellite image applications.

Chapter "Improved Human Skin Segmentation Using Fuzzy Fusion Based on Optimized Thresholds by Genetic Algorithms" uses genetic algorithm for obtaining the optimal thresholding of images. These optimal thresholds, in turn, are used with fuzzy in fusion for segmentation of human skin image information.

In Chapter "Uncertainty Based Spatial Data Clustering Algorithms for Image Segmentation", the authors have discussed about various clustering approaches used in image segmentation. In addition, they have established the efficiency of hybridization in overcoming the limitations in crisp as well as fuzzy-based clustering techniques.

The authors of Chapter "Coronary Artery Segmentation and Width Estimation Using Gabor Filters and Evolutionary Computation Techniques" have shown the efficacy of hybridization of Gabor filtering with evolutionary approach for achieving the very important task of estimation of the width of coronary artery after its appropriate segmentation.

In Chapter "Hybrid Intelligent Techniques for Segmentation of Breast Thermograms", the authors use hybridization techniques to identify and classify affected regions (out of abnormal growth) due to breast carcinoma.

Finally, Chapter "Modeling of High-Dimensional Data for Applications of Image Segmentation in Image Retrieval and Recognition Tasks" illustrates a hybridization approach in the form of Probabilistic Features Combination (PCF) method for multidimensional data modeling, extrapolation, and interpolation using the set of high-dimensional feature vectors.

The design of different hybrid soft computing algorithms might evolve over the years, and more and more efficient algorithms are likely to come. But one issue must be mentioned at this point. This book will be one of the stepping-stones in the forward direction enticing the readers to develop and apply indigenous and robust hybrid soft computing algorithms for image segmentation.

I herewith would like to invite the readers to enjoy the book and take most of its benefits. One could join the team of hybrid soft computing algorithm designers and bring new insights into this developing and challenging enterprise.

Finland
July 2016

Xiao-Zhi Gao

Preface

The field of image segmentation has assumed paramount importance in the computer vision research community given the vast amount of uncertainty involved therein. Proper segmentation of real-life images plays a key role in many real-life applications. Traditional applications include image processing, image mining, video surveillance, intelligent transportation systems, to name a few.

With the shortcomings and limitations of classical platforms of computation, particularly for handling uncertainty and imprecision prevalent in the challenging thoroughfare of image processing, soft computing as an alternative along with extended computation paradigm has been making its presence felt. Accordingly, a phenomenal growth of research initiative in this field is being witnessed. Soft computing techniques include (i) the elements of fuzzy mathematics, primarily used for handling various real-life problems engrossed with uncertainty, (ii) the ingredients of artificial neural networks, usually applied for cognition, learning and subsequent recognition by machine inducing thereby the flavor of intelligence in a machine through the process of its learning and (iii) components of evolutionary computation mainly used for search, exploration, efficient exploitation of contextual information and knowledge useful for optimization.

There has been ample research reporting based on such soft computing techniques applied effectively to solve various real-life problems. The spectrum of applications is practically all pervading. These techniques individually have their points of strength as well as of limitation. On the several real-life contexts, it is being observed that they play supplementary role to one another. Naturally, this has given rise to a serious research initiative for exploring avenues of hybridization of the above-mentioned soft computing techniques. Present day research initiative finds more orientation towards hybridization as an alternative to individual soft computing methods. Moreover, it is observed that hybrid approaches in the form of neuro-fuzzy, fuzzy genetic, rough-neuro, rough-fuzzy, neuro-fuzzy-genetic, neuro-fuzzy-rough, quantum neuro-fuzzy architectures usually offer more robust and intelligent solutions. Interestingly the scope of such hybridization is gradually being found all encompassing.

In this backdrop, the editors, in the present scope, became motivated to invite people from research community to share their latest findings. As a result, the present edited volume may be viewed as a formidable platform particularly aimed at accommodating problems pertaining to image segmentation. There are 12 chapters reported with each representing a self-contained and complete individual contribution.

In the chapter entitled "Hybrid Swarms Optimization Based Image Segmentation", Mohamed Abd El Aziz, Ahmed A. Ewees, Aboul Ella Hassanien have demonstrated as to how an efficient admixture of Firefly Algorithm (FA) and Social Spider Optimization (SSO) could achieve multilevel thresholding in image segmentation. The authors could justify their finding in terms of less CPU time consumption.

The chapter entitled "Grayscale Image Segmentation Using Multilevel Thresholding and Nature-Inspired Algorithms" by Genyun Sun, Aizhu Zhang and Zhenjie Wang demonstrates how an effective hybridization, namely GSA-GA, of Gravitational Search Algorithm (GSA) with Genetic Algorithm (GA) could be used for achieving multilevel image thresholding. The authors also substantiated results through extensive numerical means.

The focal point of the chapter entitled "A Novel Hybrid CS-BFO Algorithm for Optimal Multilevel Image Thresholding Using Edge Magnitude Information" is utilization of optimal edge magnitude information (second-order statistics) of an image to obtain multilevel threshold values on the basis of the Gray-Level Co-occurrence Matrix (GLCM) of the image. Sanjay Agrawal, Leena Samantaray, and Rutuparna Panda use a novel hybrid cuckoo search bacterial foraging optimization (CS-BFO) algorithm, which plays a very crucial role for obtaining optimal threshold values.

The chapter entitled "REFII Model and Fuzzy Logic as a Tool for Image Classification Based on Image Example" by Goran Klepac is a very informative article where he tries to establish the effectiveness of REFII model coupled with fuzzy logic for the purpose of image classification.

In the chapter entitled "Microscopic Image Segmentation Using Hybrid Technique for Dengue Prediction", Pramit Ghosh, Ratnadeep Dey, Kaushiki Roy, Debotosh Bhattacharjee and Mita Nashipuri take up an important practical problem. They offer a hybrid methodology capable of providing an automated platelet counting system for efficient, easy, and fast detection of dengue infection as well as treatment through segmentation of platelets from microscopic images of a blood smear.

The chapter entitled "Extraction of Knowledge Rules for the Retrieval of Mesoscale Oceanic Structures in Ocean Satellite Images" deals with rule extraction for the oceanic structures of mesoscale dimension where the imaging modality is ocean satellite. The authors Eva Vidal-Fernández, Jesús M. Almendros-Jiménez, José A. Piedra, and Manuel Cantón also propose a comprehensive tool for this.

B.K. Tripathy and P. Swarnalatha, in the chapter entitled "Hybrid Uncertainty Based Techniques for Segmentation of Satellite Imagery and Applications", try to establish the effectiveness of hybridization in comparison to the earlier techniques, both classical and fuzzy, towards achieving segmentation of satellite image applications

In the chapter entitled "Improved Human Skin Segmentation Using Fuzzy Fusion based on Optimized Thresholds by Genetic Algorithms" Anderson Santos, Jônatas Paiva, Claudio Toledo, and Helio Pedrini consider genetic algorithm for ensuring optimal thresholding. These thresholds, in turn, are used in fuzzy fusion for achieving segmentation of human skin image information.

Chapter "Uncertainty Based Spatial Data Clustering Algorithms for Image Segmentation" happens to be of enormous importance from the survey point of view. It will be particularly useful for young researchers. The authors Deepthi P. Hudedagaddi and B.K. Tripathy in this work discuss the pros and cons of various clustering approaches used in image segmentation and try to establish as to how hybridization is capable in overcoming the limitations inherent in crisp as well as fuzzy-based clustering techniques.

Fernando Cervantes-Sanchez, Ivan Cruz-Aceves, and Arturo Hernandez-Aguirre in their Chapter "Coronary Artery Segmentation and Width Estimation Using Gabor Filters and Evolutionary Computation Techniques" try to establish as to how effective it could be to hybridize Gabor filtering with evolutionary approach for achieving the very important task of estimation of the width of coronary artery after its appropriate segmentation.

In Chapter "Segmentation and Analysis of Breast Thermograms for Abnormality Prediction Using Hybrid Intelligent Techniques", Sourav Pramanik, Mrinal Kanti Bhowmik, Debotosh Bhattacharjee, and Mita Nasipuri through their contribution entitled "Segmentation and Analysis of Breast Thermograms for Abnormality Prediction Using Hybrid Intelligent Techniques" try to demonstrate and thereby establish the effectiveness of thermal imaging modality and technique towards the identification of abnormal growth, possibly due to breast carcinoma, by classifying the affected region.

Dariusz Jakóbczak in Chapter "Modeling of High-Dimensional Data for Applications of Image Segmentation in Image Retrieval and Recognition Tasks" offers a hybridized approach called Probabilistic Features Combination (PCF) method for multidimensional data modeling, extrapolation, and interpolation using the set of high-dimensional feature vectors.

The editors of the present treatise aimed to bring out some of the latest findings in the field of hybrid soft computing applied to proper segmentation of images. Their mission has met success with a number of quality chapters reported. The editors want to make use of this opportunity to express their sincere gratitude to the authors of the chapters for extending their wholehearted support in sharing some of their latest findings. Without their significant contribution, this volume could not have fulfilled its mission. The editors also extend their heartiest congratulations to the specific team members who took the trouble to make the present endeavor a success, not to mention Springer for providing the editors with an opportunity to work with them. The editors would also like to take this opportunity to extend their heartfelt thanks to Mr. Ronan Nugent, Senior Editor, Springer, for his constructive support during the tenure of the book project. The editors feel encouraged to make further efforts to explore and address other areas of research significance in the days to come. The editors would also like to thank, in anticipation, graduate students and researchers in computer science, electronics communication engineering, electrical engineering, and information technology who will read this as a reference book and as an advanced textbook for their active feedback; their suggestions will be of utmost academic importance to the editors.

Kolkata, India — Siddhartha Bhattacharyya
Santiniketan, India — Paramartha Dutta
Cooch Behar, India — Sourav De
Zagreb, Croatia — Goran Klepac
July 2016

Contents

Hybrid Swarms Optimization Based Image Segmentation

Mohamed Abd El Aziz, Ahmed A. Ewees and Aboul Ella Hassanien

Abstract This chapter proposed multilevel thresholding hybrid swarms optimization algorithm for image segmentation. The proposed algorithm is inspired by the behavior of fireflies and real spider. It uses Firefly Algorithm (FA) and Social Spider Optimization (SSO) algorithm (FASSO). The objective function used for achieving multilevel thresholding is the maximum between class variance criterion. The proposed algorithm uses the FA to optimize threshold, and then uses this thresholding value to partition the images through SSO algorithm of a powerful global search capability. Experimental results demonstrate the effectiveness of the FASSO algorithm of image segmentation and provide faster convergence with relatively lower CPU time.

Keywords Swarms optimization · Firefly Algorithm · Social Spider Optimization · Multilevel thresholding · Image segmentation

1 Introduction

Image segmentation is an elementary and essential phase in digital image analysis. It is defined as the process of splitting an image into homogeneous areas (segments) with comparable features (i.e., color, contrast, brightness, texture, and gray level) based on a predefined criterion. It is also a method used to distinguish between

M.A. El Aziz (✉)
Department of Mathematics, Faculty of Science, Zagazig University, Zagazig, Egypt
e-mail: abd_el_aziz_m@yahoo.com

A.A. Ewees
Department of Computer, Damietta University, Damietta, Egypt
e-mail: a.ewees@hotmail.com; ewees@du.edu.eg

A.E. Hassanien
Faculty of Computers and Information, Cairo University, Giza, Egypt
e-mail: aboitcairo@gmail.com

A.E. Hassanien
Scientific Research Group in Egypt (SRGE), Giza, Egypt

S. Bhattacharyya et al. (eds.), *Hybrid Soft Computing for Image Segmentation*, DOI 10.1007/978-3-319-47223-2_1

foreground and background [1]. It is a necessary issue since it is the initial phase for image recognition and understanding [2]. In recent years, the image segmentation has more attention in many applications such as face recognition [3], medical diagnosis [4], optical character recognition [5], and satellite image [6]. The images in different applications can be corrupted by noise from equipment or environment and the variability of the background, these challenges make the image segmentation a complex problem. There are several techniques for image segmentation such as edge detection [7], region extraction [8], histogram thresholding, clustering algorithms [9], and threshold segmentation [10].

Thresholding is considered as one of the popular methods for image segmentation, which is essentially a pixel classification problem; it can be divided into two categories: bi-level and multilevel [1, 11]. From a grayscale image, bi-level thresholding can be applied to produce two classes of objects and backgrounds, while multilevel thresholding is required to segment complex images. Also, the multilevel thresholding determines multiple thresholds such as tri-level or quad-level, which separates pixels into multiple homogeneous classes (regions) based on intensity. The main purpose of using bi-level or multilevel thresholding is to locate the best threshold value [12, 13]. Bi-level thresholding techniques has an advantage, it can produce adequate outcomes in cases where the image includes two gray levels only, but the foremost restriction is that the time-consuming computation is often high when these techniques are used to determine the optimal multilevel thresholding [14].

The thresholds also can be obtained at a global or local level. In a local level, different thresholds are allocated for each segment of the image, whereas in a global level, a single threshold is given to the entire image [15]. The reason to use global thresholding is that foreground and background areas in an image can be calculated by taking its histogram with probabilities for each gray scale. The results of bi-level thresholding are not suitable to real application images. So, there is a robust requirement of multilevel thresholding [11]. The essential purpose of multilevel thresholding is to locate threshold values which split pixels into various classes that can be determined by the Eq. (1) [16].

$$\begin{aligned} C_0 &= \{(i,j)\,\varepsilon\, I \mid 0 \le g\,(i,j) \le t_1 - 1\} \\ C_1 &= \{(i,j)\,\varepsilon\, I \mid t_1 \le g\,(i,j) \le t_2 - 1\}, \ldots \\ C_n &= \{(i,j)\,\varepsilon\, I \mid t_k \le g\,(i,j) \le L - 1\} \end{aligned} \tag{1}$$

where C_n are the classes of multilevel thresholding, $g(i,j)$ is the gray level of the pixel (i,j), t_k ($k = 1, \ldots, n$) is the kth threshold value, n is the number of thresholds, and L is the gray level in image I.

Multilevel thresholding segmentation may have some inconveniences such as [1, 2, 14, 16]:

- Selection of the threshold number corresponding to the number of the segments in the image.

- It shows a slow convergence and requires significantly large computational time since computational complexity of the deterministic techniques increases with the number of thresholds.
- There is no analytic solution when the number of segments to be detected increases.

In addition, usually, the optimal thresholding techniques search for the thresholds by optimizing some criterion functions that are determined from images and they use the chosen thresholds as parameters. So, the selection of optimal thresholds in multilevel thresholding is an NP-hard problem [17], and it has been considered as a challenge over decades. To define the optimal thresholds, most techniques analyze the histogram of the image through either maximizing or minimizing an objective function with regard to the thresholds values.

Classical techniques are suitable when the number of thresholds is small; so, in order to avoid these drawbacks, many swarm intelligence (SI) techniques, inspired by the behaviors of natural systems, have been applied. These techniques include particle swarm optimization (PSO), Gray Wolf Optimization (GWO), genetic algorithm (GA) and ant colony optimization (ACO). There are several SI techniques for image segmentation combined with thresholding algorithm to improve the cluster centers [17, 18].

Jie et al. (2013) [19] presented multi-threshold segmentation based on firefly optimization algorithm (FA) and k-means. It could effectively overcome the problem that K-means algorithm is sensitive to the initial center. It is compared with the K-means algorithm; the results show that the proposed technique achieved a higher performance, low run-time, and better segmentation result; also, it gets a better peak signal-to-noise ratio (PSNR) than the traditional fast FCM algorithm and PSO-FFCM. Whereas, Chaojie et al. (2013) [20] applied a proposed technique based on FA to solve multi-threshold searching when applying to image segmentation. The results of this technique are better than GA.

Li Fang et al. (2014) [21] presented an algorithm called FA based on Otsus method and applied it to segmentation of mineral belt image of shaking table. The results demonstrate that the proposed algorithm can accurately segment different mineral. Whereas, Vishwakarma et al. (2014) [22] presented a new meta-heuristic technique for image segmentation using FA. The results show that this technique produces higher results than the traditional K-means clustering algorithm. On the other hand, Zhou et al. (2015) [10] presented a method for image segmentation that integrates FA with two-dimensional Otsu to overcome the problems of low accuracy, time consuming, and giving false segmentation image.

Rajinikanth et al. (2015) [23] presented multilevel segmentation techniques based on RGB histogram using classic FA, Lvy search based FA (LFA), and Brownian search based FA (BFA). The results show that the FA and LFA techniques represent a faster convergence than BFA.

Erdmann et al. [24] presented an application of FA for multilevel thresholding segmentation. The experiments show that the results of FA are close to the exhaustive search. In [25] kai et al. presented a multilevel image thresholding model based on the improved FA (IFA). The IFA used Cauchy mutation and neighborhood strategy to increase the global search ability and accelerate the convergence. Also, IFA is applied to search multilevel global best thresholds. Seven images are used to verify the IFA. The results show that the IFA based model can efficiently search the multilevel thresholds and segment images into background and objects.

The PSO and its variant [26–31] are used to determine the multilevel thresholding for image segmentation. Also, the spider algorithm is used for image segmentation as proposed by Richard [32].

However, each of SI techniques has its own drawbacks; to solve this problem, they should be hybridized with other swarm techniques. The aim of using the hybridization concept is to combine different SI techniques to avoid the drawbacks of the individual swarm. Also, the hybrid techniques achieve global solution with short time.

In this chapter, we present a new segmentation technique based on combining two swarm techniques the Firefly Algorithm (FA) and Social Spider Optimization (SSO) called FASSO to solve the problem of thresholding for image segmentation. Where, the characteristics of these two algorithms are used to improve the performance of thresholding. The properties of the FA are the ability of local search, fast convergence. Where the population of FA algorithm is divided into multiple subgroups and each group searches its own local best value. There is a global best solution over all the local values. Unlike FA, the SSO is characterized by global search ability, but its convergence is poor and lack of local search ability. To achieve the global solution, the SSO algorithm uses random walk scheme. To determine the next position of food source, each spider receives the vibration that propagates over the web (the search space of optimization problems). Also, this new hybrid technique is used to avoid the problem of time complexity in other swarm techniques such as PSO.

2 Preliminaries

In this section, a brief review of Firefly and Social Spider Optimization algorithms is presented, which are used in the proposed approach.

2.1 Firefly Algorithm (FA)

Firefly algorithm (FA) was presented by Yang [33]. FA has been implemented in many areas of optimization problems such as continuous, dynamic, combinatorial, noisy, and constrained optimization. It has been applied in clustering and classification problems in neural networks, machine learning and data mining, image processing, industrial optimization, semantic web, civil engineering, robotics, business optimization, chemistry, [16, 34–39], image compression [40, 41], feature selection, classifications, and clustering [35, 42]. The FA uses three rules as follows [43, 44]:

- Fireflies will be attracted to each other regardless of their type because all of them are unisex.
- Attractiveness is proportional to their brightness, thus, the less bright one will move to the brighter one. If there is no brighter one than a specific firefly, it will move randomly.
- The brightness of a firefly is determined and affected by the landscape of the cost function.

The two factors in FA are the brightness and attractiveness and these factors will be constantly updated. The attractiveness β between i and j is defined as [45, 46]:

$$\beta = \beta_0 \times e^{(-\gamma m^2)} \tag{2}$$

where β_0 indicates the attractiveness at $m = 0$ that can be set as 1 for many problems, γ is the coefficient of the light absorption that can be used as a constant, and m is a space distance between individual i and j, as follows:

$$m_{ij} = ||u_i - u_j|| = \sqrt{\sum_{t=1}^{d} (u_{i,k} - u_{j,t})^2} \tag{3}$$

where $u_{(i,t)}$ is the tth component of the spatial coordinate d_i of ith firefly.

The movement of a firefly i is attracted to another attractive firefly j, and is calculated by:

$$u_i = u_i + \beta \times (u_i - u_j) + \alpha \times \varepsilon_i \tag{4}$$

where $\alpha \in [0, 1]$ is a random parameter and $\varepsilon_i \in N(\mu, \sigma)$ is a random number in a vector.

The basic structure of the FA is shown in Algorithm 1 [43, 47]. For a maximization problem, the brightness is proportional to the value of cost function. Other forms of the brightness could be determined as GA fitness function [47].

2.2 Social Spider Optimization Algorithm (SSO)

The Social Spider Optimization (SSO) algorithm is a new swarm technique provided by Cuevas [48]. This algorithm emulates the behavior of social spiders to perform optimization. There are many applications that use SSO such as automatic text [49], image segmentation [32], and wireless sensor networks [50]. Where the solution of an optimization problem is the position of spiders through a spider web that is responsible for translating the information between spiders. The communication between the spiders is performed using vibration transmitted when the spiders move to a new position. The vibration between spider i and j is generated as [48]:

Algorithm 1 Firefly Algorithm

1: Cost function $f(u)$, $u = (u_1, u_2, u_d)^t$
2: Generation a population of fireflies $u_i (i = 1, 2, \ldots, N)$
3: Light intensity I_i at u_i is calculated by $f(u_i)$
4: Define light absorption coefficient γ
5: while ($t < MaximumGenerations$)
6: **for** $i = 1 : n(allnfireflies)$ **do**
7: **for** $j = 1 : i$ **do**
8: **if** $I_i < I_j$ **then**
9: Move firefly i to j in $d-$ dimensions
10: **end if**
11: Attractiveness changes with distance r by $exp[-\gamma r]$
12: Evaluate new solutions and update light intensity
13: **end for**
14: **end for**
15: Sort the fireflies based on light intensity and locate the current best
16: end while
17: Post process results

$$Vib_{ij} = w_j e^{-d_{ij}^2},\ d_{ij} = \left\| x_i - x_j \right\| \tag{5}$$

where w_j is the weight spider j which is defined as:

$$w_j = \frac{F\left(x_j\right) - worst_x}{\text{best}_x - worst_x},$$

$$best_x = \max_{k=1,\ldots,N} F\left(x_k\right), worst_x = \min_{k=1,\ldots,N} F\left(x_k\right) \tag{6}$$

where $F(x_j)$ is the cost function value. There are two strategies to update the position of spiders depending on their gender. Where there are two types of gender males and females (The females represent 60–90 % from the total number of spiders). The female spider updates its position as follows:

$$x_{F_i}^{k+1} = \begin{cases} x_{F_i}^k + \alpha \times Vibc_i \times \left(x_c - x_{F_i}^k\right) + \beta \times Vibb_i \times \left(x_b - x_{F_i}^k\right) + \delta \times (r - .5), & PF \\ x_{F_i}^k - \alpha \times Vibc_i \times \left(x_c - x_{F_i}^k\right) - \beta \times Vibb_i \times \left(x_b - x_{F_i}^k\right) + \delta \times (r - .5), & 1 - PF \end{cases} \tag{7}$$

where α, β, δ, and r are random numbers in the range [0, 1], while k denotes the repetition number. The x_c represents the nearest member to i that holds a higher values and x_b is the best individual. $Vibb_i$ is the vibration transmitted by the best spider b,

$$Vibb_i = w_b e^{-d_{ib}^2}, \quad w_b = \max_{k=1,..,N} w_k \tag{8}$$

$$Vibc_i = w_c e^{-d_{ic}^2} \quad w_c > w_i \tag{9}$$

In Eq. (7), PF represents the probability of movement of the female toward the source of vibrations. Unlike the female strategy, the males update their position as follows [48]:

$$x_{m_i}^{k+1} = \begin{cases} x_{m_i}^k + \alpha \times Vibf_i \times \left(x_f - x_{m_i}^k\right) + \delta \times (r - .5), \\ x_{m_i}^k + \alpha \left(\eta - x_{m_i}^k\right), \end{cases} \tag{10}$$

where $\eta = \sum_{j=1}^{N_m} x_{m_i}^k . w_{N_f+j} / \sum_{j=1}^{N_m} w_{N_f+j}$ and $Vibf_i = w_f e^{-d_{if}^2}$. If male is dominated the first branch of Eq. (10) is used otherwise (nondominated) the second branch is used.

The final step is the offspring generation by mating between the dominant males, and females laying in the neighbor with mating radius defined as:

$$r = \frac{\sum_{j=1}^{n} \left(x_j^{high} - x_j^{low}\right)}{2n} \tag{11}$$

The cost function of offspring is computed and compared with the worst solutions. If any offspring is better a spider, it is replaced with this spider. The SSO steps are shown in Algorithm 2.

Algorithm 2 Social-Spider (SSO)

1: Output: x_{best} Best Position.
2: Initial value to N; // number of spider,
3: Generate an initial population of n spiders $\mathbf{x}_i$, $i = 1, 2, \ldots, n$
4: $G = 1$ //generation number
5: **repeat**
6:   **for** all $\mathbf{x}_i$ do // parallel techniques **do**
7:     Compute the Fitness function
8:     **if** $(F_i < F_{best})$ **then**
9:       $x_{best} = x_i$, $F_{best} = F_i$
10:       $x_{best} = \mathbf{x}_i$ such that $i = \min_i F_i$
11:       Break;
12:     **end if**
13:   **end for**
14:   Determine the best and worst value of the fitness F_i
15:   Determine $w_i = \frac{F(x_i) - worstx}{best_x - worst_x}$
16:   Determine the index of the max w_i, where index = arg max w_i
17:   Update female position using Eq. (7)
18:   Update male position using Eq. (10)
19:   Perform the mating operation using Eq. (11) to generate offspring's
20:   Compute the Fitness function for offspring (x_{offSp})
21:   Replace the offspring (x_{offSp}) with the worst spider.
22:   $G = G + 1$
23: **until** $G <$ itern

3 The Proposed Hybrid Segmentation Algorithm

According to the analysis above, FA behavior improves the search for an optimal solution, while SSO is better at reaching a global region. Therefore, the new hybrid algorithm called FASSO is proposed. It employs FA for generation jumping to avoid SSO getting stuck in the local optima problem. FASSO integrates SSO global optimization and FA fast convergence. The FASSO steps are illustrated in Algorithm 3.

Algorithm 3 FASSO

1: Population-based initializing (pop, size N), Parameters of SSO and FA algorithms.
2: **repeat**
3: Pass the agents to the Algorithm 1
4: Evaluating fitness of each agent in FA.
5: Find the best solution using the Algorithm 1.
6: Pass the new solution to the Algorithm 2.
7: Find the best solution using SSO.
8: **until** $iter < maxiter$

4 Experiments and Discussion

The FASSO algorithm was programmed in MATLAB and run in Windows environment with 64-bit support. The proposed algorithm is tested on a six common images from the Berkeley University database [51] as in Fig. 1. The parameters of the proposed algorithm were: the maximum iterative time was 100, the size of the swarm was 20, $Vmax = 3$, and $c1 = c2 = 1.49$.

We evaluate the proposed algorithm's performance by computing the time, fitness function value, Jaccard and then comparing its results with other algorithms, including SSO, FA, PSO [30], FPSO, and DPSO [27]. The results are illustrated in Tables 1, 2, 3 and 4.

In Table 1, the fitness values are computed at thresholds 2, 3, and 4. However, there are small differences; all algorithms find the optimal solution. The FA, PSO, DPSO, and FPSO algorithms have higher fitness value than the SSO and FASSO algorithms.

Tables 2 and 3 give the selected thresholds and the computation times of algorithms, respectively. The results in Table 3 and Fig. 2 show that the FASSO method is better in term convergence speed. Where the computation time of the FASSO is less than computation times of the other algorithms.

Figures 3, 4, 5, 6, 7 and 8 show the results of the FASSO algorithm for segmentation using six images and the segmented images with different threshold levels. We can conclude from these figures that images with higher level contain more detail than the other.

To evaluate the quality of techniques to select multi-thresholds, the Jaccard measure is used,

Fig. 1 The original six images

$$Jac = \frac{|I_{orginalimage} \bigcap J_{segmentedimage}|}{|I_{orginalimage} \bigcup J_{segmentedimage}|} \quad (12)$$

where the value of $Jac \in [0, 1]$. The best technique is the one that has higher value of *Jac* Obviously, From Table 4, we can conclude that the results of the proposed technique have higher Jaccard than those of the other techniques.

From all previous results, we concluded that the threshold value determined by the FASSO technique is the best value. Also, our technique FASSO is better than others in terms of execution time.

Table 1 Average of fitness values of algorithms using six images

Images	Thresholds	FASSO	SSO	PSO	DPSO	FPSO	FA
Image A	2	3023.63	3023.90	3036.23	3030.56	3029.19	3042.09
	3	3600.27	3620.50	3680.15	3639.59	3630.55	3634.17
	4	3815.70	3844.18	3851.14	3845.91	3846.49	3846.81
Image B	2	691.64	696.16	715.24	697.34	691.64	700.91
	3	887.13	900.83	932.21	901.71	902.10	889.74
	4	962.11	971.80	998.82	974.17	971.27	999.70
Image C	2	1143.81	1149.38	1167.81	1148.64	1145.61	1152.11
	3	1370.89	1377.43	1392.37	1374.50	1372.05	1374.12
	4	1448.12	1459.21	1570.85	1510.01	1510.56	1506.94
Image D	2	1826.56	1829.63	1834.83	1828.96	1829.65	1829.62
	3	2010.13	2013.23	2043.50	2019.45	2021.50	2014.05
	4	2111.45	2136.86	2148.75	2140.72	2142.34	2141.05
Image E	2	3153.34	3154.84	3167.11	3155.93	3154.21	3158.37
	3	3633.59	3637.29	3645.72	3637.61	3635.02	3631.35
	4	3855.98	3873.02	3893.97	3873.19	3863.14	3875.55
Image F	2	3820.70	3830.67	3852.72	3836.18	3829.89	3825.23
	3	4160.50	4161.47	4171.59	4162.78	4161.04	4163.57
	4	4340.90	4360.70	4367.40	4361.34	4357.95	4358.48

Table 2 Selected thresholds of techniques

Images	Thresholds	SSO	FASSO	PSO	DPSO	FPSO	FA
Image A	2	55 146	76 163	73 150	74 151	75 158	110 198
	3	71 115 181	73 140 205	72 138 197	70 137 195	73 120 182	74 142 207
	4	66 107 171 229	67 119 156 225	63 121 158 220	64 122 181 230	63 112 158 209	82 98 144 207
Image B	2	104 145	108 158	101 146	101 144	100 143	79 157
	3	90 121 182	103 136 175	89 122 159	95 113 160	90 123 160	70 147 213
	4	90 111 134 176	79 120 152 173	82 109 135 169	83 111 137 171	83 110 136 170	69 116 160 196
Image C	2	143 194	147 201	145 196	146 197	151 201	103 143
	3	132 155 218	95 149 212	101 151 223	126 168 230	116 161 203	81 118 153
	4	99 140 179 215	104 153 167 210	108 147 177 208	110 149 179 210	109 148 178 209	91 126 170 211
Image D	2	105 145	102 158	97 148	108 158	98 149	140 190
	3	70 97 140	53 111 153	67 117 155	61 101 154	68 107 153	118 144 210

(continued)

Table 2 (continued)

Images	Thresholds	SSO	FASSO	PSO	DPSO	FPSO	FA
	4	28 81 125 154	4 4 51 103	58 97 134 165	60 98 135 167	59 108 145 201	133 169 196 227
Image E	2	113 198	116 195	114 193	115 194	112 186	103 155
	3	80 138 182	69 141 203	75 132 197	76 133 198	79 145 209	47 115 150
	4	68 124 168 204	66 122 176 199	71 117 158 203	72 118 159 204	71 117 158 202	64 91 131 165
Image F	2	111 182	109 185	108 181	109 182	109 182	82 164
	3	85 132 182	79 127 181	73 122 186	74 123 187	74 123 187	76 131 182
	4	74 112 126 184	72 87 124 203	71 108 146 195	71 108 146 195	72 109 147 196	65 118 168 209

Table 3 The average CPU process time of different segmentation techniques

Images	Thresholds	SSO	FASSO	PSO	DPSO	FPSO	FA
Image A	2	2.57	3.46	4.11	3.48	3.90	2.63
	3	3.17	3.08	5.61	5.60	5.91	3.87
	4	4.43	3.89	6.01	6.33	6.28	3.34
Image B	2	2.24	2.46	3.15	3.26	3.89	2.60
	3	3.63	3.65	5.93	4.42	5.10	3.28
	4	3.86	3.76	5.94	6.11	6.47	3.93
Image C	2	2.22	2.26	3.20	2.19	2.82	2.45
	3	3.01	3.06	4.64	5.12	5.88	3.27
	4	4.59	4.29	7.95	6.72	5.64	3.97
Image D	2	3.01	2.24	3.24	2.27	2.87	3.10
	3	3.01	2.97	4.52	3.80	4.71	3.26
	4	3.73	3.72	5.89	5.75	6.06	5.46
Image E	2	2.25	2.23	3.40	2.98	4.08	2.55
	3	3.10	3.01	5.28	5.53	5.49	3.18
	4	4.78	4.72	7.24	6.38	5.83	4.99
Image F	2	2.18	2.22	3.37	3.11	2.45	2.68
	3	2.98	3.04	4.54	4.10	5.79	3.16
	4	3.76	3.71	5.83	5.24	6.69	4.13

Table 4 The average Jaccard measures of different segmentation techniques

Images	Thresholds	SSO	FASSO	PSO	DPSO	FPSO	FA
Image A	2	0.49	0.53	0.47	0.45	0.46	0.51
	3	0.76	0.79	0.73	0.74	0.74	0.73
	4	0.80	0.84	0.75	0.75	0.79	0.79
Image B	2	0.66	0.67	0.46	0.59	0.60	0.65
	3	0.79	0.81	0.80	0.79	0.79	0.72
	4	0.85	0.86	0.79	0.76	0.69	0.82
Image C	2	0.65	0.69	0.66	0.65	0.67	0.65
	3	0.73	0.78	0.75	0.72	0.76	0.76
	4	0.80	0.94	0.88	0.92	0.89	0.89
Image D	2	0.55	0.59	0.56	0.52	0.54	0.57
	3	0.56	0.63	0.61	0.58	0.59	0.62
	4	0.87	0.92	0.90	0.91	0.89	0.89
Image E	2	0.49	0.54	0.51	0.53	0.49	0.53
	3	0.54	0.59	0.57	0.55	0.53	0.56
	4	0.71	0.75	0.73	0.72	0.72	0.74
Image F	2	0.67	0.68	0.63	0.67	0.67	0.67
	3	0.75	0.76	0.74	0.72	0.76	0.75
	4	0.81	0.87	0.82	0.84	0.87	0.85

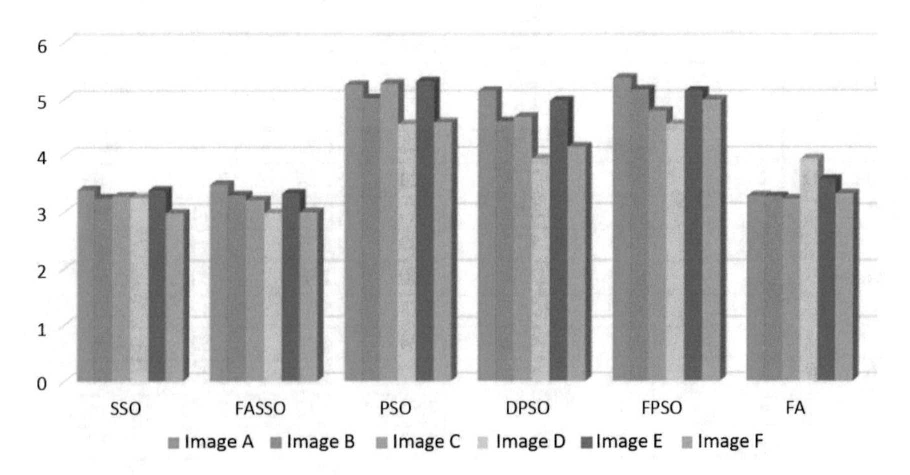

Fig. 2 Average of time for each image

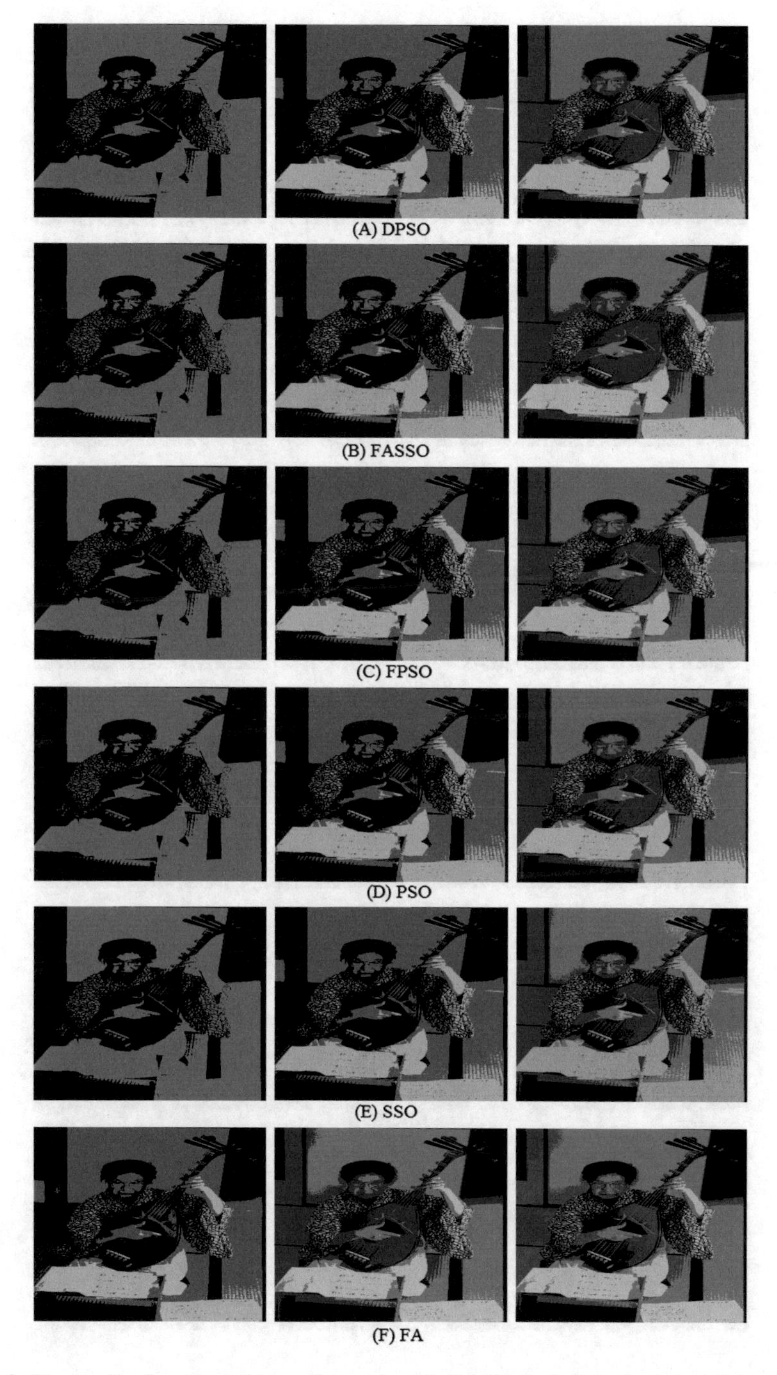

Fig. 3 The result of segmentation on Image **A** with 2 (at *left*), 3 (at *center*), and 4 (at *right*) thresholds

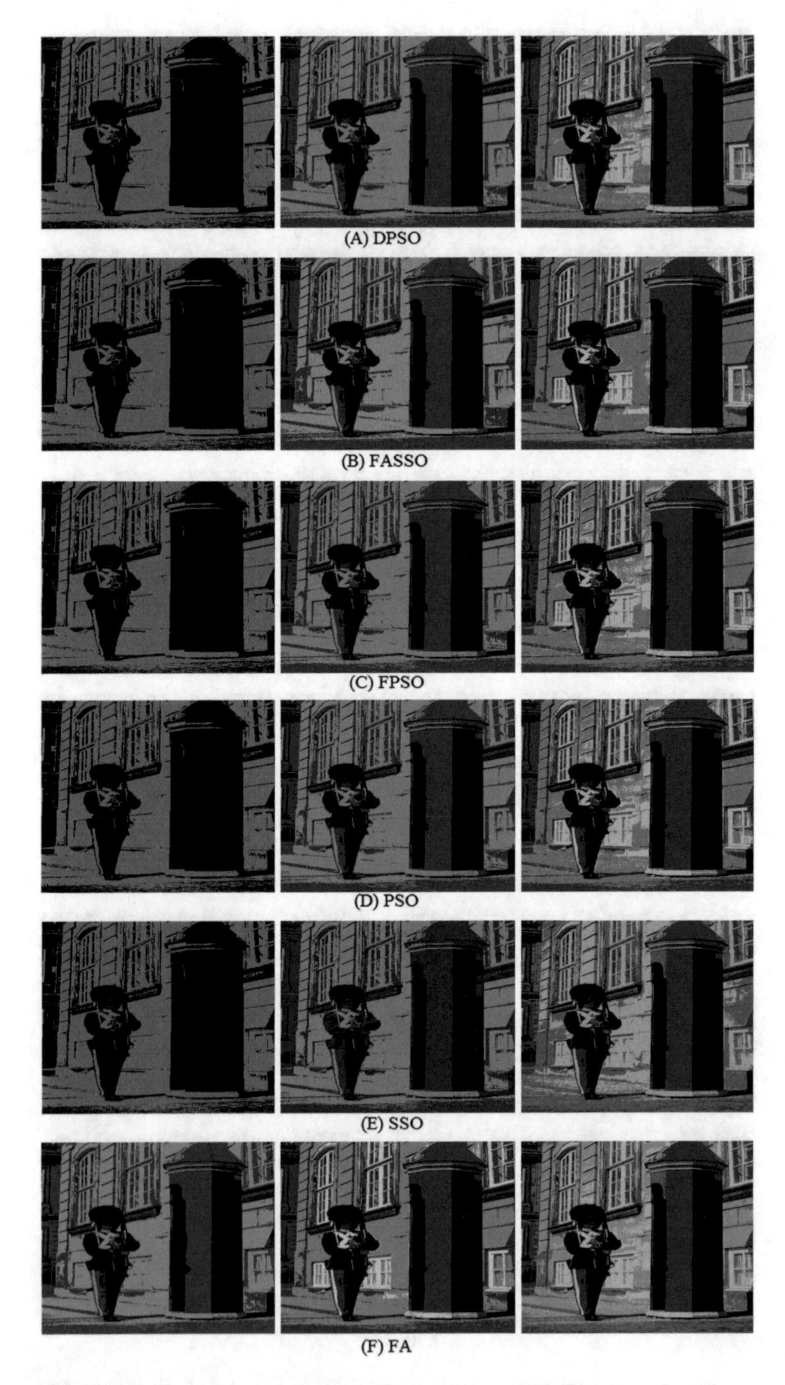

Fig. 4 The result of segmentation on Image **B** with 2 (at *left*), 3 (at *center*), and 4 (at *right*) thresholds

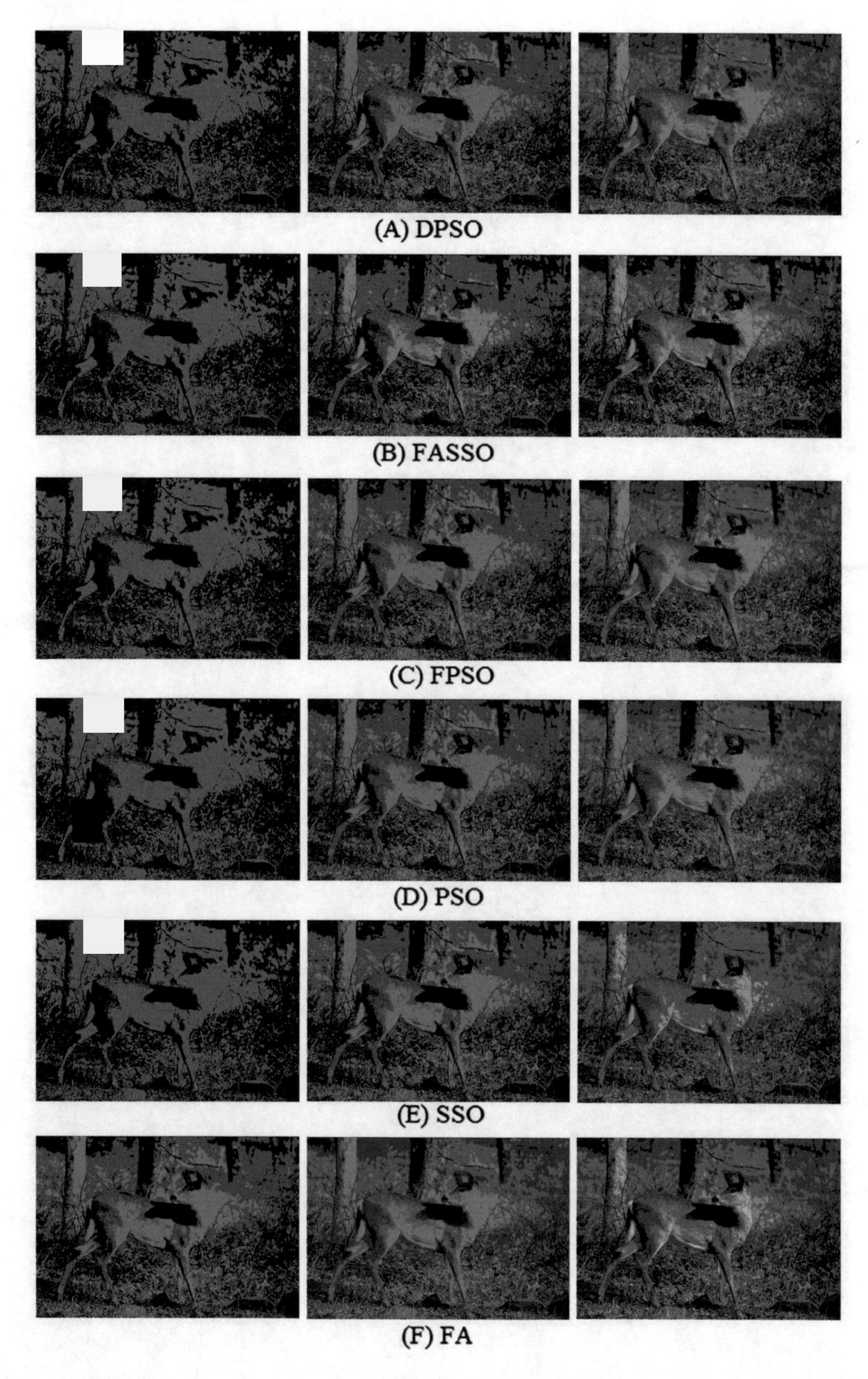

Fig. 5 The result of segmentation on Image **C** with 2 (at *left*), 3 (at *center*), and 4 (at *right*) thresholds

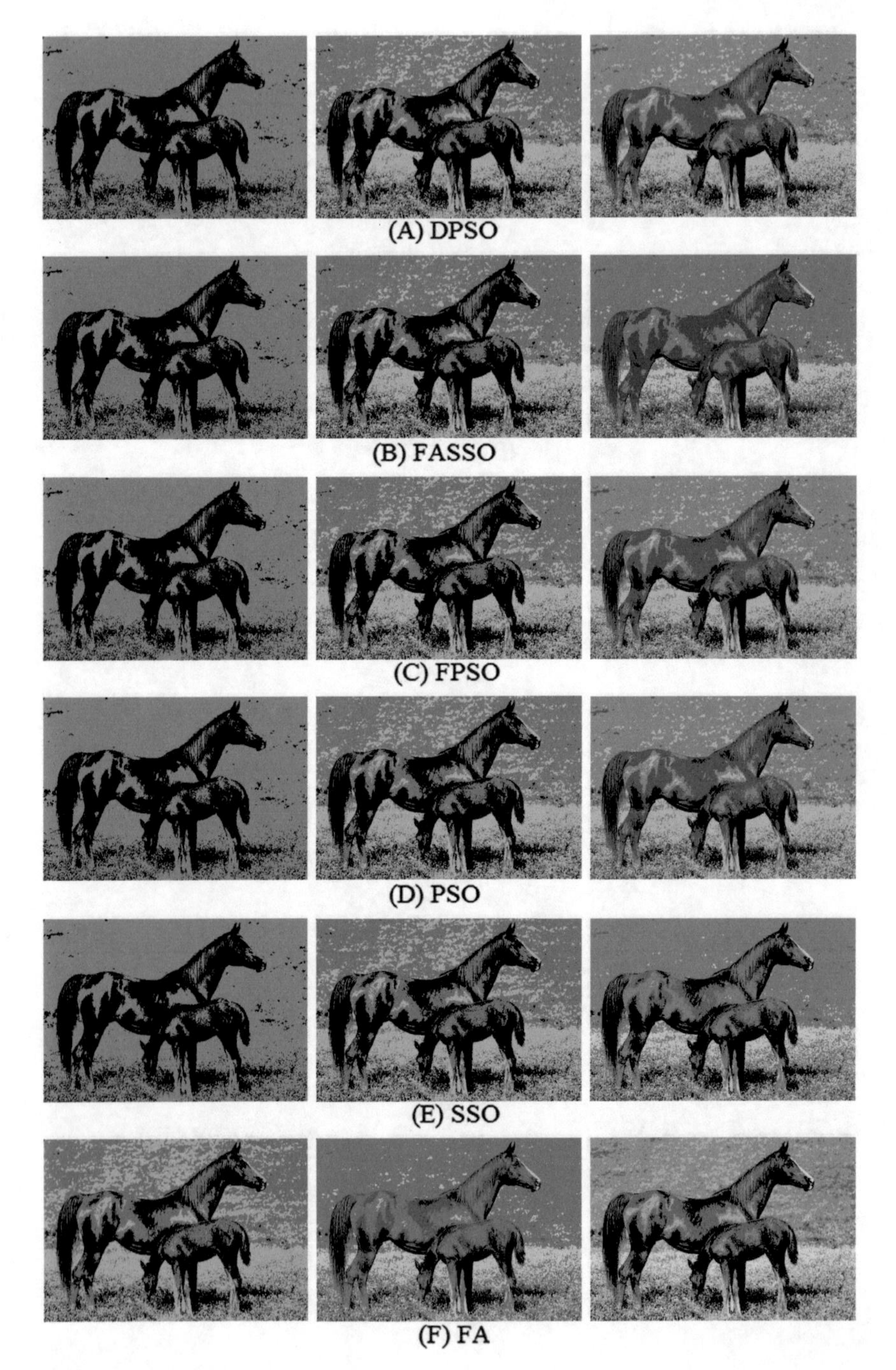

Fig. 6 The result of segmentation on Image **D** with 2 (at *left*), 3 (at *center*), and 4 (at *right*) thresholds

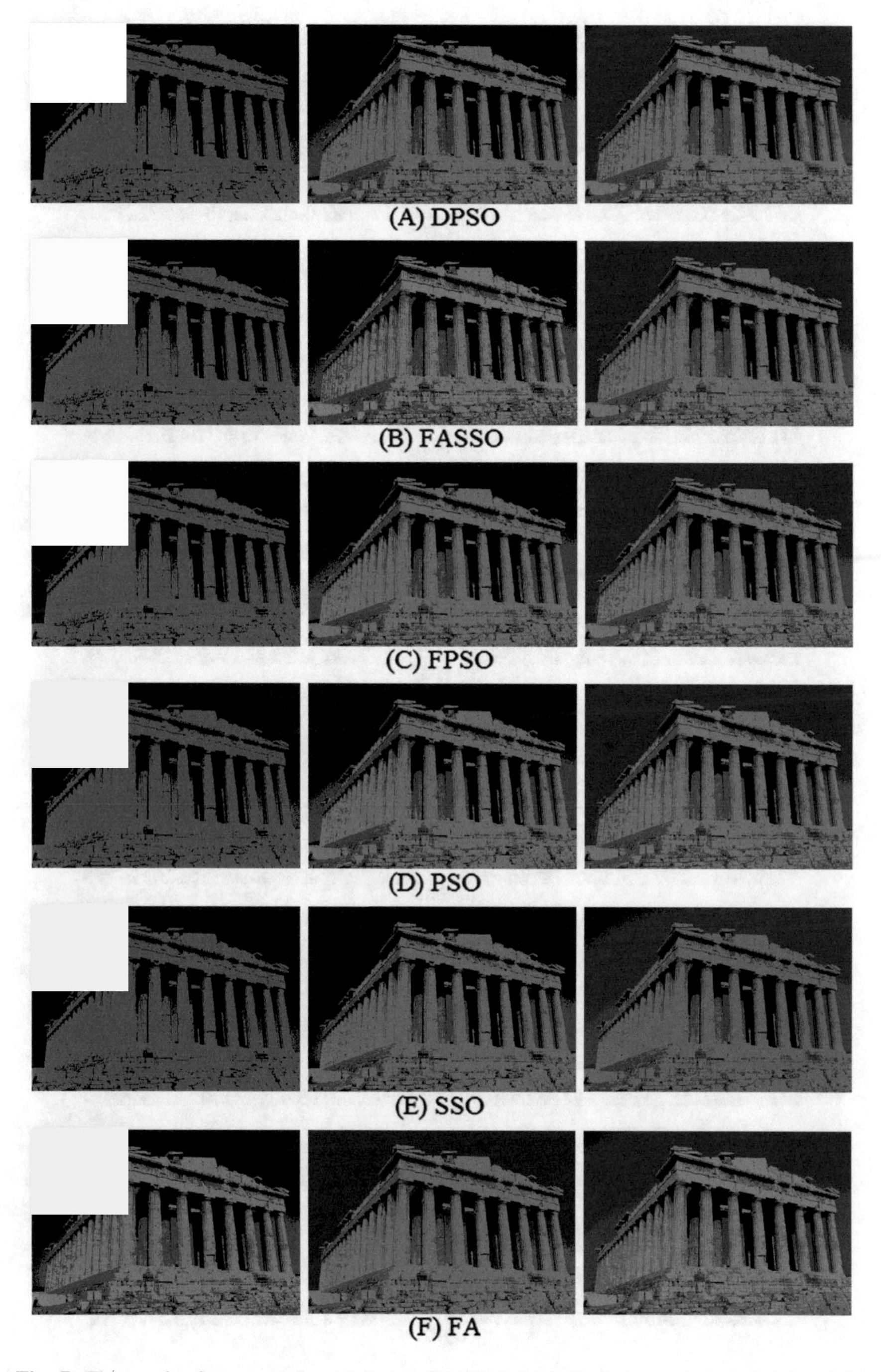

Fig. 7 The result of segmentation on Image **E** with 2 (at *left*), 3 (at *center*), and 4 (at *right*) thresholds

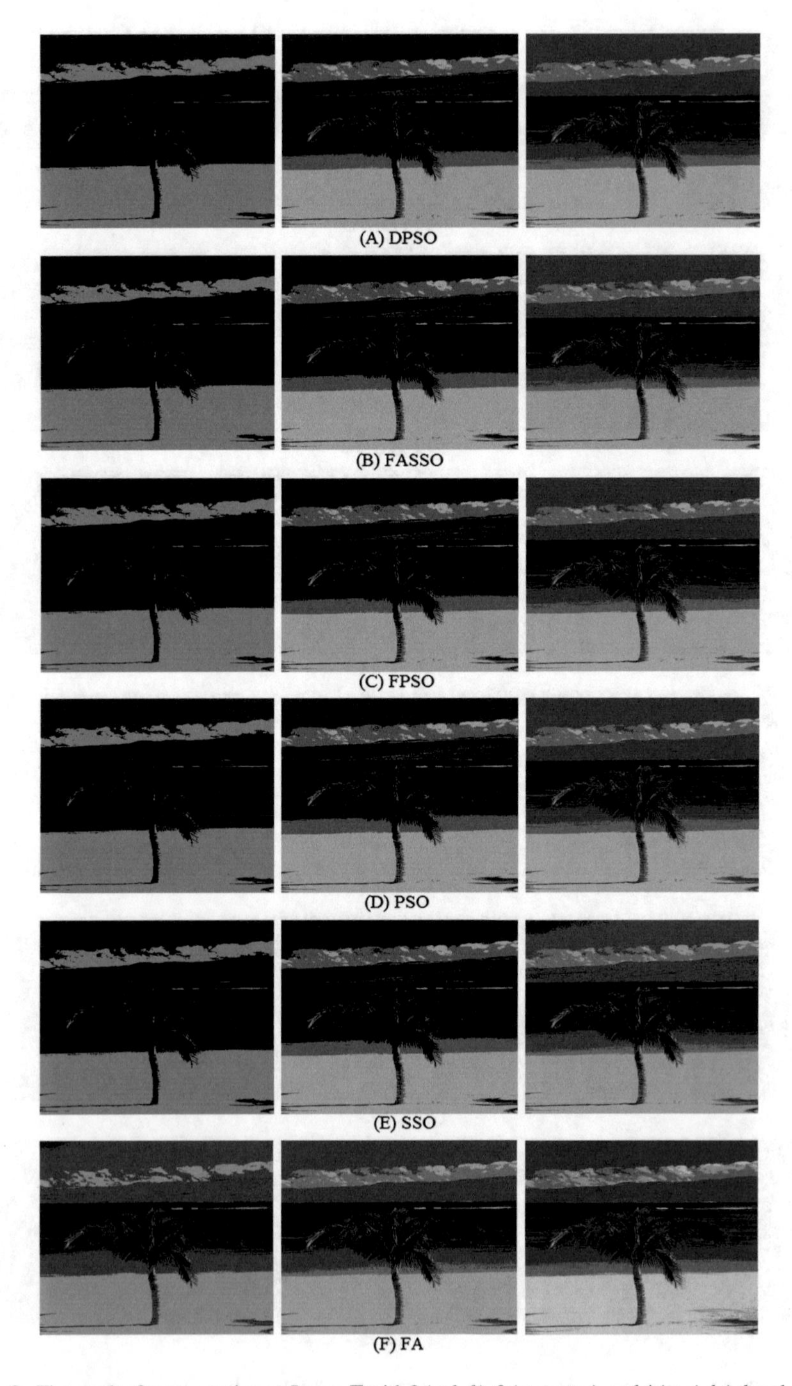

Fig. 8 The result of segmentation on Image **F** with 2 (at *left*), 3 (at *center*), and 4 (at *right*) thresholds

5 Conclusion and Future Work

Image segmentation is an essential part of image processing. In this chapter, a new hybrid Firefly Algorithm (FA) and Social Spider Optimization (SSO) algorithm (FASSO) is proposed for multi-thresholding image segmentation. The proposed algorithm employs the FA algorithm to optimize threshold, and then uses this thresholding value to partition the images through SSO. So, the FASSO integrates SSO global optimization and FA fast convergence. The experiment results were compared against four algorithms, namely, SSO, PSO, DPSO, and FPSO. The FASSO achieved better results than SSO, PSO, DPSO, and FPSO; also, the FASSO provided a faster convergence with relatively lower CPU time. In future, the FASSO can be applied to other complex image segmentation problems.

References

1. Sarkar, S., Sen, N., Kundu, A., Das, S., Chaudhuri, S.S.: A differential evolutionary multilevel segmentation of near infra-red images using Renyi's entropy. In: Proceedings of the International Conference on Frontiers of Intelligent Computing: Theory and Applications (FICTA), pp. 699–706. Springer Berlin (2013)
2. Cuevas, E., Sossa, H.: A comparison of nature inspired algorithms for multi-threshold image segmentation. Expert Syst. Appl. **40**(4), 1213–1219 (2013)
3. Ngambeki, S.S., Ding, X., Nachipyangu, M.D.: Real time face recognition using region-based segmentation algorithm. Int. J. Eng. Res. Technol. **4**(4) (2015). ESRSA Publications
4. Zhao, F., Xie, X.: An overview of interactive medical image segmentation. Ann. BMVA **7**, 1–22 (2013)
5. Kim, S.H., An, K.J., Jang, S.W., Kim, G.Y.: Texture feature-based text region segmentation in social multimedia data. Multimed. Tools Appl. pp. 1–15 (2016)
6. Pare, S., Bhandari, A.K., Kumar, A., Singh, G.K., Khare, S.: Satellite image segmentation based on different objective functions using genetic algorithm: a comparative study. In: 2015 IEEE International Conference on Digital Signal Processing (DSP), pp. 730–734. IEEE (2015)
7. Ju, Z., Zhou, J., Wang, X., Shu, Q.: Image segmentation based on adaptive threshold edge detection and mean shift. In: 2013 4th IEEE International Conference on Software Engineering and Service Science (ICSESS), pp. 385–388. IEEE (2013)
8. Li, Z., Liu, C.: Gray level difference-based transition region extraction and thresholding. Comput. Electr. Eng. **35**(5), 696–704 (2009)
9. Tan, K.S., Isa, N.A.M.: Color image segmentation using histogram thresholding. Fuzzy C-means hybrid approach. Pattern Recognit. **44**(1), 1–15 (2011)
10. Zhou, C., Tian, L., Zhao, H., Zhao, K.: A method of two-dimensional Otsu image threshold segmentation based on improved firefly algorithm. In: Proceeding of IEEE international conference on cyber technology in automation, control, and intelligent systems 2015, Shenyang, pp. 1420–1424 (2015)
11. Bhandari, A.K., Singh, V.K., Kumar, A., Singh, G.K.: Cuckoo search algorithm and wind driven optimization based study of satellite image segmentation for multilevel thresholding using Kapur's entropy. Expert Syst. Appl. **41**(7), 3538–3560 (2014)
12. Guo, C., Li, H.: Multilevel thresholding method for image segmentation based on an adaptive particle swarm optimization algorithm. In: AI 2007: Advances in Artificial Intelligence, pp. 654–658. Springer, Berlin (2007)
13. Zhang, Yudong, Lenan, Wu: Optimal multi-level thresholding based on maximum tsallis entropy via an artificial bee colony approach. Entropy **13**(4), 841–859 (2011)

14. Dirami, A., Hammouche, K., Diaf, M., Siarry, P.: Fast multilevel thresholding for image segmentation through a multiphase level set method. Sig. Process. **93**(1), 139–153 (2013)
15. Akay, B.: A study on particle swarm optimization and artificial bee colony algorithms for multilevel thresholding. Appl. Soft Comput. **13**(6), 3066–3091 (2013)
16. Yang, X.-S.: Cuckoo search and firefly algorithm: overview and analysis. Stud. Comput. Intell. **516**, 1–26 (2013)
17. Marciniak, A., Kowal, M., Filipczuk, P., Korbicz, J.: Swarm intelligence algorithms for multilevel image thresholding. In: Intelligent Systems in Technical and Medical Diagnostics, pp. 301–311. Springer, Berlin (2014)
18. Ayala, H.V.H., dos Santos, F.M., Mariani, V.C., dos Santos Coelho, L.: Image thresholding segmentation based on a novel beta differential evolution approach. Expert Syst. Appl. **42**(4), 2136–2142 (2015)
19. Yang, J., Yang, Y., Yu, W., Feng, J.: Multi-threshold Image Segmentation based on K-means and Firefly Algorithm, Atlantis Press, pp. 134–142 (2013)
20. Yu, C., Jin, B., Lu, Y., Chen, X., et al.: Multi-threshold image segmentation based on firefly algorithm. In: Proceedings of ninth international conference on IIH-MSP 2013, Beijing, pp. 415–419 (2013)
21. He, L.F., Tong, X., Huang, S.W.: Mineral belt image segmentation using firefly algorithm. Adv. Mater. Res. **989–994**, 4074–4077 (2014)
22. Vishwakarma, B., Yerpude, A.: A Meta-heuristic approach for image segmentation using firefly algorithm. Int. J. Comput. Trends Technol. (IJCTT) **11**(2), 69–73 (2014)
23. Rajinikantha, V., Couceirob, M.S.: RGB histogram based color image segmentation using firefly algorithm. Procedia Comput. Sci. **46**, 1449–1457 (2015)
24. Erdmann, H., Wachs-Lopes, G., Gallão, C., Ribeiro, M.P., Rodrigues, P.S.: A Study of a Firefly Meta-Heuristics for Multithreshold Image Segmentation, Developments in Medical Image Processing and Computational Vision. Lecture Notes in Computational Vision and Biomechanics, vol. 19, pp. 279–295. Springer, Berlin (2015)
25. Chen, K., Zhou, Y., Zhang, Z., Dai, M., Chao, Y., Shi, J.: Multilevel image segmentation based on an improved firefly algorithm. Math. Probl. Eng. **2016**, 1–12 (2016)
26. Djerou, L., Khelil, N., Dehimi, H. E., & Batouche, M.: Automatic multilevel thresholding using binary particle swarm optimization for image segmentation. In: International Conference of Soft Computing and Pattern Recognition, 2009. SOCPAR'09, pp. 66–71. IEEE (2009)
27. Ghamisi, P., Couceiro, M.S., Benediktsson, J.A., Ferreira, N.M.: An efficient method for segmentation of images based on fractional calculus and natural selection. Expert Syst. Appl. **39**(16), 12407–12417 (2012)
28. Nakib, A., Roman, S., Oulhadj, H., Siarry, P.: Fast brain MRI segmentation based on two-dimensional survival exponential entropy and particle swarm optimization. In: 29th Annual International Conference of the IEEE in Engineering in Medicine and Biology Society, 2007. EMBS 2007, pp. 5563–5566 (2007)
29. Wei, C., Kangling, F.: Multilevel thresholding algorithm based on particle swarm optimization for image segmentation. In: 27th Chinese Conference in Control, 2008. CCC 2008, pp. 348–351. IEEE (2008)
30. Yin, P.Y.: Multilevel minimum cross entropy threshold selection based on particle swarm optimization. Appl. Math. Comput. **184**(2), 503–513 (2007)
31. Zhiwei, Y., Zhengbing, H., Huamin, W., Hongwei, C.: Automatic threshold selection based on artificial bee colony algorithm. In: The 3rd International Workshop on Intelligent Systems and Applications (ISA), 2011, pp. 1–4 (2011)
32. Richard, M., Marie, B.-A., Guilhelm, S., Pascal, D.: Image Segmentation Using Socials Agents. 21 p. (2008)
33. Yang, X.-S.: Engineering Optimization: An Introduction with Metaheuristic Applications. Wiley, Hoboken (2010)
34. Gandomi, A.H., Yang, X.-S., Talatahari, S., Alavi, A.H.: Firefly algorithm with chaos. Commun. Nonlinear Sci. Numer. Simul. **18**, 89–98 (2013)

35. Fister, I., Fister Jr., I., Yang, X.S., Brest, J.: A comprehensive review of firefly algorithms. Swarm Evol. Comput. **13**, 34–46 (2013)
36. Su, H., Cai, Y.: Firefly algorithm optimized extreme learning machine for hyperspectral image classification. In: 2015 23rd International Conference on Geoinformatics, Wuhan, pp. 1–4 (2015)
37. Senthilnath, J., Omkar, S.N., Mani, V.: Clustering using firefly algorithm: Performance study. Swarm Evol. Comput. **1**(3), 164–171 (2011)
38. Kanimozhi, T., Latha, K.: An integrated approach to region based image retrieval using firefly algorithm and support vector machine. Neurocomputing **151**(3), 1099–1111 (2015)
39. Yang, X.-S. Firefly Algorithm, Lvy Flights and Global Optimization, Research and Development in Intelligent Systems XXVI, pp. 209–218 (2010)
40. Horng, M.H.: Vector quantization using the firefly algorithm for image compression. Expert Syst. Appl. **39**(1), 1078–1091 (2012)
41. Horng, M.H., Lee, M.C., Liou, R.J., Lee, Y.X.: Firefly meta-heuristic algorithm for training the radial basis function network for data classification and disease diagnosis, pp. 115–132. INTECH Open Access Publisher (2012)
42. Rajini, A., David, V.K.: A hybrid metaheuristic algorithm for classification using micro array data. Int. J. Sci. Eng. Res. **3**(2), 1–9 (2012)
43. Yang, Xin-She: Firefly algorithms for multimodal optimization. Stoch. Algorithms: Found. Appl. **5792**, 169–178 (2009)
44. Yang, X.S., He, X.: Firefly algorithm: recent advances and applications. Int. J. Swarm Intell. **1**(1), 36–50 (2013)
45. Zhou, Z., Zhu, S., Zhang, D.: A Novel K-harmonic means clustering based on enhanced firefly algorithm. In: Intelligence Science and Big Data Engineering. Big Data and Machine Learning Techniques, pp. 140–149, Springer International Publishing (2015)
46. Yang, X.-S.: Nature-inspired Metaheuristic Algorithms, Luniver Press, pp. 84–85 (2010)
47. Arora, S., Singh, S.: The firefly optimization algorithm: convergence analysis and parameter selection. Int. J. Comput. Appl. **69**(3), 48–52 (2013)
48. Cuevas, E., Cienfuegos, M., Zald'ivar, D., Prez-Cisneros, M.: A swarm optimization algorithm inspired in the behavior of the social-spider. Expert Syst. Appl. **40**(16), 6374–6384 (2013)
49. Boudia, M.A., Hamou, R.M., Amine, A., Rahmani, M.E., Rahmani, A.: A new multilayered approach for automatic text summaries mono-document based on social spiders. Computer Science and Its Applications, pp. 193–204. Springer International Publishing, Berlin (2015)
50. Benahmed, K., Merabti, M., Haffaf, H.: Inspired social spider behavior for secure wireless sensor networks. Int. J. Mob. Comput. Multimed. Commun. (IJMCMC) **4**(4), 1–10 (2012)
51. Martin, D., Fowlkes, C., Tal, D., Malik, J.: A database of human segmented natural images and its application to evaluating segmentation algorithms and measuring ecological statistics. In: Proceedings of 8th IEEE International Conference on Computer Vision, vol. 2, pp. 416–423. IEEE, Chicago (2001)

Grayscale Image Segmentation Using Multilevel Thresholding and Nature-Inspired Algorithms

Genyun Sun, Aizhu Zhang and Zhenjie Wang

Abstract Multilevel image thresholding plays a crucial role in analyzing and interpreting the digital images. Previous studies revealed that classical exhaustive search techniques are time consuming as the number of thresholds increased. To solve the problem, many nature-inspired algorithms (NAs) which can produce high-quality solutions in reasonable time have been utilized for multilevel thresholding. This chapter discusses three typical kinds of NAs and their hybridizations in solving multilevel image thresholding. Accordingly, a novel hybrid algorithm of gravitational search algorithm (GSA) with genetic algorithm (GA), named GSA-GA, is proposed to explore optimal threshold values efficiently. The chosen objective functions in this chapter are Kapur's entropy and Otsu criteria. This chapter conducted experiments on two well-known test images and two real satellite images using various numbers of thresholds to evaluate the performance of different NAs.

Keywords Image segmentation · Multilevel thresholding · Nature-inspired algorithms (NAs) · Gravitational search algorithm (GSA) · Genetic algorithm (GA) · Kapur's entropy · Otsu

1 Introduction

For the analysis, interpretation, and understanding of digital images, image segmentation is one of the most essential and fundamental technologies [1]. It is useful for partitioning a digital image into multiple regions/objects with distinct gray-levels [2]. Over the several decades, three main categories in image segmentation have been

G. Sun (✉) · A. Zhang · Z. Wang
School of Geosciences, China University of Petroleum,
Qingdao 266580, Shandong, China
e-mail: genyunsun@163.com

A. Zhang
e-mail: zhangaizhu789@163.com

Z. Wang
e-mail: sdwzj@upc.edu.cn

S. Bhattacharyya et al. (eds.), *Hybrid Soft Computing for Image Segmentation*, DOI 10.1007/978-3-319-47223-2_2

exploited in the literatures: (1) edge-based image segmentation, (2) region-based image segmentation, and (3) special theory-based image segmentation. Detailed introduction of their properties are presented in [3]. Particularly, thresholding technique of the region-based segmentation is regard as the most popular one out of others [4]. The main purpose of image thresholding is to determine one (bi-level thresholding) or m (multilevel thresholding) appropriate threshold values for an image to divide pixels of the image into different groups [5, 6]. In the recent years, increasing complexity of digital images, such as intensity inhomogeneity, makes multilevel thresholding approaches drawn much more attention. This is mainly due to its easy implementation and low storage memory characteristic [7].

Generally, the multilevel thresholding transforms the image segmentation to an optimization problem where the appropriate threshold values are found by maximizing or minimizing a criterion. For example, in the popular Otsu's method [8], thresholds are determined by maximizing the between-class variance. In the Kapur's entropy [9], the optimum thresholds are achieved by maximizing the entropy of different classes. A fuzzy entropy measure is applied for picking the optimum thresholds in [10] while Qiao et al. [11] formulated the thresholding criterion by exploring the knowledge in terms of intensity contrast. Researches have also developed some other preferable criteria, including fuzzy similarity measure [12], Bayesian error [13], cross entropy [14], Tsallis entropy [15], and so on.

Exhaustive search algorithms achieved optimization of the aforementioned criteria. However, the methods will become quiet time consuming if the number of desired thresholds is increased [16]. Moreover, the exhaustive search algorithms prone to premature convergence when dealing with complex real life images [17–19]. Nature-inspired algorithms (NAs) possess the ability for searching the optimal solutions on the basis of any objective function [7]. Furthermore, the population-based nature of NAs allows the generation of several candidate thresholds in a single run [20]. The population-based nature of NAs thus remarkably reduces the computational time of multilevel thresholding. Consequently, many NAs have been preferred in finding optimum thresholds for image thresholding.

The NAs can be categorized into three typical kinds: swarm intelligence algorithms (SIAs), evolutionary algorithms (EAs), and physics-based algorithms (PAs) in accordance with their original inspirations. Popular NAs chosen for multilevel thresholding include genetic algorithm (GA) [8, 21–24], differential evolution (DE) [18, 25–27], simulated annealing (SA) [16, 28], ant colony optimization (ACO) [29, 30], artificial bee colony optimization (ABC) [17, 31], differential search algorithm (DS) [16, 32], particle swarm optimization (PSO) [17, 33–35], and so on. Generally speaking, all these algorithms have achieved certain successes and have showed different advantages. For example, DE achieves higher quality of the optimal thresholds than GA, PSO, ACO, and SA whereas PSO converges the most quickly when comparing with ACO, GA, DE, and SA [21]. Besides, the DS consumes the shortest running time for multilevel color image thresholding when comparing with DE, GA, PSO, ABC, etc. [16].

Moreover, to further improve the optimization ability of these NAs, large amounts of variants, such as hybrid algorithms have been proposed. Many of the hybrid algorithms have been utilized in solving multilevel thresholding problems, including GAPSO (hybrid GA with PSO) [36, 37], ACO/PSO (hybrid ACO with PSO) [38], SA/PSO (hybrid SA with PSO) [39], BBO-DE (hybrid DE with biogeography-based optimization (BBO)) [40] etc. It is demonstrated that these hybrid algorithms can always perform more effective search ability than the original ones. But simple hybrid model easily causes high-computational complexity.

Therefore, in this chapter, we developed a novel hybrid algorithm of GSA with GA (GSA-GA) for multilevel thresholding. The proposed GSA-GA is expected to rapidly obtain the high-quality optimal thresholds. In GSA-GA, the discrete mutation operator [41] is introduced to promote the population diversity when premature convergence occurred. Moreover, for selecting the particles for mutation, the roulette selection [42] is also introduced. The introduction of these operators therefore could promote GSA-GA to perform faster and more accurate multilevel image thresholding. Both Kapur's entropy and Otsu are considered as evaluation criteria for GSA-GA.

Section 2 formulates the multilevel thresholding problem as an optimization problem first and then introduces two criteria briefly. In Sect. 3, typical NAs and hybrid NAs based multilevel image thresholding are overviewed. Section 4 describes the proposed GSA-GA algorithm and states its application in multilevel thresholding. In Sect. 5, experiments are conducted over two standard test images and two real satellite images to evaluate the effectiveness of the GSA-GA. Finally, the chapter is concluded in Sect. 6.

2 Formulation of Multilevel Thresholding Problem

Optimal thresholding methods search for the optimum thresholds by minimizing or maximizing an objective function. Parameters in the utilized objective function are made up by the selected thresholds in each iteration. Specially, the purpose of multilevel thresholding is to classify the *Num* pixels of an image into K classes using a set of thresholds $Th = (td_1, td_2, \ldots, td_m)$ where $m = K - 1$. Obviously, different sets of thresholds will produce different image segmentation results. Researchers have formulated some criteria for constructing efficient objective functions [8, 9, 12, 13, 43, 44]. Two most preferred criteria are Kapur's entropy and Otsu methods as they are relatively easy to use. They are also chosen as criteria in this chapter.

2.1 Kapur's Entropy Criterion

The Kapur's entropy criterion, proposed by Kapur et al. [9], is originally developed for bi-level image thresholding. In this method, proper segmentation of an image is achieved by maximizing the Kapur's entropy. For a bi-level thresholding, the Kapur's entropy can be described as follows:

$$HE_0 = -\sum_{i=0}^{td_0-1} \frac{prob_i}{\omega_0} ln \frac{prob_i}{\omega_0}, \omega_0 = \sum_{i=0}^{td_0-1} prob_i$$
$$HE_1 = -\sum_{i=td_0}^{L-1} \frac{prob_i}{\omega_1} ln \frac{prob_i}{\omega_1}, \omega_1 = \sum_{i=td_0}^{L-1} prob_i \quad (1)$$

where HE_i is the entropy of class i ($i \in 0, 1$) and ω_i is the probability of class i, td_0 is the optimum threshold of the bi-level thresholding. Assume that there are L gray-levels in a given image, the probability of the gray-level i can be defined as follows:

$$prob_i = \frac{hist_i}{\sum_{i=0}^{L-1} hist_i}, \quad (2)$$

where $hist_i$ denotes the number of pixels with gray-level i.

As shown in (1) and (2), td_0 is the selected threshold. To obtain the optimum threshold td_0, the fitness function (3) needs to be maximized as follows:

$$f(td_0) = HE_0 + HE_1, \quad (3)$$

To solve complex multilevel thresholding problems, the Kapur's entropy has been extended to determine m thresholds for a given image, i.e., $[td_1, td_2, \ldots, td_m]$. The objective of the Kapur's entropy-based multilevel thresholding is to maximize the fitness function (4) as follows:

$$f([td_1, td_2, \ldots, td_m]) = HE_0 + HE_1 + HE_2 + \cdots + HE_m, \quad (4)$$

where

$$H_E0 = -\sum_{i=0}^{td_1-1} \frac{prob_i}{\omega_0} ln \frac{prob_i}{\omega_0}, \omega_0 = \sum_{i=0}^{td_1-1} prob_i$$
$$HE_1 = -\sum_{i=td_1}^{td_2-1} \frac{prob_i}{\omega_1} ln \frac{prob_i}{\omega_1}, \omega_1 = \sum_{i=td_1}^{td_2-1} prob_i$$
$$HE_2 = -\sum_{i=td_2}^{td_3-1} \frac{prob_i}{\omega_2} ln \frac{prob_i}{\omega_2}, \omega_2 = \sum_{i=td_2}^{td_3-1} prob_i \quad (5)$$
$$HE_m = -\sum_{i=td_m}^{L-1} \frac{prob_i}{\omega_m} ln \frac{prob_i}{\omega_m}, \omega_m = \sum_{i=td_m}^{L-1} prob_i.$$

2.2 Otsu Criterion

The Otsu criterion [8] is another histogram-based multilevel image thresholding algorithm. The method divides the given image into m classes so that the total variance of different classes is maximized. The object function of Otsu is defined as the sum of sigma functions of each class as follows:

$$f(td_0) = \sigma_0 + \sigma_1, \quad (6)$$

The sigma functions are as follows:

$$\begin{aligned}\sigma_0 &= \omega_0(\mu_0 - \mu_T)^2,\ \mu_0 = \frac{\sum_{i=0}^{td_0-1} i \cdot prob_i}{\omega_0},\\ \sigma_1 &= \omega_1(\mu_1 - \mu_T)^2,\ \mu_1 = \frac{\sum_{i=td_0}^{L-1} i \cdot prob_i}{\omega_1},\end{aligned} \tag{7}$$

where μ_i is the mean gray-level of class i ($i \in 0, 1$) and μ_T is the mean intensity of the original image.

Similar to the entropy method, the Otsu criterion has already been extent to multi-level thresholding problems. The objective in the Otsu-based multilevel thresholding is to maximize the following function:

$$f([td_1, td_2, \ldots, td_m]) = \sigma_0 + \sigma_1 + \sigma_2 + \cdots + \sigma_m, \tag{8}$$

where

$$\begin{aligned}\sigma_0 &= \omega_0(\mu_0 - \mu_T)^2,\ \mu_0 = \frac{\sum_{i=0}^{td_1-1} i \cdot prob_i}{\omega_0},\\ \sigma_1 &= \omega_1(\mu_1 - \mu_T)^2,\ \mu_1 = \frac{\sum_{i=td_1}^{td_2-1} i \cdot prob_i}{\omega_1},\\ \sigma_2 &= \omega_2(\mu_2 - \mu_T)^2,\ \mu_2 = \frac{\sum_{i=td_2}^{td_3-1} i \cdot prob_i}{\omega_2},\\ \sigma_m &= \omega_m(\mu_m - \mu_T)^2,\ \mu_m = \frac{\sum_{i=td_m}^{L-1} i \cdot prob_i}{\omega_m}.\end{aligned} \tag{9}$$

3 Nature-Inspired Algorithms Based Multilevel Thresholding

In this chapter, we categorized the NAs utilized for multilevel thresholding into three kinds: SIAs, EAs, and PAs based on their original inspirations. Since different kinds of NAs possess unique information sharing mechanisms, certain successes have been produced by each kind of NA. Sections 3.1–3.3 give a detailed review of the three aforementioned kinds of NAs based multilevel thresholding, respectively. In Sect. 3.4, we briefly reviewed the hybrid NAs based multilevel thresholding.

3.1 Swarm Intelligence-Based Optimization Algorithms Based Multilevel Thresholding

SIAs are mainly proposed by mimicking the foraging or hunting behaviors of different species like ants, bees, birds, cuckoos, bats, wolf, and so on [45]. Well-known SIAs include ACO [29], ABC [31], PSO [46], cuckoo search algorithm (CS) [47], Bat Algorithm (BA) [48], and Gray Wolf Optimizer (GWO) [49], etc. In these algorithms, one special particle/individual with the best performance is usually chosen as exemplar/leader for others. The guidance of the exemplar ensures the population evolves directionally in each iteration. The directional search property equips the SIAs with the fast convergence performance [50].

Application of swarm intelligence thereby makes a success in decreasing the computational complexity problem of multilevel thresholding. For example, in [30], Tao et al. utilized ACO to optimize the fuzzy entropy criterion while Zhang et al. search for the optimum multilevel thresholding on the basis of the maximum Tsallis entropy [51]. Similarly, ABC, BA, and GWO also obtained certain successes in multilevel thresholding [6, 17, 51–54]. Besides, as one of the most popular swarm intelligence-based optimization algorithm, PSO and its various variants have contributed to multilevel thresholding a lot [2, 17, 33–35, 55]. Moreover, Hammouche et al. have found that PSO converges more quickly than ACO, ABC, and SA because of all particles can directly learn from the global best particle [21].

Although SIAs contribute a lot to image thresholding, their fast convergence property caused rapid decrease of population diversity. That is, exploration of SIAs is insufficient. This defective exploration of SIAs may result in temperature convergence of multilevel thresholding.

3.2 Evolutionary Algorithms Based Multilevel Thresholding

EAs are motivated by the principle of evolution through selection and mutation. In the past decade, many EAs, such as GA [8], DE [27], and DS [32], have been widely applied. In conventional EAs, evolutionary processes are conducted through reproduction, crossover, mutation, and selection operators. Particularly, only two individuals are randomly selected to exchange information in a single crossover operation [50]. Similarly, in a traditional mutation operator, only two genes in the individuals are randomly varied [50]. Obviously, evolutionary performance of EAs is not as directional as SIAs do. It is omnidirectional to some extent. Thereby EAs have better exploration than SIAs.

Naturally, a number of researchers have paid close attention to obtain optimal thresholds by EAs, especially the GA [21–24] and DE [18, 25, 26]. The favorable exploration ability of EAs makes them achieve remarkable progress in the research area of multilevel thresholding. However, the omnidirectional characteristic of EAs leads to their slow convergence. Moreover, the success of some EAs in solving specific problems highly rely on the user set trial vector generation strategies and control parameters [16].

3.3 Physics-Based Algorithms Based Multilevel Thresholding

The PAs, as the name suggested, are constructed by simulating the phenomenon of physics. For instance, SA [28] imitates the annealing process of melted iron;

gravitational search algorithm (GSA) [56] inspired by the gravitational kinematics; electromagnetism-like algorithm (EM) [57] and charged search system (CSS) algorithm [58] are proposed on the basis of electrodynamics. More detailed survey of the PAs can be found in [59]. Individuals in PAs explore the search space follows several exemplars. Accordingly, interactions among individuals result in iterative improvement of solutions quality over time. Search diversity of PAs is better than that of SIAs and thus PAs usually perform better exploration. On the other hand, their more directional search property keep higher convergence speed of them. Nevertheless, none of the PAs can solve all the optimization problems in accordance with the "No Free Lunch Theorems" [60].

Comparing with the SIAs and EAs, few researchers have focused on their application on multilevel thresholding. GSA, which possesses simple concept, high-computational efficiency, and few parameters, has proven its promising efficiency in solving complex problems [56, 61]. Furthermore, numerical experiments has confirmed the superiority of GSA with respect to search ability and speed over many other NAs, such as PSO, GA, ACO etc. [56, 62–64]. These advantages make GSA a potential choice for solving multilevel thresholding.

3.4 *Hybrid NAs Based Multilevel Thresholding*

As discussed in Sects. 3.1–3.3, each kind of NAs has separate and unique advantages and disadvantages. To incorporate different advantages of various NAs and improve the optimization ability of them, large amounts of hybrid algorithms have been proposed over the past decades. For instance, in GAPSO [9], the mutation operator of GA is introduced to PSO to tackle the premature convergence problem while the memory property of PSO is preserved. BBO-DE [40] incorporate both the crossover operator and the selection operator of DE algorithm to BBO to promote exploration ability of BBO while the exploitation ability of BBO is kept. Experiments reported in the literatures demonstrated that these hybrid algorithms can always perform more effective search ability than the original ones.

Many of the hybrid algorithms have already been utilized to solving multilevel thresholding problems, including GAPSO [9], ACO/PSO [38], SA/PSO [39], and BBO-DE [40], etc. The hybrid algorithms usually possess advantages of each composed algorithm. Due to simple hybrid model easily causes high-computational complexity, new NAs, or efficient hybrid NAs are desirable.

4 GSA-GA Based Multilevel Image Thresholding

This chapter introduces a novel hybrid GSA-GA for multilevel image thresholding [65]. Although many researches have paid close attention to the improvement of GSA and presented some GSA variants in the past few years [1, 61, 66, 67], very few of the algorithms have focused on the application on multilevel thresholding. When applying GSA into multilevel image thresholding, especially when the desired number of thresholds is large, the premature convergence and high time-consuming problems become more serious.

The lack of diversity is one important reason for the premature convergence [68]. In GSA-GA, the discrete mutation operator of GA was introduced to promote the population diversity when premature convergence occurred. Moreover, for decreasing computational time, adaptive method is utilized to judge whether the hybridization of GSA and GA are performed. The details of GSA-GA and its application for multilevel image thresholding are given in Sects. 4.1 and 4.2, respectively.

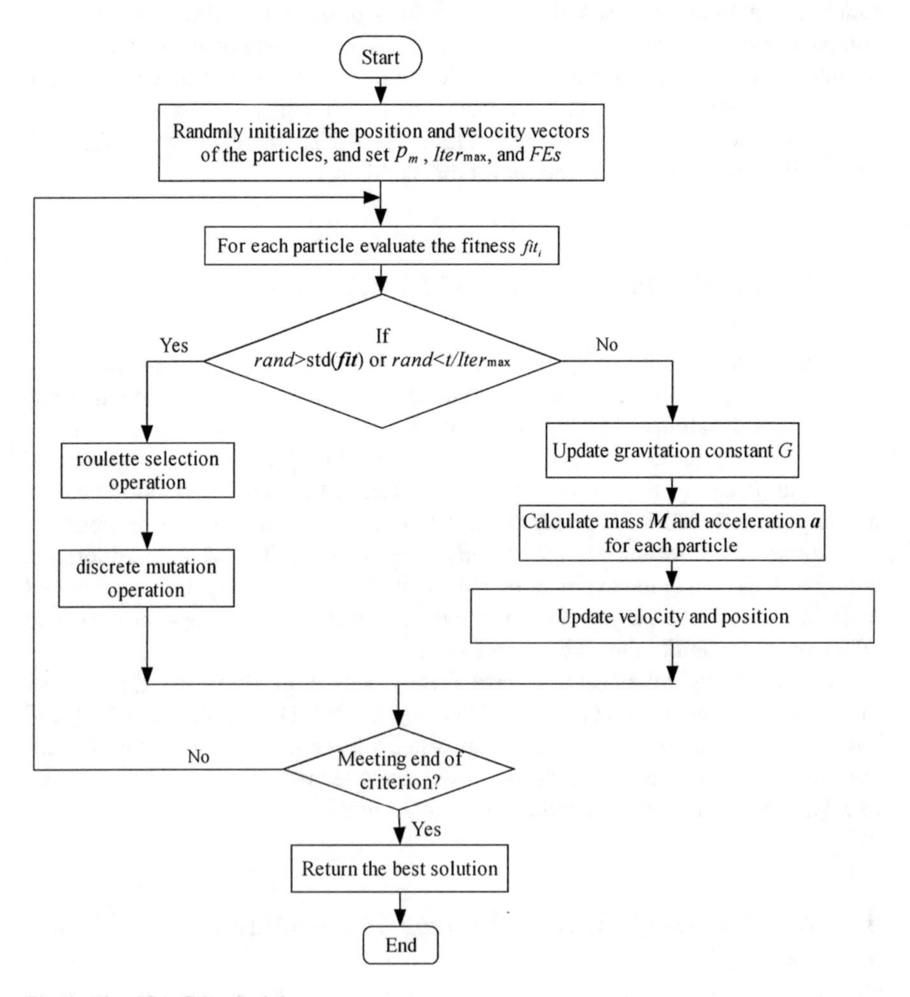

Fig. 1 The GSA-GA principle

4.1 GSA-GA

In GSA-GA, to adaptively identify whether the hybridization of GSA and GA is necessary, the standard deviations of objective functions and the ratio of the current iteration t to the maximum iterations *maxiterations* are calculated before conducting the selection and mutation operators of GA. More specifically, when particles are trapped (judging by $rand < std(fit)$ or $rand < t/maxiterations$), the roulette selection and discrete mutation operators are carried out to update position of each particle and thus to diversify the population. Due to the population diversity is crucial for a

1. Input test image ***IMG***.
2. Set P_m, $Iter_{\max}$, N, G_0, β, m, and FEs.
3. Set the thresholding criterion F_{index}='entropy' or 'between-class variance'.
4. Initialize positions $\boldsymbol{x}$ and velocities $\boldsymbol{v}$ of particles in the population.
5. **For** $t=1:Iter_{\max}$
6.   **If** F_{index}='entropy'
7.   Evalutate the fitness functions values ***fit*** by Eq. (4);
8.   **Else**
9.   Evalutate the fitness functions values ***fit*** by Eq. (8);
10.   **End**
11.   **End**
12.   **If** $(rand>std\,(\boldsymbol{fit})\,\|\,rand<(t/Iter_{\max})$
13.   Calculate the selection probability $p_s^t(i)$ of each particle i;
14.   Calculate cumulative probability $p_{cs}^t(i)$ of each particle i;
15.   **For** $i=1:N$
16.   Generate *rand* randomly;
17.   **If** $\mathrm{p}_{cs}^t(j) < rand < \mathrm{p}_{cs}^t(j+1) \quad j \in [1,2,\ldots,N]$
18. $\boldsymbol{Newx}_i^t = \boldsymbol{x}_j^t$;
19.   **End**
20.   **End**
21.   Randomly generate basic vector ***BaseV*** for discrete mutation in the gray range of ***IMG*** ;
22.   Create mutation mask matrix by $\boldsymbol{BaseM} = \boldsymbol{BaseV}(ones(N,1),:)$;
23. $\boldsymbol{x}^{t+1} = rem(\boldsymbol{Newx}^t + (rand < p_m).*ceil(rand.*(BaseM-1)), BaseM)$;
24.   **Else**
25.   Update the gravitation constant by $G(t) = G_0 \times \exp(-\beta \times \frac{t}{Iter_{\max}})$;
26.   **For** $i=1:N$
27.   Calculate the mass by $Mass_i^t = \frac{nmfit_i^t}{\sum_{j=1}^{N} nmfit_j^t}$ and $nmfit_i^t = \frac{fit_i^t - worst^t}{best^t - worst^t}$;
28.   Calculate the force acting on particle i from other parrticles in demension d
29.   by $F_{ij}^d(t) = G(t)\frac{Mass_i^t \times Mass_j^t}{R_{ij}^t + \varepsilon}(x_{jd}^t - x_{id}^t)$ and $F_i^d(t) = \sum_{j=1, j\neq i}^{N} rand \cdot F_{ij}^d(t)$;
30.   Calculate the acceleration by $a_{id}^t = \frac{F_i^d(t)}{Mass_i^t}$;
31.   Update the position and velocity of particle i by:
32. $v_{id}^{t+1} = rand \times v_{id}^t + a_{id}^t$;
33. $x_{id}^{t+1} = x_{id}^t + v_{id}^{t+1}$;
34.   **End**
35.   **End**
36. **End**

Fig. 2 Pseudocode of the GSA-GA algorithm for multilevel image thresholding

health search, combination of GA, and GSA can help the algorithm escape from local optima. Besides, because of satisfaction of the condition '$rand < t/maxiterations$' becomes easier by the lapse of time, GSA-GA can utilize selection and mutation operators to accelerate convergence in the last iterations. The principle of GSA-GA is shown in Fig. 1. The detailed introduction of the method can be found in [65].

4.2 *Implementation of GSA-GA for Multilevel Thresholding*

The application of GSA-GA approach to the multilevel image thresholding problem depends on the criterion used for optimization. In this chapter, the Kapur's entropy and Otsu criteria were implemented. To start the GSA-GA for multilevel thresholding, initial population $\boldsymbol{X} = [\boldsymbol{X}_1, \boldsymbol{X}_2, \ldots, \boldsymbol{X}_i, \ldots, \boldsymbol{X}_N]$ where $\boldsymbol{X}_i = [x_{i1}, x_{i2}, \ldots, x_{iD}]$ should be randomly generated first. Each particle $\boldsymbol{X}_i$ is comprised of a set of gray values, which stands for a candidate solution of the required threshold values. The size of the population is N (can be set by users), and dimension of each particle is D, which is also the number of desired thresholds: $D = m$. In the iteration process, the fitness value of each particle is calculated from the Kapur's entropy or Otsu criterion using Eqs. (4)–(5) and (8)–(9), respectively. The pseudocode of the GSA-GA algorithm for multilevel image thresholding is shown in Fig. 2.

5 Experimental Results and Discussion

In this section, the experimental results are presented. The performance of GSA-GA has been evaluated and compared with the other 6 NAs: the standard GSA [56], GGSA (Adaptive gbest-guided GSA) [67], PSOGSA (hybrid PSO and GSA) [66], DS (Differential Search algorithm) [32], BBO-DE [40], and GAPSO [36] by conducting experiments on two benchmark images that has been tested in many related literatures. Furthermore, to evaluate the ability of GSA-GA based thresholding on more complex and difficulty images, we conducted experiments on two real satellite images adopted from NASA (http://earthobservatory.nasa.gov/Images/?eocn=topnav046eoci=images). The obtained experimental results of the 7 algorithms are also compared. For the satellite images, the low spectral and spatial resolution usually makes objects hard to be segmented, while the high resolution always leads to highly computational complexity during the image segmentation. The four experimental images and corresponding histogram of each image are given in Fig. 3 as shown in Sect. 5.1.

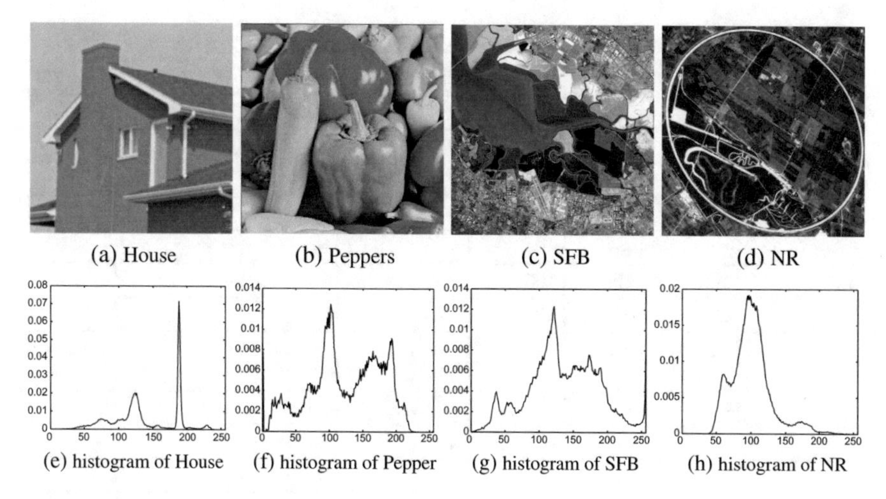

Fig. 3 Test images used in the experiments

5.1 *Test Images*

The two standard benchmark test images are two widely utilized images: House and pepper, as shown in Fig. 3a and b, respectively. Size of every tested benchmark images is 256×256. The two tested satellite images are named SFB and NR respectively as shown in Fig. 3c and d. The image SFB is acquired by the Thematic Mapper on Landsat 5 on September 9, 2002 while image NR is an astronaut photograph taken by Kodak 760 C digital camera on November 8, 2006. Size of the image SFB and NR are 512×512 and 539×539, respectively. Due to the two satellite images are multi-spectral images, we transformed the two images using a famous orthogonal transformation method, i.e., principal component analysis (PCA). As the most informative component, the first principal component is selected for multilevel image thresholding in this chapter.

5.2 *Experimental Settings*

The relative parameter settings of algorithms utilized in this chapter are shown in Table 1. As illustrated in Table 1, to perform a fair experiment, in all of the seven algorithms, the population size (N) and maximum number of fitness evaluations (FEs) were set to 30 and 3000, respectively. For GSA-GA, GSA, PSOGSA, and GGSA, the initial gravitational constant G_0 and decrease coefficient β were set to 100 and 20 as the default values in original GSA [56]. For PSOGSA, the two acceleration coefficients (c_1 and c_2) were set to 0.5 and 1.5, respectively as suggested in [66]. For GGSA, the two acceleration coefficients (c_1 and c_2) were set to $-2t^3/Iter_{\max}^3 + 2$ and

Table 1 Parameter settings of the compared algorithms in this chapter

Parameters	GSA-GA	GSA	PSOGSA	GGSA	DS	BBO-DE	GAPSO
FEs	300	300	300	300	300	300	300
N	30	30	30	30	30	30	30
p_m	0.1	–	–	–	–	DE/$rand$/1	0.1
p_c/CR	–	–	–	–	–	0.9	1
F	–	–	–	–	–	0.5	–
G_0	100	100	100	100	–	–	–
β	20	20	20	20	–	–	–
c_1	–	–	0.5	$-2t^3/Iter_{\max}^3+2$	–	–	2
c_2	–	–	1.5	$2t^3/Iter_{\max}^3$	–	–	2
p_1	–	–	–	–	0.3*$rand$	–	–
p_2	–	–	–	–	0.3*$rand$	–	–
n_{elit}	–	–	–	–	–	2	–

$2t^3/Iter_{\max}^3$, respectively as recommended in [67]. The other parameters in DS, BBO-DE, and GAPSO were all adopted as the suggested values in the original papers. To be specific, in DS, the two control parameters (i.e., p_1 and p_2) were both set to 0.3*$rand$; in BBO-DE, the mutation scheme was the DE/$rand$/1, the crossover probability (CR), the scaling factor (F), and the elitism parameter (n_{elit}) were set to 0.9, 0.5, and 2 [40]; in GAPSO, the mutation probability (p_m) was set to 0.1, the crossover probability (denoted by p_c) was set to 1, and the two acceleration coefficients were both set to 2 [36]. For GSA-GA, the mutation probability p_m was set to 0.1 as it is recommended in many GA based algorithms [23, 37, 40].

5.3 Performance Metrics

In this chapter, the uniformity measure [69] is adopted to qualitatively judge the segmentation results based on different thresholds. The uniformity measure (u) is calculated as follows:

$$u = 1 - 2 \times m \times \frac{\sum_{j=1}^{m+1} \sum_{i \in Re_j} (f_i - g_j)^2}{Num \times (f_{max} - f_{min})^2}, \tag{10}$$

where m is the number of desired thresholds, Re_j is the j-th segmented region, Num is the total number of pixels in the given image $\boldsymbol{IMG}$, f_i is the gray-level of pixel i, g_j is the mean gray-level of pixels in j-th region, f_{max} and f_{min} are the maximum and minimum gray-level of pixels in the given image, respectively. Typically,

$u \in [0, 1]$ and a higher value of uniformity indicates that there is better uniformity in the thresholding image.

5.4 Experiment 1: Maximizing Kapur's Entropy

The Kapur's entropy criterion was first maximized to construct an optimally segmented image for all the seven NAs. Table 2 presented the best uniformity (u) and the corresponding threshold values (Th) on all the test images produced by the algorithms after 30 independent runs. Moreover, to test the stability of thresholding methods, average uniformity of 30 independent runs are reported in Table 3. Besides, to test the computation complexity, the mean of CPU times are reported in Table 4.

As shown in Table 2, for image House, GSA-GA outperformed GSA, PSOGSA, GGSA, DS, and GAPSO on all the different number of thresholds. Although BBO-DE gained the best uniformity when $m = 4$, GSA-GA has a better mean uniformity as illustrated in Table 3. For image Pepper, SFB and NR, GSA-GA performed as well as or was better than the other GSA-based segmentation methods on the best uniformity reported in Table 2. Meanwhile, mean uniformity of the three images shown in Table 3 also confirmed the superiority of GSA-GA in most cases. This may be due to that the introduced mutation operator prevents GSA from getting stuck to local optima. Furthermore, when comparing with DS, BBO-DE, and GAPSO, GSA-GA displayed the best on both the best and mean uniformity as presented in Tables 2 and 3. That is to say, GSA variant-based multilevel image thresholding methods are more efficient than other NAs based techniques. That is because particles in GSA variants can learn from other particles fuller than other NAs. Besides, it is worth noting that the mean CPU times of GSA-GA are visible smaller than the comparison algorithms as reported in Table 4.

For a visual interpretation of the segmentation results, the segmented images of the two satellite images obtained based on the optimum thresholds shown in Table 2 were presented in Figs. 4 and 5 with $m = 2 - 5$. For the two images, it is hard to say which number of thresholds is more suitable for due to the complexity of these images. Their inherent uncertainty and ambiguity make it a challenging task for these tested algorithms to determine the proper number of classes in a given image.

5.5 Experiment 2: Maximizing Otsu

This section devotes to maximize the Otsu criterion to construct an optimally segmented image for all of the seven algorithms. Experiments were performed on all the test images shown in Fig. 3. The best uniformity (u) and the corresponding threshold values (Th) produced by the algorithms were given in Table 5 after 30 independent runs. Similar to Sect. 5.4, to test the stability of the algorithms, Table 6 presented the mean of uniformity. In addition, Table 7 provided the mean of CPU times.

Table 2 Comparison of algorithms taking the Kapur's entropy as evaluation criterion in terms of best uniformity and corresponding thresholds

Image	m	GSA		PSOGSA		GGSA	
		Th	*u*	*Th*	*u*	*Th*	*u*
House	2	103, 197	0.9302	98, 194	0.9265	98, 193	0.9268
	3	64, 116, 206	0.9146	65, 112, 194	0.9109	65, 112, 193	0.9113
	4	59, 97, 132, 193	0.9728	59, 88, 115, 194	0.8932	61, 101, 137, 183	0.9844
	5	66, 101, 134, 164, 201	0.9886	61, 101, 137, 183, 195	0.9868	61, 101, 137, 182, 194	0.9869
Pepper	2	71, 138	**0.9783**	71, 138	**0.9783**	71, 138	**0.9783**
	3	64, 121, 169	**0.9817**	65, 122, 169	**0.9817**	64, 121, 168	**0.9817**
	4	49, 89, 129, 171	**0.9843**	71, 138, 152, 253	0.9656	49, 88, 129, 172	**0.9843**
	5	52, 79, 116, 151, 183	0.9856	65, 122, 152, 169, 252	0.9742	62, 111, 141, 174, 219	0.9777
SFB	2	92, 156	**0.9719**	92, 156	**0.9719**	92, 156	**0.9719**
	3	83, 138, 187	**0.9760**	83, 139, 191	**0.9760**	82, 140, 189	0.9759
	4	76, 125, 163, 202	0.9785	92, 102, 156, 195	0.9687	72, 114, 152, 200	**0.9788**
	5	60, 108, 136, 171, 208	0.9817	38, 83, 102, 139, 191	0.9723	89, 94, 138, 162, 197	0.9714
NR	2	89, 137	**0.9865**	89, 137	**0.9865**	89, 138	0.9864
	3	78, 105, 145	0.9885	79, 106, 146	**0.9886**	79, 106, 146	**0.9886**
	4	73, 97, 118, 155	**0.9896**	89, 131, 137, 254	0.9752	89, 126, 137, 253	0.9772
	5	67, 87, 104, 124, 158	**0.9902**	79, 106, 131, 146, 253	0.9841	74, 103, 131, 140, 253	0.9824
DS		BBO-DE		GAPSO		GSA-GA	
Th	*u*	*Th*	*u*	*Th*	*u*	*Th*	*u*
112, 196	0.9351	68, 99	0.9170	111, 194	0.9354	67, 123	**0.9446**
63, 110, 200	0.9077	104, 184, 196	0.9785	64, 133, 195	0.9630	109, 152, 196	**0.9830**
63, 104, 137, 185	0.9828	51, 96, 163, 205	**0.9834**	52, 82, 145, 197	0.9786	51, 106, 161, 198	**0.9834**
36, 78, 107, 149, 201	0.9844	46, 98, 127, 179, 199	0.9839	66, 90, 126, 180, 192	0.9839	70, 103, 139, 177, 202	**0.9887**

(continued)

Table 2 (continued)

DS		BBO-DE		GAPSO		GSA-GA	
Th	*u*	*Th*	*u*	*Th*	*u*	*Th*	*u*
75, 142	0.9781	76, 146	0.9776	68, 136	0.9782	71, 138	**0.9783**
60, 133, 176	0.9800	57, 122, 173	0.9810	66, 114, 172	0.9803	64, 123, 170	**0.9817**
49, 111, 134, 182	0.9789	45, 99, 121, 152	0.9771	53, 115, 153, 185	0.9806	38, 83, 128, 171	0.9835
57, 98, 126, 177, 222	0.9781	44, 83, 131, 161, 172	0.9827	55, 115, 148, 171, 249	0.9758	42, 81, 117, 153, 183	**0.9860**
88, 153	0.9718	84, 155	0.9715	93, 170	0.9697	93, 157	**0.9719**
76, 126, 185	0.9740	79, 151, 187	0.9724	100, 144, 190	0.9731	83, 138, 189	**0.9760**
93, 141, 181, 206	0.9742	64, 110, 160, 201	0.9770	56, 98, 145, 226	0.9705	73, 115, 154, 202	**0.9788**
60, 98, 120, 164, 194	0.9767	45, 97, 132, 175, 212	0.9786	85, 108, 133, 174, 237	0.9757	68, 107, 140, 176, 218	**0.9825**
88, 139	0.9864	82, 135	0.9857	85, 138	0.9861	89, 137	**0.9865**
81, 108, 136	0.9875	75, 97, 142	0.9873	93, 118, 166	0.9846	79, 106, 146	**0.9886**
85, 118, 154, 191	0.9847	67, 99, 117, 160	0.9877	75, 97, 145, 204	0.9850	80, 108, 127, 157	0.9882
77, 102, 121, 156, 191	0.9890	80, 84, 103, 143, 183	0.9854	76, 98, 141, 150, 215	0.9832	77, 89, 109, 123, 159	**0.9891**

Table 3 Comparison of algorithms taking the Kapur's entropy as evaluation criterion in terms of the mean of uniformity

Image	*m*	GSA	PSOGSA	GGSA	DS	BBO-DE	GAPSO	GSA-GA
House	2	0.9185	0.9265	0.9267	0.9109	0.9103	0.9162	**0.9300**
	3	0.9084	0.9109	0.9108	0.8882	0.9366	0.9084	**0.9370**
	4	0.9064	0.8869	0.9129	0.9306	0.9218	0.9288	**0.9311**
	5	0.9870	0.9366	0.9846	0.9322	0.9191	0.9805	**0.9877**
Pepper	2	**0.9783**	**0.9783**	**0.9783**	0.9764	0.9757	0.9778	**0.9783**
	3	0.9816	0.9700	**0.9817**	0.9755	0.9779	0.9782	0.9816
	4	0.9793	0.9643	0.9707	0.9748	0.9740	0.9769	**0.9809**
	5	0.9769	0.9669	0.9694	0.9768	0.9786	0.9718	**0.9803**
SFB	2	**0.9719**	**0.9719**	0.9718	0.9692	0.9700	0.9682	**0.9719**
	3	0.9757	0.9652	0.9751	0.9688	0.9709	0.9703	**0.9760**
	4	0.9773	0.9556	0.9704	0.9716	0.9737	0.9689	**0.9785**
	5	0.9740	0.9648	0.9658	0.9717	0.9732	0.9731	**0.9766**
NR	2	**0.9865**	**0.9865**	0.9864	0.9859	0.9845	0.9845	**0.9865**
	3	0.9885	0.9749	0.9877	0.9838	0.9818	0.9835	**0.9886**
	4	**0.9891**	0.9752	0.9732	0.9824	0.9827	0.9823	0.9869
	5	0.9860	0.9833	0.9771	0.9857	0.9817	0.9808	**0.9862**

Table 4 Comparison of algorithms taking the Kapur's entropy as evaluation criterion in terms of the mean of CPU times (in seconds)

Image	*m*	GSA	PSOGSA	GGSA	DS	BBO-DE	GAPSO	GSA-GA
House	2	1.0488	1.0765	1.0323	1.1322	2.3825	1.0707	**0.9144**
	3	1.1113	1.0908	1.1569	1.1852	2.5713	1.1439	**0.9246**
	4	1.0932	1.1090	1.1449	2.0981	2.8502	1.1865	**0.9645**
	5	1.2845	1.1854	1.1890	1.9882	2.9229	1.2625	**1.0268**
Pepper	2	0.7239	0.7881	0.7153	0.6398	1.5238	1.7950	**0.4939**
	3	0.7545	0.7495	0.7318	0.6501	1.5997	0.7175	**0.5024**
	4	0.7661	0.7739	0.7559	0.6829	1.6234	0.6883	**0.5288**
	5	0.7680	0.7942	0.7950	0.6981	1.7266	0.7097	**0.5375**
SFB	2	1.4644	1.4473	1.4620	1.4138	2.9996	1.4216	**1.0782**
	3	1.4855	1.4803	1.4838	1.4334	3.0790	1.5437	**1.0633**
	4	1.4879	1.4911	1.5245	1.4626	3.7202	1.4634	**1.0861**
	5	1.5132	1.5120	1.5095	1.4603	3.2803	1.6632	**1.0911**
NR	2	1.5570	1.5690	1.5323	1.5154	3.2308	1.5084	**1.1575**
	3	1.5604	1.5729	1.5594	1.5209	3.2722	1.5922	**1.2417**
	4	1.5937	1.5997	1.5696	1.5484	3.3852	1.5545	**1.1933**
	5	1.5952	1.5852	1.6206	1.5784	3.4412	1.5969	**1.2306**

Fig. 4 Segmented images for the SFB image by Kapur's entropy with different optimization techniques

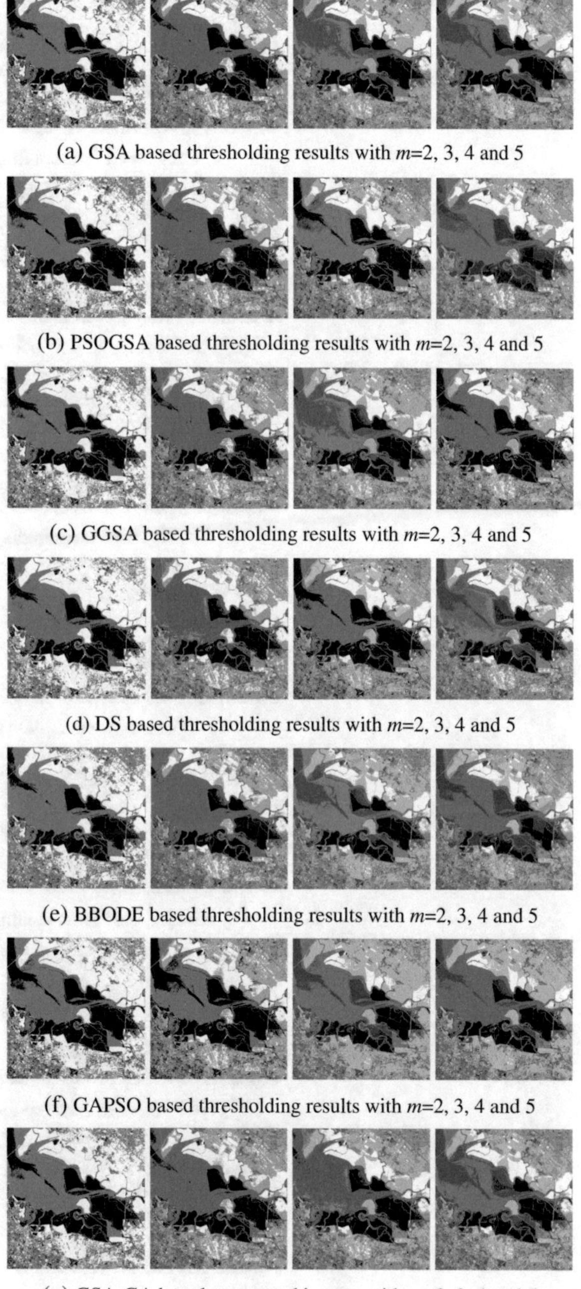

(a) GSA based thresholding results with m=2, 3, 4 and 5

(b) PSOGSA based thresholding results with m=2, 3, 4 and 5

(c) GGSA based thresholding results with m=2, 3, 4 and 5

(d) DS based thresholding results with m=2, 3, 4 and 5

(e) BBODE based thresholding results with m=2, 3, 4 and 5

(f) GAPSO based thresholding results with m=2, 3, 4 and 5

(g) GSA-GA based segmented images with m=2, 3, 4 and 5

Fig. 5 Segmented images for the NR image by Kapur's entropy with different optimization techniques

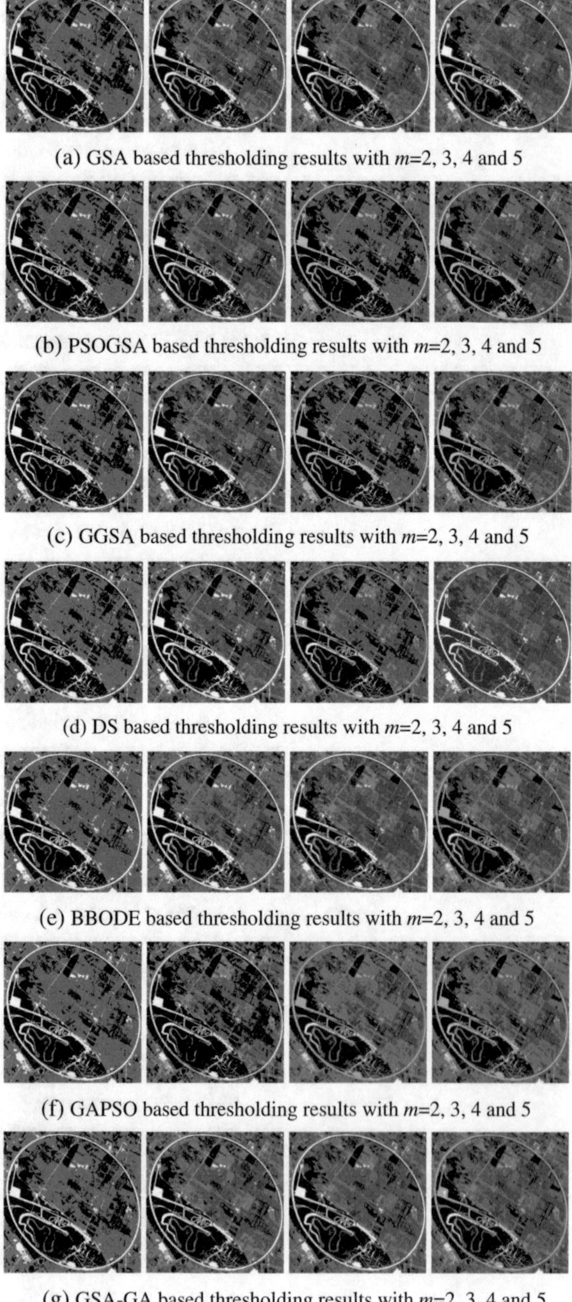

(a) GSA based thresholding results with m=2, 3, 4 and 5

(b) PSOGSA based thresholding results with m=2, 3, 4 and 5

(c) GGSA based thresholding results with m=2, 3, 4 and 5

(d) DS based thresholding results with m=2, 3, 4 and 5

(e) BBODE based thresholding results with m=2, 3, 4 and 5

(f) GAPSO based thresholding results with m=2, 3, 4 and 5

(g) GSA-GA based thresholding results with m=2, 3, 4 and 5

Table 5 Comparison of algorithms taking the Otsu as evaluation criterion in terms of best uniformity and corresponding thresholds

Image	m	GSA		PSOGSA		GGSA	
		Th	*u*	*Th*	*u*	*Th*	*u*
House	2	96, 155	**0.9875**	96, 155	**0.9875**	96, 155	**0.9875**
	3	79, 110, 154	0.9862	90, 109, 147	0.9847	81, 112, 158	**0.9864**
	4	81, 109, 153, 201	0.9875	96, 109, 155, 197	0.9851	78, 93, 109, 149	0.9823
	5	80, 102, 115, 154, 194	0.9812	96, 108, 152, 199, 204	0.9815	80, 110, 148, 164, 238	0.9814
Pepper	2	77, 147	0.9774	77, 148	0.9772	77, 148	0.9772
	3	63, 113, 164	0.9809	63, 113, 163	0.9809	63, 113, 163	0.9809
	4	50, 88, 131, 173	**0.9843**	39, 87, 132, 174	0.9834	49, 87, 128, 172	**0.9843**
	5	43, 78, 118, 162, 197	0.9828	45, 82, 113, 145, 180	**0.9860**	48, 85, 118, 157, 197	0.9831
SFB	2	87, 175	0.9662	86, 176	0.9653	86, 176	0.9653
	3	82, 145, 201	0.9754	80, 140, 199	0.9755	80, 140, 199	0.9755
	4	66, 109, 155, 202	**0.9782**	68, 109, 155, 200	**0.9782**	57, 99, 149, 199	0.9765
	5	50, 86, 128, 165, 206	0.9794	52, 89, 128, 165, 204	0.9799	51, 87, 127, 164, 204	0.9795
NR	2	131, 193	0.9684	132, 193	0.9681	132, 193	0.9681
	3	86, 133, 193	0.9821	84, 133, 193	0.9818	84, 133, 193	0.9818
	4	88, 136, 181, 221	0.9774	82, 122, 156, 196	0.9830	84, 132, 178, 219	0.9779
	5	71, 106, 148, 192, 220	0.9822	77, 106, 133, 163, 196	**0.9873**	80, 116, 147, 191, 223	0.9815
DS		BBO-DE		GAPSO		GSA-GA	
Th	*u*	*Th*	*u*	*Th*	*u*	*Th*	*u*
103, 157	0.9871	91, 145	0.9867	99, 156	0.9874	96, 155	**0.9875**
81, 109, 156	0.9863	81, 106, 165	0.9856	74, 110, 169	0.9853	80, 111, 158	**0.9864**
76, 106, 162, 204	0.9869	85, 109, 133, 164	0.9855	83, 125, 160, 216	0.9850	83, 113, 155, 202	**0.9876**

(continued)

Table 5 (continued)

DS		BBO-DE		GAPSO		GSA-GA	
Th	*u*	*Th*	*u*	*Th*	*u*	*Th*	*u*
78, 91, 127, 175, 213	0.9848	66, 104, 137, 177, 199	0.9883	90, 115, 152, 184, 212	0.9870	76, 102, 132, 167, 199	**0.9888**
65, 136	**0.9781**	72, 122	0.9754	59, 127	0.9769	77, 148	0.9772
65, 134, 166	0.9797	59, 113, 183	0.9766	53, 109, 156	0.9794	61, 121, 166	**0.9816**
72, 109, 145, 181	0.9804	50, 71, 136, 169	0.9791	67, 107, 145, 184	0.9811	51, 88, 130, 173	**0.9843**
42, 62, 94, 128, 175	0.9821	77, 113, 141, 160, 185	0.9770	56, 99, 143, 186, 192	0.9772	40, 83, 117, 161, 197	**0.9827**
79, 150	**0.9710**	92, 174	0.9678	83, 162	0.9705	83, 161	0.9707
89, 139, 193	0.9756	73, 134, 193	0.9749	90, 154, 199	0.9735	81, 138, 195	**0.9758**
76, 110, 165, 204	0.9750	79, 103, 132, 204	0.9707	60, 110, 166, 218	0.9734	64, 107, 151, 200	**0.9782**
37, 95, 144, 160, 198	0.9723	58, 104, 119, 154, 208	0.9770	69, 102, 127, 190, 205	0.9703	72, 102, 136, 177, 213	**0.9816**
126, 199	0.9692	120, 171	0.9737	90, 136	**0.9864**	127, 194	0.9692
92, 131, 200	0.9815	85, 126, 175	**0.9843**	84, 145, 185	0.9802	84, 131, 193	0.9819
91, 125, 153, 203	0.9817	81, 117, 141, 177	**0.9850**	89, 117, 147, 201	0.9841	81, 117, 141, 177	**0.9850**
89, 99, 129, 156, 194	0.9832	80, 107, 155, 180, 230	0.9827	86, 121, 162, 189, 224	0.9792	77, 106, 133, 163, 196	**0.9873**

Table 6 Comparison of algorithms taking the Otsu as evaluation criterion in terms of the mean of uniformity

Image	m	GSA	PSOGSA	GGSA	DS	BBO-DE	GAPSO	GSA-GA
House	2	**0.9875**	**0.9875**	**0.9875**	0.9830	0.9864	0.9867	**0.9875**
	3	0.9848	0.9758	0.9837	0.9849	0.9843	0.9847	**0.9863**
	4	0.9811	0.9695	0.9791	0.9840	0.9837	0.9839	**0.9866**
	5	0.9812	0.9757	0.9802	0.9834	**0.9859**	0.9833	0.9853
Pepper	2	0.9768	0.9768	**0.9772**	0.9714	0.9722	0.9728	**0.9772**
	3	0.9808	**0.9809**	**0.9809**	0.9723	0.9723	0.9775	**0.9809**
	4	0.9803	0.9830	0.9838	0.9752	0.9750	0.9769	**0.9839**
	5	0.9820	**0.9836**	0.9824	0.9750	0.9699	0.9753	0.9814
SFB	2	0.9657	0.9653	0.9653	0.9634	0.9628	0.9683	**0.9664**
	3	0.9753	**0.9755**	0.9754	0.9701	0.9717	0.9580	0.9746
	4	0.9760	0.9726	0.9761	0.9675	0.9656	0.9594	**0.9763**
	5	0.9765	0.9768	0.9765	0.9621	0.9738	0.9661	**0.9792**
NR	2	**0.9682**	0.9681	0.9681	0.9657	0.9685	0.9756	0.9679
	3	**0.9818**	0.9813	0.9817	0.9763	0.9755	0.9748	**0.9818**
	4	0.9752	0.9770	0.9762	0.9754	0.9763	0.9733	**0.9775**
	5	0.9768	0.9803	0.9802	0.9666	0.9710	0.9745	**0.9809**

Table 7 Comparison of algorithms taking the Otsu as evaluation criterion in terms of the mean of CPU times (in seconds)

Image	m	GSA	PSOGSA	GGSA	DS	BBO-DE	GAPSO	GSA-GA
House	2	1.1504	0.9398	0.8722	1.0113	2.2981	0.9848	**0.7026**
	3	0.9435	0.9785	0.8744	0.9876	2.3724	1.0608	**0.7657**
	4	1.0936	1.0273	1.1108	1.1089	2.5155	1.0065	**0.7762**
	5	1.0838	1.0335	1.0464	1.2391	2.6006	1.0453	**0.7325**
Pepper	2	0.7789	0.7764	0.7861	0.7115	1.5719	0.6825	**0.6266**
	3	0.8076	0.8006	0.8307	0.7262	1.8429	0.7527	**0.6437**
	4	0.9038	0.8315	0.9399	0.7397	2.0726	0.7486	**0.6428**
	5	0.8507	0.8810	0.8743	0.7648	1.8911	0.7573	**0.6644**
SFB	2	1.5234	1.4966	1.4985	1.5176	3.5485	1.4641	**1.3279**
	3	1.5384	1.5032	1.5652	1.5036	3.2132	1.4819	**1.2740**
	4	1.5683	1.5484	1.6148	1.6120	3.6203	1.5370	**1.3008**
	5	1.5844	1.5753	1.6248	1.5562	3.3839	1.4967	**1.1065**
NR	2	1.5666	1.5938	1.6228	1.5742	3.2607	1.6392	**1.4363**
	3	1.5838	1.6257	1.6352	1.6902	4.0370	1.6036	**1.4148**
	4	1.6289	1.7799	1.6631	1.6329	3.4892	1.7265	**1.4367**
	5	1.6597	1.6547	1.7577	1.6487	3.6371	1.6850	**1.4455**

Generally speaking, with the Otsu criterion, GSA-GA yielded better results than the other six comparison algorithms as illustrated in Table 5. Moreover, similar to the results given in Table 4, the mean uniformity reported in Table 6 also confirms the superiority of GSA-based methods comparing with other NAs. Mean CPU times illustrated in Table 7 showed that the computational time of GSA-GA is also the lowest. Especially, time consuming of GSA-GA is much lower than that of DS. In Figs. 6 and 7, the segmentation images of SFB and NR by all the seven NAs used the Otsu criterion are presented.

Furthermore, comparing the mean uniformity shown in Tables 3 and 6, we can concluded that for a given image, thresholds determined based on different criteria can be different. For images Pepper, SFB, and NR, the thresholding results using Otsu criterion is better than the results on the basis of Kapur's entropy. While for the image House, we observed completely opposite conclusions. Thereby, comprehensive analysis for choosing the most appropriate objective function are desired in the future real-world application.

5.6 Running Time Analysis Using Student's t-test

To statistically analyze the time-consuming shown in Tables 4 and 7, a parametric significance proof known as the Student's t-test was conducted in this section [55]. This test allows assessing result differences between two related methods (one of the methods is chosen as control method). In this chapter, the control algorithm is GSA-GA. Results of GSA-GA were compared with other six comparison NAs in terms of the mean CPU times. The null hypothesis in this chapter is that there is no significance difference between the mean CPU times achieved by two compared algorithms for a test image over 30 independent runs. The significance level is $\alpha = 0.05$. Accordingly, if the produced t-value of a test is smaller than or equal to the critical value "-2.045", the null hypothesis for the paired t-test should be rejected [70, 71].

Tables 8 and 9 reported the t-values produced by Student's t-test for the pairwise comparison of six groups. These groups were formed by GSA-GA (with Kapur's entropy and Otsu, respectively) versus other comparison algorithms. In Tables 8 and 9, if the null hypothesis is rejected, we define a symbol "$h = 0$" which means the difference between the results obtained by two algorithms are not different. By contrast, "$h = 1$" indicates that the difference between GSA-GA and the compared algorithm is significant.

As illustrated in Tables 8 and 9, for all the test images, the produced t-values for the experiments on the Kapur's entropy and Otsu criteria were smaller than the critical value. The results indicated that GSA-GA performed significantly fast image segmentation compared with the other six algorithms.

Fig. 6 Segmented images for the SFB image by Otsu with different optimization techniques

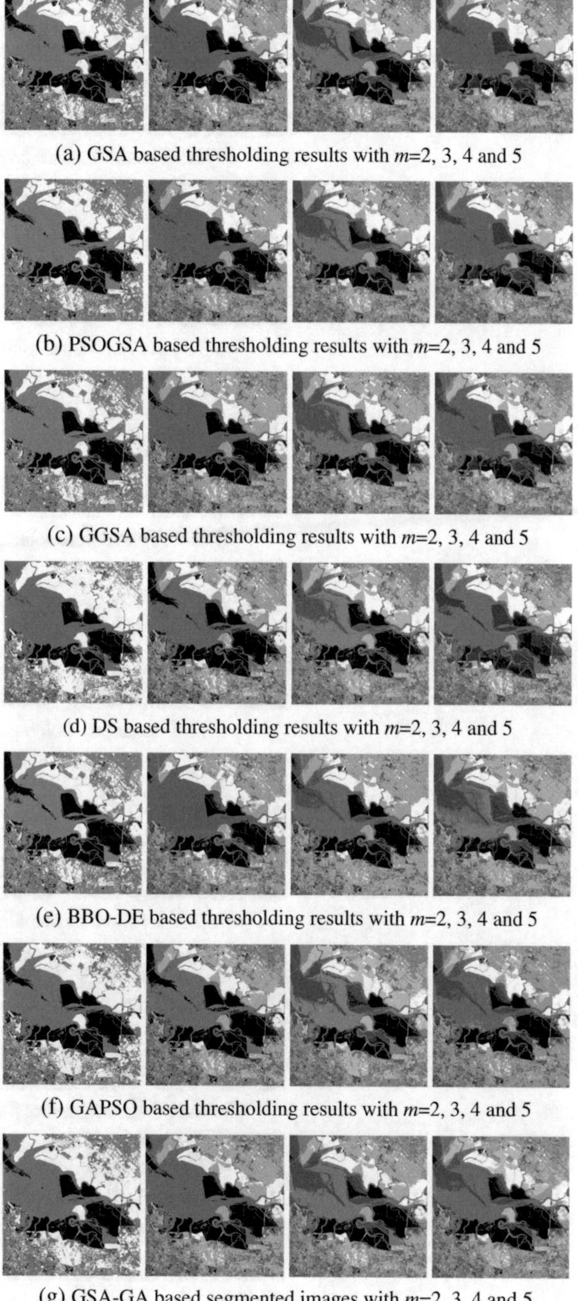

(a) GSA based thresholding results with m=2, 3, 4 and 5

(b) PSOGSA based thresholding results with m=2, 3, 4 and 5

(c) GGSA based thresholding results with m=2, 3, 4 and 5

(d) DS based thresholding results with m=2, 3, 4 and 5

(e) BBO-DE based thresholding results with m=2, 3, 4 and 5

(f) GAPSO based thresholding results with m=2, 3, 4 and 5

(g) GSA-GA based segmented images with m=2, 3, 4 and 5

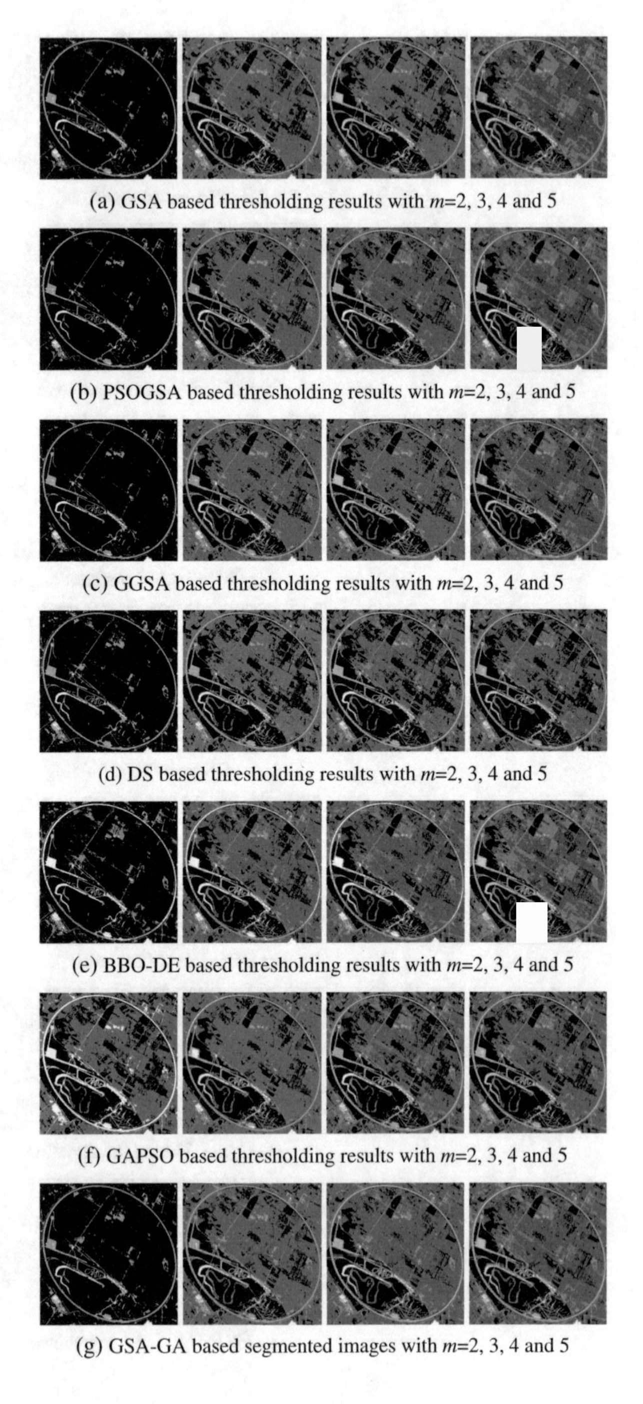

(a) GSA based thresholding results with m=2, 3, 4 and 5

(b) PSOGSA based thresholding results with m=2, 3, 4 and 5

(c) GGSA based thresholding results with m=2, 3, 4 and 5

(d) DS based thresholding results with m=2, 3, 4 and 5

(e) BBO-DE based thresholding results with m=2, 3, 4 and 5

(f) GAPSO based thresholding results with m=2, 3, 4 and 5

(g) GSA-GA based segmented images with m=2, 3, 4 and 5

Fig. 7 Segmented images for the NR image by Otsu with different optimization techniques

Table 8 Statistical analysis of 30 runs for the experimental results on the Kapur's entropy criterion

Image	m	GSA-GA VS											
		GSA		PSOGSA		GGSA		DS		BBO-DE		GAPSO	
House	2	−15.1156	1	−19.1332	1	−14.2639	1	−24.7401	1	−107.8601	1	−24.4811	1
	3	−32.6476	1	−19.9512	1	−26.7523	1	−44.5436	1	−109.2128	1	−29.0906	1
	4	−39.5957	1	−24.2611	1	−20.0202	1	−44.2350	1	−102.7793	1	−46.5004	1
	5	−34.7283	1	−30.5527	1	−43.5916	1	−26.5975	1	−222.9597	1	−39.2664	1
Pepper	2	−10.7081	1	−10.4253	1	−11.1125	1	−5.9857	1	−65.2916	1	−3.8069	1
	3	−64.8463	1	−57.2600	1	−55.2033	1	−29.4831	1	−94.3847	1	−30.4911	1
	4	−36.0189	1	−42.7815	1	−26.6089	1	−23.4535	1	−33.6639	1	−24.6679	1
	5	−2.5627	1	−2.9435	1	−2.8870	1	−1.3815	1	−16.6481	1	−1.2784	1
SFB	2	−27.6977	1	−24.3816	1	−29.6109	1	−12.1633	1	−45.1326	1	−23.9737	1
	3	−22.4245	1	−21.4709	1	−23.8949	1	−21.7750	1	−151.1436	1	−19.7951	1
	4	−14.6253	1	−13.5423	1	−17.3410	1	−11.0828	1	−79.4278	1	−10.7647	1
	5	−75.6004	1	−57.9269	1	−54.1767	1	−66.5839	1	−341.5912	1	−50.8013	1
NR	2	−21.8825	1	−22.3767	1	−23.0353	1	−22.0319	1	−278.2518	1	−12.4801	1
	3	−10.3605	1	−12.4493	1	−13.3727	1	−11.0882	1	−71.6375	1	−11.6034	1
	4	−13.2716	1	−11.2340	1	−14.4755	1	−13.4430	1	−144.0898	1	−14.9473	1
	5	−12.9729	1	−12.7915	1	−16.0183	1	−11.9100	1	−111.5124	1	−13.4166	1

Table 9 Statistical analysis of 30 runs for the experimental results on the Otsu criterion

Image	m	GSA-GA VS											
		GSA		PSOGSA		GGSA		DS		BBO-DE		GAPSO	
		t-value	h	t-value	h	t-value	h	t-value	h	t-value	h	t-value	h
House	2	−9.5851	1	−11.7822	1	−9.4314	1	−14.6542	1	−64.9566	1	−11.9011	1
	3	−20.6436	1	−15.8776	1	−18.2415	1	−20.7056	1	−133.0959	1	−22.5299	1
	4	−14.1373	1	−19.2192	1	−27.5020	1	−11.2347	1	−45.3072	1	−28.2547	1
	5	−35.3391	1	−22.4042	1	−27.7047	1	−49.8531	1	−120.2696	1	−38.4260	1
Pepper	2	−65.6607	1	−36.0491	1	−70.6444	1	−41.2547	1	−58.9202	1	−3.5054	1
	3	−80.3597	1	−73.7771	1	−84.5025	1	−56.4896	1	−97.0729	1	−53.7705	1
	4	−51.9612	1	−44.2863	1	−44.7699	1	−21.8072	1	−205.1454	1	−18.4475	1
	5	−50.4190	1	−56.5007	1	−52.2792	1	−35.4027	1	−92.6841	1	−18.5975	1
SFB	2	−40.5912	1	−40.1226	1	−40.2393	1	−35.7284	1	−213.5478	1	−39.6544	1
	3	−40.0638	1	−47.2556	1	−45.0096	1	−39.0400	1	−202.4002	1	−21.6851	1
	4	−37.4041	1	−36.5145	1	−41.7963	1	−36.6461	1	−92.7530	1	−35.9711	1
	5	−55.9229	1	−51.7381	1	−53.8157	1	−55.1532	1	−170.0324	1	−36.5197	1
NR	2	−81.1067	1	−70.7878	1	−73.9369	1	−57.9793	1	−21.2059	1	−77.3845	1
	3	−51.7581	1	−78.3780	1	−66.3253	1	−77.0030	1	−390.4326	1	−27.7027	1
	4	−41.9257	1	−48.8776	1	−45.0429	1	−40.2674	1	−25.7615	1	−44.0133	1
	5	−43.5523	1	−43.6292	1	−46.6667	1	−42.4596	1	−231.6469	1	−35.1170	1

6 Conclusion

Multilevel thresholding is a fundamental and important technology for image segmentation and a fundamental process in image interpretation. An overview of the grayscale image segmentation using multilevel thresholding and nature-inspired algorithms is discussed in this chapter. On one hand, the NAs based multilevel image thresholding can reduce the computational consuming of exhaustive search algorithms based techniques. On the other hand, the global optimization property of NAs makes them as preferable choices for multilevel thresholding. Although some drawbacks of the NAs limited their application, hybrid algorithms can promote their performances. Therefore, a novel multilevel thresholding algorithm for image segmentation by employing an improved GSA variant, called GSA-GA is proposed in this chapter. In GSA-GA, the roulette selection and discrete mutation operators of GA were introduced into GSA to improve the search accuracy and speed of GSA. Experiments on benchmark images and real-word satellite images confirmed the superiority of GSA-GA compared with GSA, PSOGSA, GGSA, DS, BBO-DE, and GAPSO. Potential future investigation can be on the analysis of different criteria. By comprehensive analysis, the most appropriate criterion for constructing objective function can be selected or developed.

Acknowledgments This work was supported by the Chinese Natural Science Foundation Projects (No. 41471353, No. 41271349), Fundamental Research Funds for the Central Universities (No. 14CX02039A, No. 15CX06001A), and the China Scholarship Council.

References

1. Zhang, A.Z., Sun, G.Y., Wang, Z.J., Yao, Y.J.: A hybrid genetic algorithm and gravitational search algorithm for global optimization. Neural. Netw. World. **25**, 53–73 (2015)
2. Ghamisi, P., Couceiro, M.S., Martins, F.M., Atli, B.J.: Multilevel image segmentation based on fractional-order Darwinian particle swarm optimization. IEEE. Trans. Geosci. Remote. Sens. **52**, 2382–2394 (2014)
3. Kang, W.X., Yang, Q.Q., Liang, R.P.: The comparative research on image segmentation algorithms. In: 2009 First International Workshop on Education Technology and Computer Science, pp. 703-707. IEEE (2009)
4. Bhandari, A.K., Singh, V.K., Kumar, A., Singh, G.K.: Cuckoo search algorithm and wind driven optimization based study of satellite image segmentation for multilevel thresholding using Kapur's entropy. Expert. Syst. Appl. **41**, 3538–3560 (2014)
5. Dey, S., Bhattacharyya, S., Maulik, U.: Quantum Behaved Multi-objective PSO and ACO Optimization for Multi-level Thresholding. IEEE, New York (2014)
6. Horng, M.H.: Multilevel thresholding selection based on the artificial bee colony algorithm for image segmentation. Expert. Syst. Appl. **38**, 13785–13791 (2011)
7. Bhandari, A.K., Kumar, A., Singh, G.K.: Tsallis entropy based multilevel thresholding for colored satellite image segmentation using evolutionary algorithms. Expert. Syst. Appl. **42**, 8707–8730 (2015)
8. Otsu, N.: A threshold selection method from gray-level histograms. IEEE. Trans. Syst. Man Cybern. **9**, 62–66 (1979)

9. Kapur, J.N., Sahoo, P.K., Wong, A.K.: A new method for gray-level picture thresholding using the entropy of the histogram. Comput. Vison Graph. **29**, 273–285 (1985)
10. Huang, L.K., Wang, M.J.J.: Image thresholding by minimizing the measures of fuzziness. Pattern Recogn. **28**, 41–51 (1995)
11. Qiao, Y., Hu, Q.M., Qian, G.Y., Luo, S.H., Nowinski, W.L.: Thresholding based on variance and intensity contrast. Pattern. Recogn. **40**, 596–608 (2007)
12. Li, X.Q., Zhao, Z.W., Cheng, H.D.: Fuzzy entropy threshold approach to breast cancer detection. Inf. Sci. Appl. **4**, 49–56 (1995)
13. Kittler, J., Illingworth, J.: Minimum error thresholding. Pattern Recogn. **19**, 41–47 (1986)
14. Li, C.H., Tam, P.K.S.: An iterative algorithm for minimum cross entropy thresholding. Pattern Recogn. Lett. **19**, 771–776 (1998)
15. de Albuquerque, M.P., Esquef, I.A., Mello, A.R.G.: Image thresholding using Tsallis entropy. Pattern. Recogn. Lett. **25**, 1059–1065 (2004)
16. Kurban, T., Civicioglu, P., Kurban, R., Besdok, E.: Comparison of evolutionary and swarm based computational techniques for multilevel color image thresholding. Appl. Soft Comput. **23**, 128–143 (2014)
17. Akay, B.: A study on particle swarm optimization and artificial bee colony algorithms for multilevel thresholding. Appl. Soft. Comput. **13**, 3066–3091 (2013)
18. Ali, M., Ahn, C.W., Pant, M.: Multi-level image thresholding by synergetic differential evolution. Appl. Soft. Comput. **17**, 1–11 (2014)
19. Chander, A., Chatterjee, A., Siarry, P.: A new social and momentum component adaptive PSO algorithm for image segmentation. Expert. Syst. Appl. **5**, 4998–500 (2011)
20. Coello, C.C., Lamont, G.B., Van Veldhuizen, D.A.: Evolutionary algorithms for solving multi-objective problems. Springer Science & Business Media, Berlin (2007)
21. Hammouche, K., Diaf, M., Siarry, P.: A multilevel automatic thresholding method based on a genetic algorithm for a fast image segmentation. Comput. Vis. Image Underst. **109**, 163–175 (2008)
22. Tao, W.B., Tian, J.W., Liu, J.: Image segmentation by three-level thresholding based on maximum fuzzy entropy and genetic algorithm. Pattern Recogn. Lett. **24**, 3069–3078 (2003)
23. Yin, P.Y.: A fast scheme for optimal thresholding using genetic algorithms. Signal Process. **72**, 85–95 (1999)
24. Zhang, J., Percy, R.G., McCarty Jr., J.C.: Introgression genetics and breeding between Upland and Pima cotton: a review. Euphytica **198**, 1–12 (2014)
25. Ayala, H.V.H., dos Santos, F.M., Mariani, V.C., dos Santos Coelho, L.: Image thresholding segmentation based on a novel beta differential evolution approach. Expert. Syst. Appl. **42**, 2136–2142 (2015)
26. Sarkar, S., Das, S., Chaudhuri, S.S.: A multilevel color image thresholding scheme based on minimum cross entropy and differential evolution. Pattern Recogn. Lett. **54**, 27–35 (2015)
27. Storn, R., Price, K.: Differential evolution-a simple and efficient heuristic for global optimization over continuous spaces. J. Global. Optim. **11**, 341–359 (1997)
28. Kirkpatrick, S.: Optimization by simulated annealing: quantitative studies. J. Stat. Phys. **34**, 975–986 (1984)
29. Drigo, M., Maniezzo, V., Colorni, A.: Ant system: optimization by a colony of cooperation agents. IEEE Trans. Syst. Man Cybern. B **26**, 29–41 (1996)
30. Tao, W.B., Jin, H., Liu, L.M.: Object segmentation using ant colony optimization algorithm and fuzzy entropy. Pattern Recogn. Lett. **28**, 788–796 (2007)
31. Karaboga, D.: An idea based on honey bee swarm for numerical optimization. In: Broy, M., Dener, E. (eds.) Software Pioneers, pp. 10–13. Springer, Heidelberg (2002)
32. Civicioglu, P.: Transforming geocentric cartesian coordinates to geodetic coordinates by using differential search algorithm. Comput. Geosci. **46**, 229–247 (2012)
33. Clerc, M., Kennedy, J.: The particle swarm-explosion, stability, and convergence in a multidimensional complex space. IEEE Trans. Evol. Comput. **6**, 58–73 (2002)
34. Nabizadeh, S., Faez, K., Tavassoli, S., Rezvanian, A.: A Novel Method for Multi-level Image Thresholding Using Particle Swarm Optimization Algorithms. IEEE, New York (2010)

35. Yin, P.Y.: Multilevel minimum cross entropy threshold selection based on particle swarm optimization. Appl. Math. Comput. **184**, 503–513 (2007)
36. Baniani, E.A., Chalechale, A.: Hybrid PSO and genetic algorithm for multilevel maximum entropy criterion threshold selection. Int. J. Hybrid Inf. Technol. **6**, 131–140 (2013)
37. Juang, C.F.: A hybrid of genetic algorithm and particle swarm optimization for recurrent network design. Proc IEEE Trans. Syst. Man Cybern. B **34**, 997–1006 (2004)
38. Patel, M.K., Kabat, M.R., Tripathy, C.R.: A hybrid ACO/PSO based algorithm for QoS multicast routing problem. Ain Shams Eng. J. **5**, 113–120 (2014)
39. Zhang, Y.D., Wu, L.N.: A robust hybrid restarted simulated annealing particle swarm optimization technique. Adv. Comput. Sci. Appl. **1**, 5–8 (2012)
40. Boussaïd, I., Chatterjee, A., Siarry, P., Ahmed-Nacer, M.: Hybrid BBO-DE Algorithms for Fuzzy Entropy-based Thresholding. Springer, Heidelberg (2013)
41. Gong, W.Y., Cai, Z.H., Ling, C.X., Li, H.: A real-coded biogeography-based optimization with mutation. Appl. Math. Comput. **216**, 2749–2758 (2010)
42. John, H.: Holland. Adaptation in Natural and Artificial Systems. MIT Press, Cambridge (1992)
43. Li, C.H., Lee, C.: Minimum cross entropy thresholding. Pattern Recogn. **26**, 617–625 (1993)
44. Tsai, W.H.: Moment-preserving thresolding: a new approach. Comput. Vison Graph. **29**, 377–393 (1985)
45. Blum, C., Li, X.: Swarm Intelligence in Optimization. Springer, Heidelberg (2008)
46. Kenndy, J., Eberhart, R.C., Labahn, G.: Particle Swarm Optimization. Kluwer, Boston (1995)
47. Yang, X.S., Deb, S.: Cuckoo Search via Lévy Flights. IEEE, New York (2009)
48. Yang, X.S., Hossein, G.A.: Bat algorithm: a novel approach for global engineering optimization. Eng. Comput. **29**, 464–483 (2012)
49. Mirjalili, S., Mirjalili, S.Z.M., Lewis, A.: Grey Wolf Optimizer. Adv. Eng. Softw. **69**, 46–61 (2014)
50. Gong Y.J., Li, J.J., Zhou, Y., Li, Y., Chung, H.S.H., Shi, Y.H., Zhang, J.: Genetic learning particle swarm optimization. IEEE Trans. Cybern., 1–14 (2015)
51. Zhang, Y.D., Wu, L.N.: Optimal multi-level thresholding based on maximum Tsallis entropy via an artificial bee colony approach. Entropy **13**, 841–859 (2011)
52. Alihodzic, A., Tuba, M.: Bat algorithm (BA) for image thresholding. Recent Res. Telecommun. Inf. Electron. Signal Process, 17–19 (2013)
53. Alihodzic, A., Tuba, M.: Improved bat algorithm applied to multilevel image thresholding. Sci World J. (2014). doi:10.1155/2014/176718
54. Ye, Z.W., Wang, M.W., Liu, W., Chen, S.B.: Fuzzy entropy based optimal thresholding using bat algorithm. Appl. Soft. Comput. **31**, 381–395 (2015)
55. Li, Y.Y., Jiao, L.C., Shang, R.H., Stolkin, R.: Dynamic-context cooperative quantum-behaved particle swarm optimization based on multilevel thresholding applied to medical image segmentation. Inf. Sci. **294**, 408–422 (2015)
56. Rashedi, E., Nezamabadi-Pour, H., Saryazdi, S.: GSA: a gravitational search algorithm. Inf. Sci. **179**, 2232–2248 (2009)
57. Birbil, S., Fang, S.C.: An electromagnetism-like mechanism for global optimization. J. Global. Optim. **25**, 263–282 (2003)
58. Kaveh, A., Talatahari, S.: A novel heuristic optimization method: charged system search. Acta. Mech. **213**, 267–289 (2010)
59. Biswas, A., Mishra, K.K., Tiwari, S., Misra, A.K.: Physics-inspired optimization algorithms: a survey. J. Optim. **2013**, 1–16 (2013)
60. Wolpert, D.H., Macready, W.G.: No free lunch theorems for optimization. IEEE Trans. Evol. Comp. **1**, 67–82 (1997)
61. Jiang, S., Wang, Y., Ji, Z.: Convergence analysis and performance of an improved gravitational search algorithm. Appl. Soft Comput. **24**, 363–384 (2014)
62. Kumar, J.V., Kumar, D.V., Edukondalu, K.: Strategic bidding using fuzzy adaptive gravitational search algorithm in a pool based electricity market. Appl. Soft Comput. **13**, 2445–2455 (2013)
63. Mirjalili, S., Hashim, S.Z.M., Sardroudi, H.M.: Training feedforward neural networks using hybrid particle swarm optimization and gravitational search algorithm. Appl. Math. Comput. **218**, 11125–11137 (2012)

64. Sabri, N.M., Puteh, M., Mahmood, M.R.: A review of gravitational search algorithm. Int. J. Adv. Soft Comput. Appl. **5**, 1–39 (2013)
65. Sun, G.Y., Zhang, A.Z., Yao, Y.J., Wang, Z.J.: A novel hybrid algorithm of gravitational search algorithm with genetic algorithm for multi-level thresholding. Appl. Soft Comput. **46**, 703–730 (2016). doi:10.1016/j.asoc.2016.01.054
66. Mirjalili, S., Hashim, S.Z.M.: A new hybrid PSOGSA algorithm for function optimization. IEEE, New York (2010)
67. Mirjalili, S., Lewis, A.: Adaptive gbest-guided gravitational search algorithm. Neural Comput. Appl. **25**, 1569–1584 (2014)
68. Herrera, F., Lozano, M., Verdegay, J.L.: Fuzzy connectives based crossover operators to model genetic algorithms population diversity. Fuzzy. Set. Syst. **92**, 21–30 (1997)
69. Sahoo, P.K., Soltani, S., Wong, A.K.C.: A survey of thresholding techniques. Neural. Comput. Appl. **41**, 233–260 (1988)
70. Derrac, J., García, S., Molina, D., Herrera, F.: A practical tutorial on the use of nonparametric statistical tests as a methodology for comparing evolutionary and swarm intelligence algorithms. Swarm Evol. Comput. **1**, 3–18 (2011)
71. Liang, J.J., Qin, A.K., Suganthan, P.N., Baskar, S.: Comprehensive learning particle swarm optimizer for global optimization of multimodal functions. IEEE Ttans. Evolut. Comput. **10**, 281–295 (2006)

A Novel Hybrid CS-BFO Algorithm for Optimal Multilevel Image Thresholding Using Edge Magnitude Information

Sanjay Agrawal, Leena Samantaray and Rutuparna Panda

Abstract Thresholding is the key to simplify image classification. It becomes challenging when the number of thresholds is more than two. Most of the existing multilevel thresholding techniques use image histogram information (first-order statistics). This chapter utilizes optimal edge magnitude information (second-order statistics) of an image to obtain multilevel threshold values. We compute the edge magnitude information from the gray-level co-occurrence matrix (GLCM) of the image. The second-order statistics uses the correlation among the pixels for improved results. Maximization of edge magnitude is vital for obtaining optimal threshold values. The edge magnitude is maximized by introducing a novel hybrid cuckoo search-bacterial foraging optimization (CS-BFO) algorithm. The novelty of our proposed CS-BFO algorithm lies in its ability to provide improved chemotaxis in BFO algorithm, which is achieved by supplementing levy flight feature of CS. Social foraging models are relatively efficient for determining optimum multilevel threshold values. Hence, CS-BFO is used for improved thresholding performance and highlighting the novelty of this contribution. We have also implemented other soft computing tools cuckoo search (CS), particle swarm optimization (PSO), and genetic algorithm (GA) for a comparison. In addition, we have incorporated constraint handling in all the above-mentioned techniques so that optimal threshold values do not cross the bounds. This study reveals the fact that CS technique provides us improved speed while the CS-BFO method shows improved results both qualitatively and quantitatively.

Keywords Multilevel thresholding · Edge magnitude · GLCM · Evolutionary computing

S. Agrawal (✉) · R. Panda
VSS University of Technology, Burla 768018, India
e-mail: agrawals_72@yahoo.com

L. Samantaray
Ajay Binay Institute of Technology, Cuttack 753014, India
e-mail: Leena_Sam@rediffmail.com

S. Bhattacharyya et al. (eds.), *Hybrid Soft Computing for Image Segmentation*, DOI 10.1007/978-3-319-47223-2_3

1 Introduction

Image segmentation is considered a vital step in image processing. Over the years, many techniques of segmentation are reported in literature [1–4]. Thresholding is used as one of the most preferred techniques for image segmentation. Most of the approaches to image thresholding are based on histogram analysis [5, 6]. For images with clearly identifiable objects and background, a single threshold value easily divides pixels of an image into two distinct classes. However, for complex images, multilevel thresholding is needed to distribute pixels into many distinct groups, where pixels belonging to a group have gray levels within a specified range. In recent times, with an increase in the number of required threshold levels, the complexity and the computation time required to solve thresholding problems have emerged as a notable challenge in this field. Due to this, many thresholding techniques are proposed. Comprehensive reviews and comparative studies on various thresholding techniques are found in literature [5–9].

Sahoo et al. [7] presented a comprehensive survey of various thresholding techniques. They used uniformity and shape measures information for automatic global thresholding. They pointed out the dependence of such techniques on first-order statistics, i.e., histogram of an image, as one of its major drawbacks. They ranked the co-occurrence matrix based method first according to the shape measure.

A comparative study of many global thresholding techniques for segmentation is presented in [10]. The authors focused on five different global thresholding techniques used for segmentation. They evaluated five different algorithms using criterion functions such as probability of error, shape, and uniformity measures. They observed that most of the techniques are well suited for images possessing bimodal histogram. They suggested for a more refined and consistent technique to obtain better threshold values that can be used for segmentation.

Sankur and Sezgin [11] presented a comprehensive survey of many thresholding techniques and their quantitative performance evaluation. The authors in their study categorized thresholding algorithms on the basis of the type of information used such as entropy, shape of the histogram, attributes of the objects under consideration, spatial correlation, and surface of the gray level. But they mainly focused on document imaging applications.

Chang et al. [12] conducted a survey and comparative analysis of thresholding techniques based on Shannons entropy and relative entropy. They investigated eight different entropy based methods and evaluated them by shape and uniformity measures. They also explored the relationship between entropy and relative entropy thresholding methods. However, they concluded that information conveyed by histograms is not sufficient to get a proper threshold value, as it does not take into account spatial correlation among the pixels of an image. Due to this reason, different images with similar histogram may result in the same threshold value.

Aforementioned methods are based on first-order statistics in the sense that they are using gray-level histogram of an image. However, they do not take into account the spatial correlation between pixels of an image. This has motivated us to exploit this spatial correlation for achieving multilevel thresholding. In this chapter, to solve the problem of multilevel thresholding, edge magnitude information from the GLCM of a given image is used to determine multiple threshold values.

Several researchers have tried using evolutionary computing techniques to solve multilevel thresholding problem. Optimal threshold values are one which is obtained by either minimizing or maximizing an objective function. Yin [13] used GA for optimal thresholding.

Cheng et al. [14] used fuzzy entropy and GA for image segmentation. Zahara et al. [15] proposed a hybrid optimization algorithm for multilevel thresholding which used the Otsus approach with Nelder–Mead simplex search technique combined with PSO. Sathya and Kayalvizhi [16] presented bacterial foraging optimization (BFO) based multilevel thresholding. They used BFO algorithm to obtain optimal thresholds by maximizing the Kapurs and Otsus fitness functions. Sarkar et al. [17] emphasized on obtaining the optimal multiple threshold values from a LISS III near-infrared (NIR) band using Renyis entropy function. They used differential evolution technique to obtain the optimal threshold values. Further, Sarkar et al. [18] presented a differential evolution based method for multilevel thresholding employing the minimum cross-entropy method. Zhang and Wu [19] proposed a global multilevel thresholding scheme for image segmentation. They employed the Tsallis entropy function. They further used artificial bee colony (ABC) approach for obtaining the optimal threshold values. Horng [20] used honey bee mating optimization technique for multilevel thresholding.

As stated above, a wide variety of algorithms are proposed for multilevel image thresholding. Bhandari et al. [21] used Kapurs entropy for multilevel thresholding of satellite images. They used wind-driven algorithm and cuckoo search optimization to optimize the Kapurs entropy function. Raja et al. [22] used Otsus function for multilevel thresholding of thermal images of breast infected with cancer [22]. They implemented an improved PSO for optimizing the Otsus function. Further, Bhandari et al. [23] presented a computationally efficient modified artificial bee colony algorithm for satellite image segmentation. They used Tsallis as well as Kapurs and Otsus criteria as fitness function. Nabizadeh et al. [24] focussed on automatic lesion detection in MR brain image using a gravitational optimization algorithm based on the histogram. Liu et al. [25] presented a modified PSO to overcome the computationally expensive component of conventional approaches to multilevel thresholding. They used the adaptive population and adaptive inertia strategy to improvise PSO. Brajevic and Tuba [26] investigated the performance of firefly and cuckoo search algorithms for multilevel thresholding. They used Kapurs entropy and Otsus criteria as the objective function. Roy et al. [27] focused on demonstrating the image segmentation scheme using cross-entropy based thresholding optimized by cuckoo search algorithm.

Use of such methods is justified for obtaining optimal threshold values as the dimension of the problem is large. Most of the above-mentioned methods optimize Otsus function or Kapurs function using various non-hybridized evolutionary computing algorithms. They mainly emphasize on entropy feature. Each evolutionary computing technique can be used separately, but a powerful advantage of it is the complementary nature of the techniques. Used together they can produce improved solutions to various problems. However, this chapter focuses on edge magnitude information which utilizes correlation among pixels giving improved thresholding performance. A new hybrid CS-BFO algorithm is proposed to maximize the edge magnitude. With increase in threshold levels, the computational complexity increases and accuracy decreases. As the search space dimension increases it is required to utilize some soft computing techniques. Hybrid approaches produce better results as compared to non-hybrid approaches. So CS-BFO is proposed. Further the fitness function defines the accuracy of the results. Hence, edge magnitude information is used for improved accuracy as it utilizes correlation among pixels in an image. Interestingly, optimal threshold values correspond to maximal edge magnitudes only. In this chapter, edge magnitude is maximized using a novel hybrid CS-BFO algorithm to get optimal threshold values. The complexity of such problem increases with increase in number of thresholds. Usually, a fixed step size for chemotaxis in BFO is considered. This does not yield accurate thresholding results. Moreover, a bacterium starts searching from the same location. Due to which, some missing nutrients do not add value to the healthy bacteria. This is improved by supplementing the levy flight feature of the CS algorithm to improvise global optimum values of the objective functions. In the context of present problem, the main feature of our proposed method is that it combines the Levy flight feature of CS algorithm with the chemotaxis property of BFO to efficiently explore the search space resulting in better fitness function value with more accurate multilevel image thresholding. Note that CS, PSO, and GA are implemented to maximize the edge magnitude for a comparison. The results obtained are compared qualitatively and quantitatively. New findings are discussed in the result section to make our idea explicitly clear.

The chapter is structured as follows: the theory of gray-level co-occurrence matrix and computation of edge magnitude is presented in Sect. 2. Section 3 provides the brief concept of evolutionary computing techniques. Section 4 presents the proposed methodology. Results and discussions are presented in Sect. 5. Finally, Sect. 6 is the conclusion.

2 Edge Magnitude Computation

Gray-level co-occurrence matrix (GLCM) has been extensively used in the area of texture segmentation [28]. Now, it has also been used in thresholding as well [29]. Let Q be an operator that defines the location of two pixels relative to each other. Let I be an image with L possible intensity levels. We can define G as a matrix, having dimension LxL whose element g_{ij} is the number of times a pixel pair with intensities z_i and z_j occur in I in the position indicated by Q, where $1 \leq i, j \leq L$. Now G is referred to as gray-level co-occurrence matrix [30]. A GLCM includes conditional joint probabilities P_{ij} of all pair-wise combinations of gray levels using two parameters, interpixel orientation θ and interpixel distance d. A different GLCM is generated for each (d, θ) pair. It is usually defined to be symmetric with respect to orientation, i.e., intensities (z_i, z_j) oriented at 0° is also considered as being oriented at 180°. Figure 1 describes the calculation of GLCM. The interpixel distance d is taken as one and interpixel orientation is taken as 0°. For the pair (1, 1), there are two instances, when pixel 1 is followed at its right, i.e., $(\theta = 0°)$ by pixel 1. Hence, the GLCM for $(\theta = 0°)$ is having a value of two in the position corresponding to $z_i = z_j = 1$. Figure 1c shows the symmetric nature of GLCM for $(\theta = 180°)$.

The GLCM is calculated for eight directions around a central pixel I_{ij} as shown in Fig. 2. However, simple relation exists among certain pair of (d, θ). Note that $d = 1$. Let $P^T(d, \theta)$ represent the transpose of $P(d, \theta)$. Then

$$\begin{aligned} P(d, 0°) &= P^T(d, 180°) \\ P(d, 45°) &= P^T(d, 225°) \\ P(d, 90°) &= P^T(d, 270°) \\ P(d, 135°) &= P^T(d, 315°) \end{aligned}$$

Thus, the evaluation of $P(d, 180°)$, $P(d, 225°)$, $P(d, 270°)$, $P(d, 315°)$ adds nothing more to the calculation of GLCM. For a given interpixel distance, d, we consider four angular GLCM only. Therefore, to avoid dependency on directions, we calculate an average matrix out of four matrices for $\theta = 0°, 45°, 90°, 135°$ as [31]

$$G = \frac{[G_1(d, 0°) + G_2(d, 45°) + G_3(d, 90°) + G_4(d, 135°)]}{4} \quad (1)$$

Let I be an image of size $M \times N$ with L gray levels $0, 1, \ldots, L-1$. Let h_i be the frequency of gray level i and $p_i = h_i/(M \times N)$ be the probability of occurrence of gray level i. A single threshold value t divides the image into two regions, namely object having gray levels from $[0, t]$ and background having gray levels from $[t +$

Fig. 1 Description of GLCM. **a** Sample Image, **b** GLCM of (**a**) for $d = 1, \theta = 0°$, **c** GLCM of (**a**) for $d = 1, \theta = 180°$

(a)

1	4	3	2	1
2	1	1	2	2
1	4	4	2	1
1	2	1	1	2
3	4	3	2	1

(b)

	1	2	3	4
1	2	3	0	2
2	5	1	0	0
3	0	2	0	1
4	3	1	2	1

(c)

	1	2	3	4
1	2	5	0	3
2	3	1	2	1
3	0	0	0	2
4	2	0	1	1

Fig. 2 Directions used for calculating GLCM

135°	90°	45°
180°	I_{ij}	0°
225°	270°	315°

$1, L-1]$. An optimum threshold t_{opt} is obtained by maximizing or minimizing an objective function

$$t_{opt} = \arg\max \ \ or \ \ \arg\min \{f(\cdot)\} \tag{2}$$

Haralick and Shapiro [32] identified fourteen features from GLCM and then suggested six of them to be most relevant for image processing applications. However, information from GLCM known as edge magnitude q is used for computation of some features. It is usually described as a difference in the gray value of the pixel pair and is defined by the position of a pixel pair in the GLCM. The contrast definition in GLCM carries the edge magnitude information defined as,

$$CON = \sum_{q=0}^{L-1} q^2 \left\{ \sum_{m=0}^{L-1} \sum_{n=0}^{L-1} G(m,n) \right\} \qquad where \quad |m-n| = q \tag{3}$$

In this chapter, edge magnitude information as defined above is used for determining the multiple threshold values. A single threshold value t is computed as [33],

$$t = \frac{1}{\eta} \sum_{m=0}^{L-1-q} \sum_{n=m+q}^{L-1} \frac{m+n}{2} G(m,n) \tag{4}$$

where

$$\eta = \sum_{m=0}^{L-1-q} \sum_{n=m+q}^{L-1} G(m,n) \tag{5}$$

The edge component q is characterized as the range in the summation operation. This range forces the equation to calculate the threshold value in an area in the GLCM restricted by $n - m \geq q$. Here, η represents the total number of pixel pairs within the GLCM with edge magnitudes greater than or equal to q. Further, the symmetrical feature of GLCM is exploited to compute the equation in the upper part of the matrix only. Based on the definition of optimum threshold as presented in Eq. (2), here it is obtained as,

$$t_{opt} = \arg\max \left\{ \sum_{q} f(\cdot) \right\} \tag{6}$$

An optimum threshold value is obtained when the summation over edge magnitude is maximized. The derivation of optimal threshold values for multilevel thresholding of an image is presented in Sect. 4. Four different evolutionary computing techniques used for maximizing edge information of an image are briefly discussed in Sect. 3.

3 Evolutionary Computing Techniques

This section presents evolutionary computing techniques used to optimize the objective function for obtaining optimal threshold values. We have defined and implemented here hybrid CS-BFO for maximizing edge magnitude. CS, PSO, and GA are also implemented for a comparison.

3.1 *Cuckoo Search Technique*

Cuckoo search technique has been proposed by Yang and Deb [34, 35] which is inspired by the parasitic behavior of the cuckoo bird. The characteristic of CS algorithm is that it uses fewer parameters for tuning. Besides, its performance is improved by Levy flights rather than random walk. The most attractive feature is its inbuilt constraint handling that helps us for rescaling of pixel values. Usually, the cuckoo birds lay their eggs in the nests of other host birds. Sometimes, they may even cleverly remove the host bird's egg to improve the hatching probability of their own eggs. As mentioned by Yang and Deb, CS is designed on the basis of three ideal rules:

1. Each of the cuckoo birds lays one egg at a time and places it in a randomly chosen nest.
2. The nests containing high quality of eggs are considered best nests and are selected for the next generation.
3. The number of available host nests is fixed. The host bird may recognize the cuckoo's egg. In such case, it may either throw the egg or discard the nest and make a new nest at a new location. This assumption is represented by a probability $p_a \in [0, 1]$, where a fraction of the total number of host nests is replaced by new nests (with new random solutions).

3.2 *Genetic Algorithm*

Genetic algorithm(GA) is a search algorithm inspired by the concepts of natural selection [36]. A chromosome in the form of string represents a solution to a given problem. It consists of a set of elements called genes that carry the value of optimizing variables. GA searches among a population of chromosomes and works with a coding of parameter set. It uses the information of the objective function without any gradient information. It is applied to a variety of function optimization, parameter estimation, and machine learning applications. A brief framework ofgenetic algorithm is given below.

begin

1. A random population of n chromosomes (which represents assumed solutions for the given problem) is generated.
2. The objective function $f(x)$ for each of the chromosomes x in the population is computed.
3. A new population is generated by repeating the steps mentioned below until the new population size is same as the original population.

   a. *Selection*: Two parent chromosomes are chosen from a population as per their fitness values (higher is the fitness value, more is the probability to be selected).
   b. *Crossover*: The parent chromosomes selected from the above step are crossed to form new offspring (children). This is performed with a predefined crossover probability. If no crossover operation is done, the offspring will be simply copy of parents.
   c. *Mutation*: This step is performed to come out of local optima. A predefined mutation probability is used to mutate new offspring at each location (position in chromosome).

4. The newly created population is used for the next run of the algorithm.
5. If the terminating condition is met, **stop**. Then the best solution in the current population is returned.
6. ***loop*** repeat from Step 2.

end

A set of solutions (characterized by chromosomes) coined as population is created using random numbers. Results from one population are selected and employed to make a new population following three main steps; selection, crossover, and mutation. It is anticipated that the new population will be improved as compared to the old one. Results (parents) which are chosen to form new solutions (offspring) are selected as per their fitness—higher the fitness, the more chances they will get to reproduce. This process is reiterated until some termination condition (for example number of population or may be an improvement over the best solution) is not met.

Iterative application of these steps on the current population creates new populations, producing solutions that are nearer to the optimum solution. The fitness function values for the individuals of the new population are again computed. These values represent the fitness of the solutions of the new generations. In each generation, if the solution is improved, it is retained as the best solution. This is repeated till a termination criterion is met.

3.3 Particle Swarm Optimization

Particle swarm optimization (PSO) is a soft computing method developed by Kennedy and Eberhart [37]. It is based on the swarming behavior of species such as fish and birds. An individual is represented as a particle in PSO with a pre-owned position. Note that the search space has a dimension D which denotes the search space of the problem. The algorithm basically involves throwing a population of particles into the search space, retaining the best (most fit) solution came across. It uses the particle's position and computes a fitness value. The position of the particles with the highest objective function value in the entire run is termed global best $(gbest)$. It is noteworthy to mention here that each particle also remembers its individual highest fitness value. Further, this location is called its personal best $(pbest)$. At the end of each iteration step, every particle updates its velocity vector. The updating depends on its momentum and the effect of both its best solution and the neighbor's best solution. Then a new point is computed to examine. The velocity and the position of each particle are updated as,

$$v^{t+1} = w * v^t + c_1 * rand * (pbest - x^t) + c_2 * rand * (gbest - x^t) \tag{7}$$

$$x^{t+1} = x^t + v^{t+1} \tag{8}$$

where c_1 and c_2 represent acceleration constant, $rand$ is the random function, w is the inertia weight, v^{t+1} is the updated velocity, and x^{t+1} is the updated current location of a particle.

3.4 Hybrid CS-BFO Algorithm

Bacteria foraging optimization (BFO) is a popular evolutionary optimization technique based on the foraging behavior of the *E. Coli* bacteria developed by Passino [38]. It is designed to solve non-gradient optimization problems. The hyperspace is searched using three main processes: chemotaxis, reproduction, and elimination-dispersal events. The chemotaxis process is carried out in two different ways; swimming (running) and tumbling. The bacterium in its lifetime alternates between these two ways of motion. In BFO, a tumble is characterized by a unit length in a random direction $\phi(j)$ which is used to define the direction of movement after a tumble. Usually, the size of the step taken in the random direction specified by the tumble is represented as $C(i)$. This step size is modified by using the Levy flight feature of cuckoo search algorithm. Here, we introduce a random walk via Levy flights as it is more efficient in exploring the search space. The reason being the step size becomes much longer in the long run. The bacteria will move to a new location by performing a Levy flight operation and the step size is now defined as,

$$C(i) = C(i) + \alpha \cdot *Levy(\lambda)$$

where $\alpha > 0$ is the factor contributing to step size. In most of the cases $\alpha = 1$ is considered [35]. Levy flights provide a random walk while their random steps are derived from a levy distribution, $Levy \sim u = i^{-\lambda}, \quad (1 < \lambda \leq 3)$ which has an infinite variance with infinite mean. The successive jumps/steps form a random walk process which follows a power law step length distribution with a heavy tail. In fact, BFO supplements Levy flight feature of CS to produce better fitness function values. With a population of S bacteria, the position of the ith bacterium of the jth chemotactic step, k^{th} reproduction step and the lth elimination/dispersal event is represented by $P_i(j, k, l) \in \Re^p$. At this position, the fitness function also known as nutrient function is denoted by $J(i, j, k, l)$. After a tumble, the location of the ith bacterium is represented as $P_i(j+1, k, l) = P_i(j, k, l) + C(i)\phi(j)$. Now if at $P_i(j+1, k, l)$, the fitness function $J(i, j+1, k, l)$ is better (lower) than $J(i, j, k, l)$, another step of size $C(i)$ is taken in the same direction. This swimming operation is continued till a lower cost is obtained, but is limited to a maximum predefined number of steps N_s. The fitness function of each bacterium in the population may also change by a type of swarming that is implemented by cell-to-cell signaling via an attractant released by the bacteria group to form swarm patterns. The cell also repels a nearby cell assuming that two cells cannot be physically present at the same location. The combined cell-to-cell attraction and repelling effects is expressed as follows:

$$\begin{aligned} J_{cc}(P, Q(j,k,l)) &= \textstyle\sum_{i=1}^{S} J_{cc_i}(P, P_i(j,k,l)) \\ &= \textstyle\sum_{i=1}^{S}\left[-d_{attract}\exp\left(-w_{attract}\sum_{m=1}^{Q}(P_m - P_{m_i})^2\right)\right] \\ &+ \textstyle\sum_{i=1}^{S}\left[-h_{repellant}\exp\left(-w_{repellant}\sum_{m=1}^{Q}(P_m - P_{m_i})^2\right)\right] \end{aligned}$$

where $d_{attract}$, $w_{attract}$, $h_{repellant}$, $w_{repellant}$ are coefficients that represent the characteristics of the attractant and repellent signals released by the cell and P_{m_i} is the mth component of ith bacterium position P_i. $Q(j, k, l)$ is the position of each member in the population of the Sbacteria and is defined as,

$$Q(j,k,l) = P_i(j,k,l) \quad for\ i = 1, 2, \ldots, S$$

The cell-to-cell signaling effect can be added to the cost function as,

$$J(i, j, k, l) + J_{cc}(P, Q)$$

A reproduction step is taken after executing a maximum number of chemotactic steps N_c. The population S is divided in two halves, which form the population members to reproduce. Out of these population members, the least healthy half bacterium perishes and the other healthiest half splits into two bacteria, which are placed at the same position. More characteristics of chemotactic and swarming behavior of bacteria are reported in [38]. The idea of using Levy flights in CS-BFO is to explore the search space more efficiently. After N_{re} reproduction steps, an elimination-dispersal event takes place for N_{ed} number of events. In this process, each bacterium could be dispersed to explore the other domain of the search space. For each elimination-dispersal event, each bacterium in the population experiences

Fig. 3 Flow chart for CS-BFO algorithm

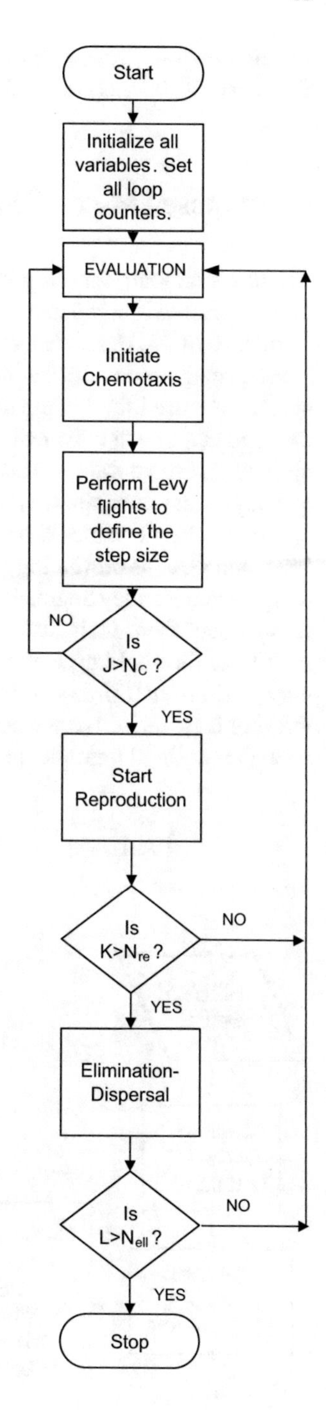

the elimination-dispersal with a predefined probability p_{ed}. For better clarity, the flow chart of the proposed CS-BFO algorithm is presented in Fig. 3.

4 Proposed Methodology

The block diagram for our proposed scheme is shown in Fig. 4 where the GLCM of the given image is computed as per Eq. 1. The edge magnitude information is obtained from the GLCM. Here, we have used a novel hybrid CS-BFO algorithm to maximize the edge magnitude information. Other evolutionary computing techniques CS, PSO, and GA are used for a comparison. The optimal threshold values are obtained from the edge magnitude information. Finally, the thresholded image is obtained by using optimal threshold values. Results obtained with different evolutionary computing techniques are compared. Here, we present theoretical formulations for obtaining optimal threshold values. Constraint handling mechanism is introduced in CS-BFO, PSO, and GA for preventing constraint violation and for violation repair. Different images from Berkley Segmentation Database [39] are considered here to experiment. It is evident from their histograms that they are all multimodal in nature. Hence, multilevel thresholding is considered here for obtaining a thresholded image. The gray levels from $0\ to\ L-1$ of the given image are divided into K classes by setting thresholds $t_1, t_2, \ldots, t_{k-1}$ where $t_0 = 0$ and $t_k = L-1$. The objective function for obtaining optimal multiple threshold values is expressed as,

$$[t_{opt1}, t_{opt2}, \ldots, t_{opt(k-1)}] = \arg \max \left\{ \sum_{q_1, q_2, \ldots, q_{k-1}} f(\cdot) \right\} \tag{9}$$

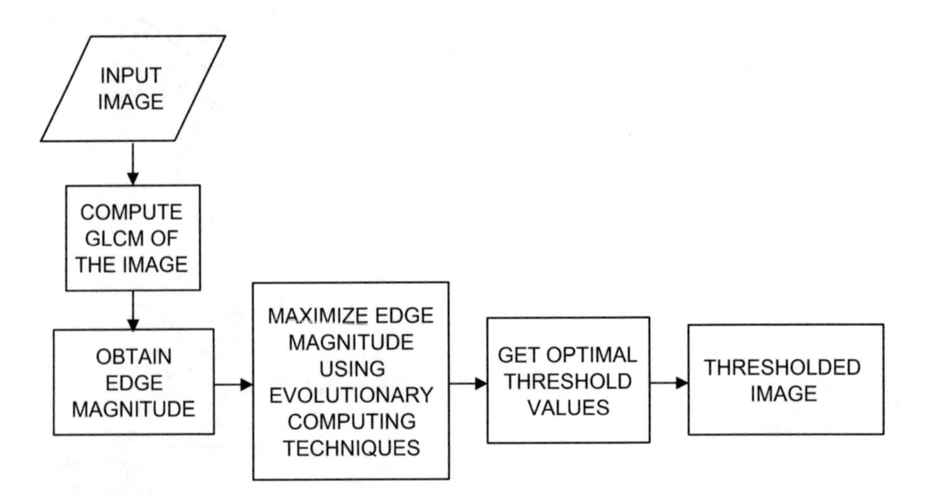

Fig. 4 Block diagram of the proposed scheme

subject to

$$0 < t_{opt1} < t_{opt2} < \cdots < t_{opt(k-1)} < L - 1 \tag{10}$$

Here, $q_1, q_2, \ldots, q_{k-1}$ represent edge magnitude. The optimal threshold values are one that maximizes Eq. (9). The optimal threshold values are computed as,

$$\begin{aligned}
t_{opt1} &= \arg\ \max\left(\frac{1}{\eta_1}\sum_{m=0}^{q_1}\sum_{n=q_1+1}^{q_2}\left(\frac{m+n}{2}\right)G(m,n)\right)\\
t_{opt2} &= \arg\ \max\left(\frac{1}{\eta_2}\sum_{m=q_1+1}^{q_2}\sum_{n=q_2+1}^{q_3}\left(\frac{m+n}{2}\right)G(m,n)\right)\\
&\vdots\\
t_{opt(k-1)} &= \arg\ \max\left(\frac{1}{\eta_{k-1}}\sum_{m=q_{k-2}+1}^{q_{k-1}}\sum_{n=q_{k-1}+1}^{L-1}\left(\frac{m+n}{2}\right)G(m,n)\right)
\end{aligned} \tag{11}$$

where

$$\begin{aligned}
\eta_1 &= \sum_{m=0}^{q_1}\sum_{n=q_1+1}^{q_2} G(m,n)\\
\eta_2 &= \sum_{m=q_1+1}^{q_2}\sum_{n=q_2+1}^{q_3} G(m,n)\\
&\vdots\\
\eta_{k-1} &= \sum_{m=q_{k-2}+1}^{q_{k-1}}\sum_{n=q_{k-1}+1}^{L-1} G(m,n)
\end{aligned} \tag{12}$$

Here, the multiple threshold values obtained are optimal when the summation over q is maximized. The multiple edge magnitude calculation area is presented in Fig. 5. The dotted line represents the symmetrical area where the edge magnitude is zero. The upper portion of the GLCM is only utilized due to its symmetrical behavior. After obtaining optimal threshold values, the given image is thresholded using the scheme as mentioned below. Let the thresholding level be 2. Then there will be two optimum threshold values. In this chapter, for a given image with gray levels, the thresholded image has gray values,

$$\begin{aligned}
&t, && 0 < t \le t_{opt1}\\
&t_{opt1}, && t_{opt1} < t \le t_{opt2}\\
&t_{opt2}, && t_{opt2} < t < L - 1
\end{aligned} \tag{13}$$

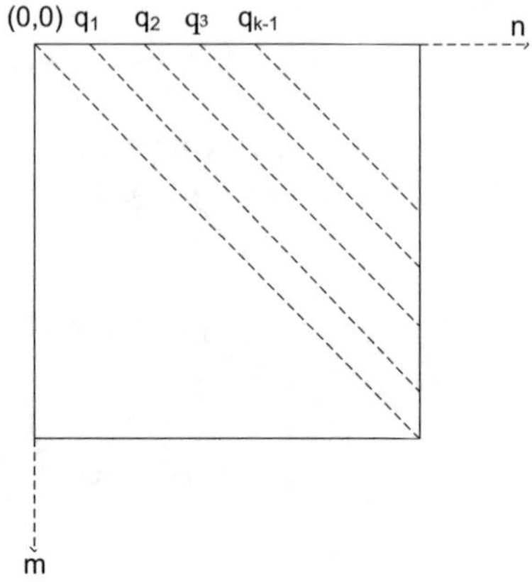

Fig. 5 Multiple edge magnitude calculation area

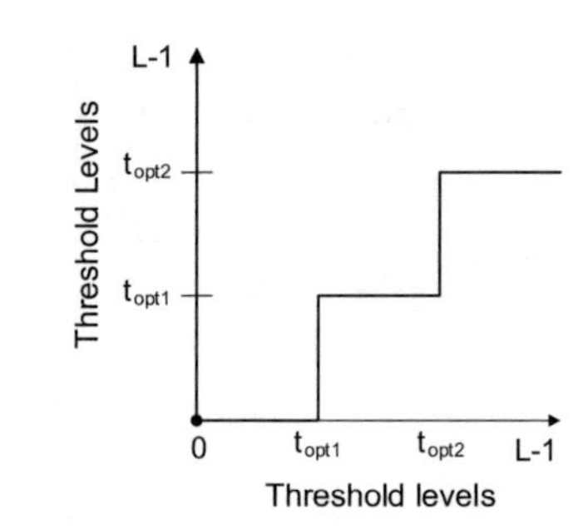

Fig. 6 Scheme for thresholding an image at level 2

Figure 6 shows the proposed scheme for thresholding an image at level 2. For higher levels of thresholding, similar scheme is used to obtain the thresholded image. The following section presents the results obtained by using four different evolutionary computing techniques for multilevel thresholding.

5 Results and Discussions

The proposed methodology is tested with standard Berkley segmentation database. Nine different images from Berkley database are considered to validate our claim. Test Images and thresholded images are displayed from Figs. 7, 8, 9, 10, 11, 12, 13, 14, 15, 16, 17, 18, 19, 20, 21, 22, 23, 24, 25 and 26. One synthetic image (square image) with six different gray levels is considered in this experiment in order to justify the use of our method for 5-level thresholding problems. Some of the images (e.g., image 5096, 118072) are having poor illumination condition while some other images (e.g., image 48017, 69007) are having good edge information. Some images have good textural information. Some images (e.g., image 289011) have a very low gray-level variation which makes it difficult to threshold at lower levels. However,

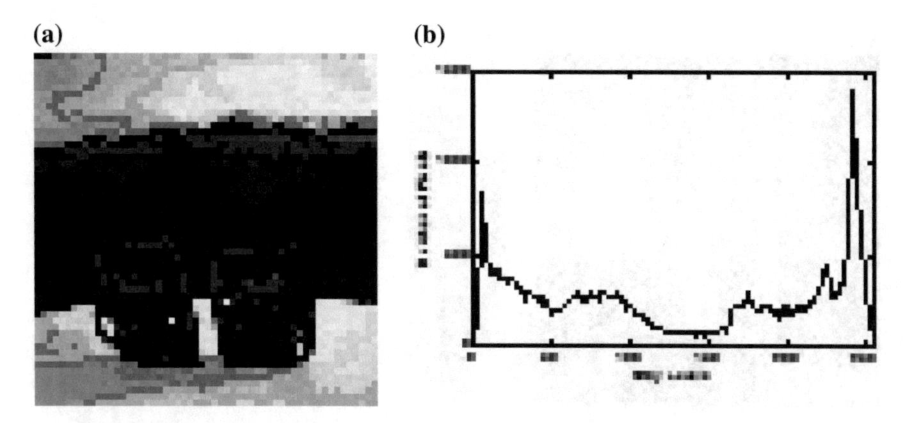

Fig. 7 **a** Original 2018 image, **b** Histogram of (**a**)

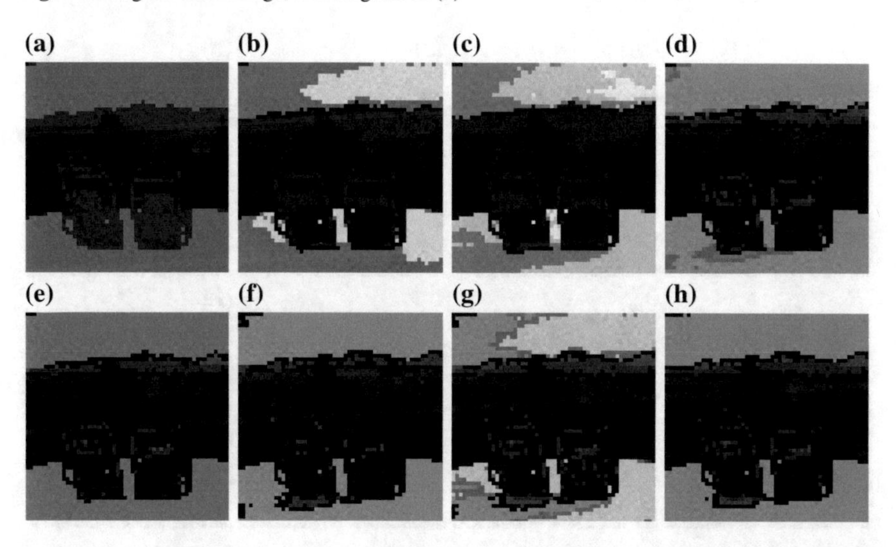

Fig. 8 Thresholded images. **a** Thresholded at level 2 with CS-BFO. **b** Thresholded at level 3 with CS-BFO. **c** Thresholded at level 4 with CS-BFO. **d** Thresholded at level 5 with CS-BFO. **e** Thresholded at level 2 with CS. **f** Thresholded at level 3 with CS. **g** Thresholded at level 4 with CS. **h** Thresholded at level 5 with CS

the thresholded image is expressive at levels 4 and above. Under the reconstruction rule in equation explained in this chapter, it is obvious that we need to consider 4 or higher level thresholding for obtaining good quality results for the square image. Here, we emphasize on the point that multilevel thresholding is very much essential for square image, which is considered to be typical having six gray-level variations. This particular synthetic square image is mostly used for validating multilevel thresholding algorithms. For 2-level thresholding, the number of classes should be 3. We can assign only two optimum threshold values (t_{opt_1}, t_{opt_2}) for two classes while value t is assigned to the other class. The fact that t refers to the original pixel value

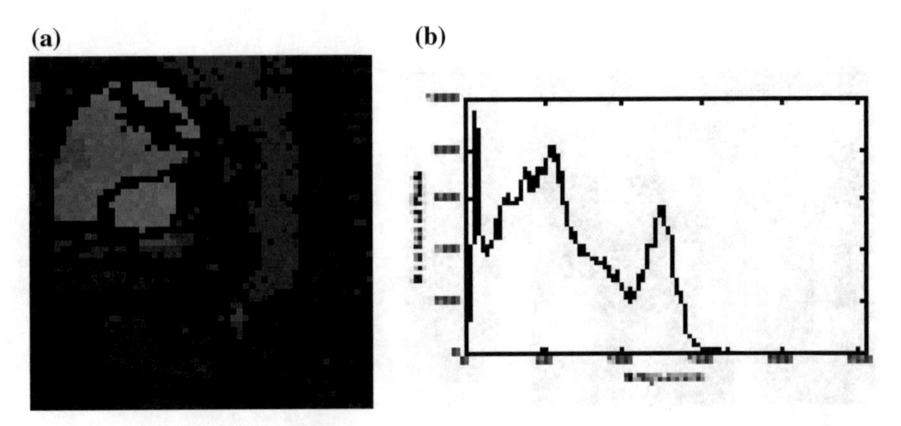

Fig. 9 **a** Original 5096 image, **b** Histogram of (**a**)

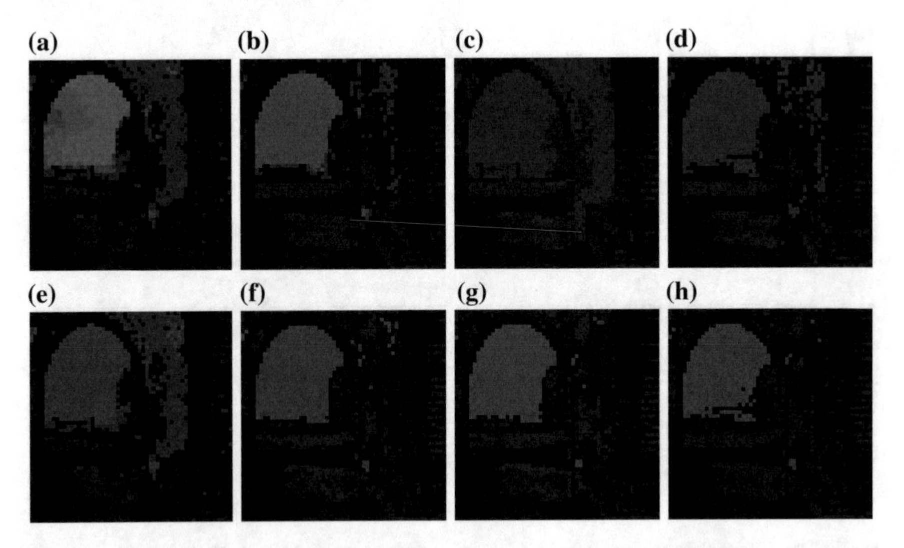

Fig. 10 Thresholded images. **a** Thresholded at level 2 with CS-BFO. **b** Thresholded at level 3 with CS-BFO. **c** Thresholded at level 4 with CS-BFO. **d** Thresholded at level 5 with CS-BFO. **e** Thresholded at level 2 with CS. **f** Thresholded at level 3 with CS. **g** Thresholded at level 4 with CS. **h** Thresholded at level 5 with CS

may be reiterated. Therefore, pixels with gray values less than t_{opt_1} will retain their original values in the output. To be precise, all gray values less than t_{opt_1} will appear in the thresholded image, causing ambiguity. But this problem can be avoided by using higher thresholding levels (4 or higher), which is clearly observed in thresholded square image. This should convince the readers that multilevel thresholding is a worthwhile subject of study. In this sense, this contribution is quite significant for the subject. Note that m number of thresholds lead to $m + 1$ classes here. Results are presented in the form of figures and tables. Experiments are carried out on Intel

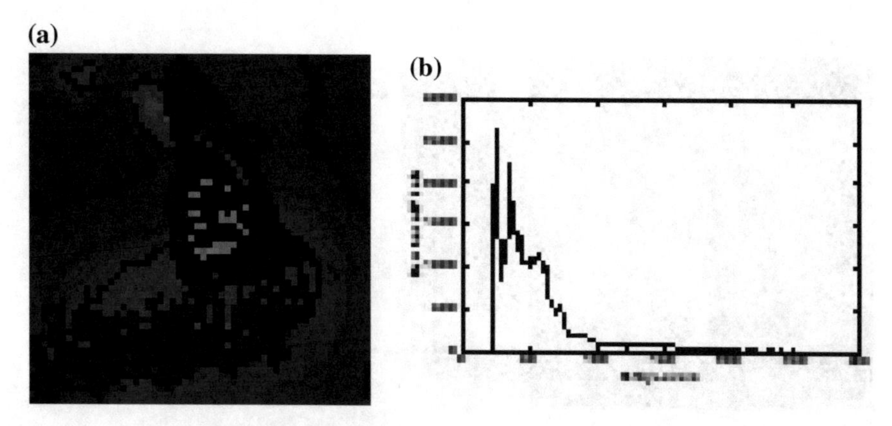

Fig. 11 a Original 35049 image, b Histogram of (a)

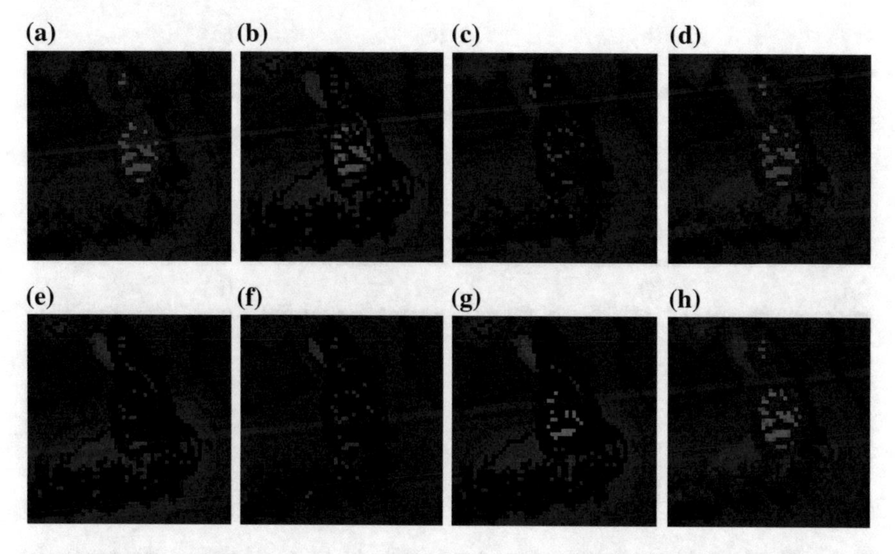

Fig. 12 Thresholded images. a Thresholded at level 2 with CS-BFO. b Thresholded at level 3 with CS-BFO. c Thresholded at level 4 with CS-BFO. d Thresholded at level 5 with CS-BFO. e Thresholded at level 2 with CS. f Thresholded at level 3 with CS. g Thresholded at level 4 with CS. h Thresholded at level 5 with CS

core i-3 processor with 4 GB RAM running under Windows 7.0 operating system. The algorithms are implemented using MATLAB. Best objective function values and optimum threshold values for test images are displayed in Table 1. Comparison of evaluation parameters, Structured Similarity Index (SSIM) and Featured Similarity Index (FSIM), for measuring thresholding performance is shown in Table 2. Comparison of evaluation parameters, Inverse Difference Moment (IDM) and Contrast (CON), for measuring thresholding performance is shown in Table 3. Peak signal-to-noise ratio (PSNR) values and CPU Time (sec) are displayed in Tables 4 and 5.

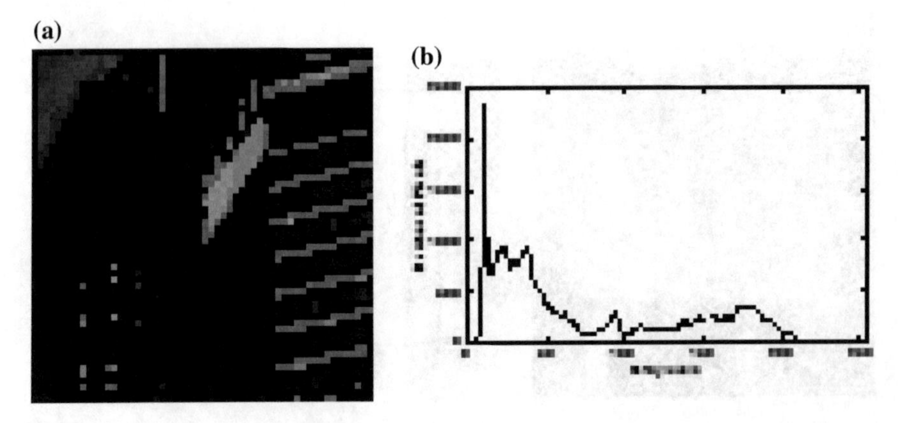

Fig. 13 **a** Original 48017 image, **b** Histogram of (**a**)

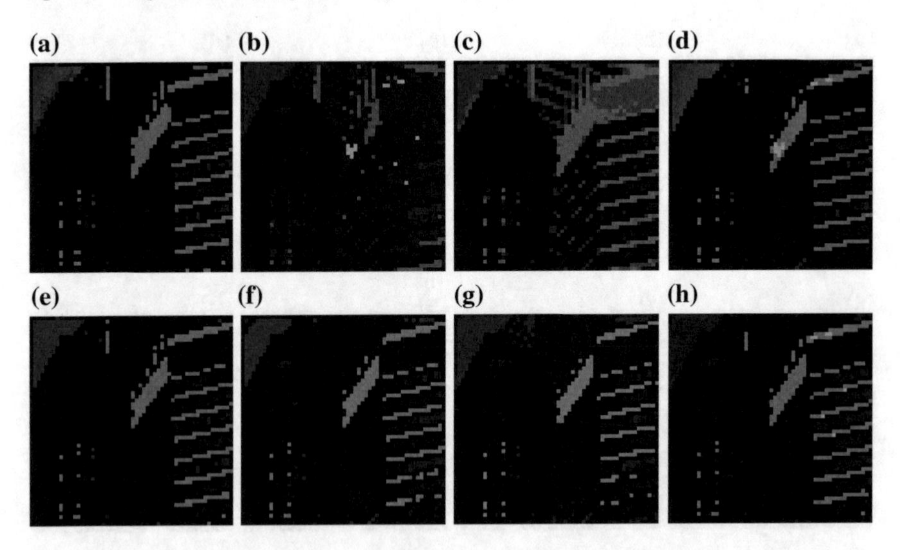

Fig. 14 Thresholded images. **a** Thresholded at level 2 with CS-BFO. **b** Thresholded at level 3 with CS-BFO. **c** Thresholded at level 4 with CS-BFO. **d** Thresholded at level 5 with CS-BFO. **e** Thresholded at level 2 with CS. **f** Thresholded at level 3 with CS. **g** Thresholded at level 4 with CS. **h** Thresholded at level 5 with CS

Test Images

The common parameter setting chosen for all techniques is: Lower Bound (LB) = 2, Upper Bound (UB) = 255, Population Size (NP) = 30, Iterations (Iter) = 30, and Image size = 256 × 256. Here we have implemented GA using OPTIM toolbox of MATLAB. The CPU time required with GA is much more than the other methods. This is the reason why we have not included it in Table 4.

The parameters chosen for cuckoo search are: mutation probability = 0.25, scale factor = 1.5.

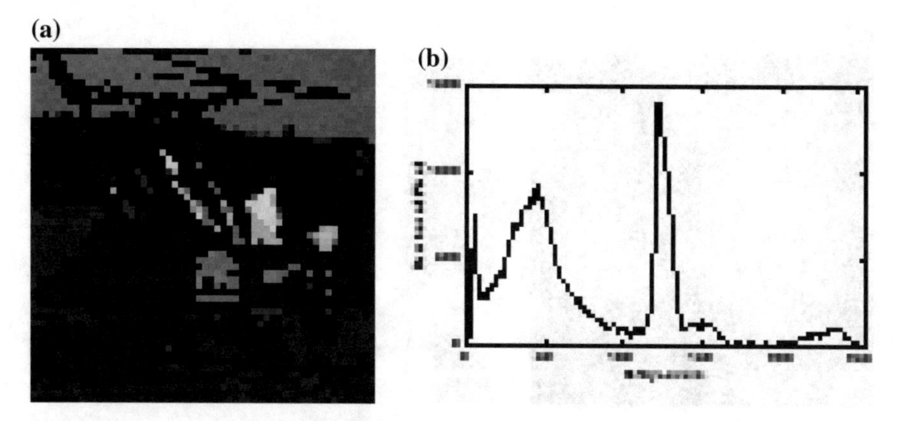

Fig. 15 **a** Original 69000 image, **b** Histogram of (**a**)

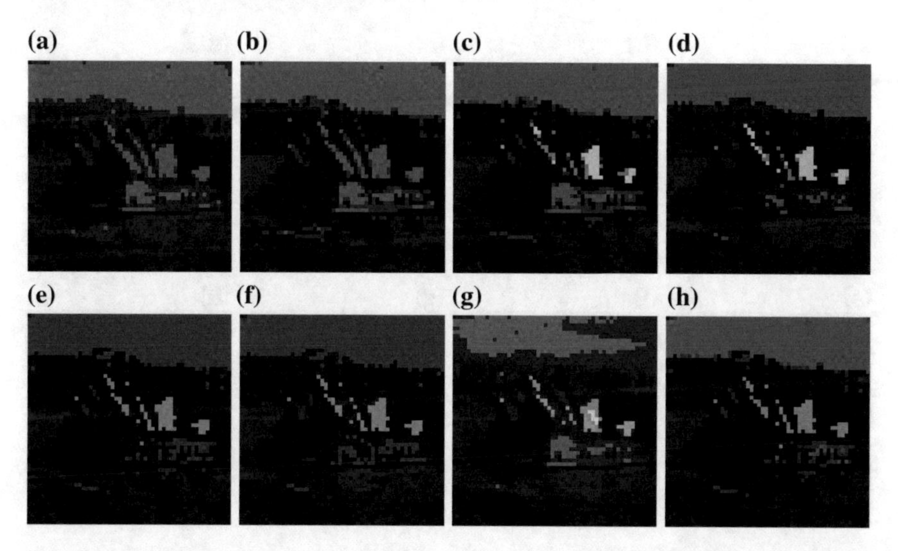

Fig. 16 Thresholded images. **a** Thresholded at level 2 with CS-BFO. **b** Thresholded at level 3 with CS-BFO. **c** Thresholded at level 4 with CS-BFO. **d** Thresholded at level 5 with CS-BFO. **e** Thresholded at level 2 with CS. **f** Thresholded at level 3 with CS. **g** Thresholded at level 4 with CS. **h** Thresholded at level 5 with CS

The parameters chosen for hybrid CS-BFO are: Number of chemotactic steps = 10, Swimming length = 10, Here $C(i) = 0.1$, $\alpha = 1$, $\lambda = 1.5$, Number of reproduction steps = 4, Number of elimination and dispersal events = 2, Probability of elimination and dispersal = 0.02, Depth of attractant = 0.1, Width of attractant = 0.2, Height of repellent = 0.1, Width of repellent = 10. The parameters chosen for particle swarm optimization are: Weight of velocity = 0.9–0.5, Coefficient of personal and social cognition = 2.

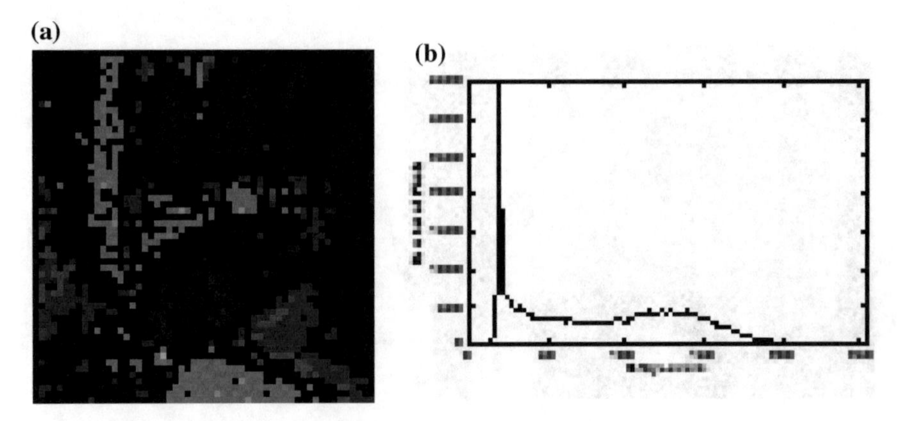

Fig. 17 **a** Original 118072 image, **b** Histogram of (**a**)

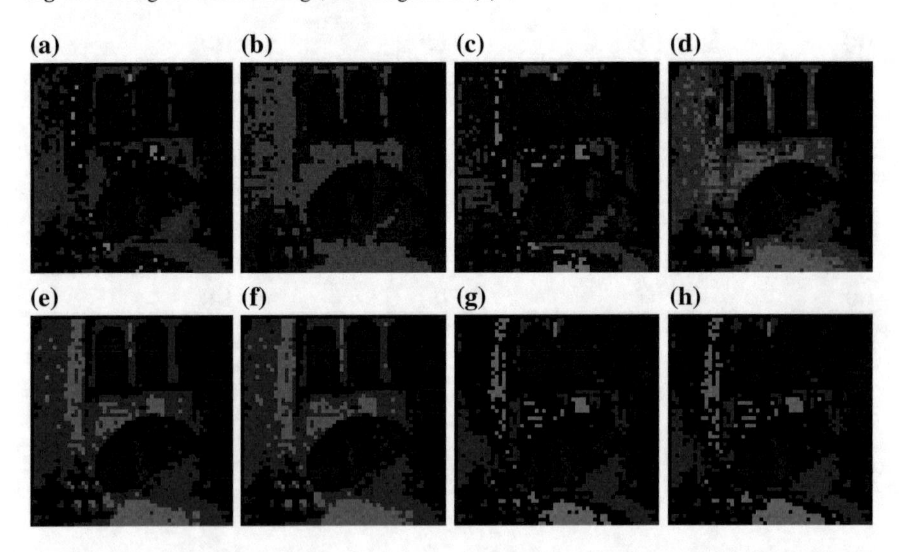

Fig. 18 Thresholded images. **a** Thresholded at level 2 with CS-BFO. **b** Thresholded at level 3 with CS-BFO. **c** Thresholded at level 4 with CS-BFO. **d** Thresholded at level 5 with CS-BFO. **e** Thresholded at level 2 with CS. **f** Thresholded at level 3 with CS. **g** Thresholded at level 4 with CS. **h** Thresholded at level 5 with CS

The parameters chosen forgenetic algorithm are: Crossover probability = 0.9, mutation probability = 0.1, Selection operator = Roulette wheel.

From Table 1 it is observed that best objective function values are obtained by using CS-BFO method. Here, we consider many parameters to measure the performance of our method. Structured Similarity Index (SSIM) computes the visual similarity between the original image and the thresholded image at a particular threshold level. It is always less than 1 and a higher value of SSIM closer to 1 exhibits better thresholding [40]. Similarly, Feature Similarity Index (FSIM) also represents the visual similarity

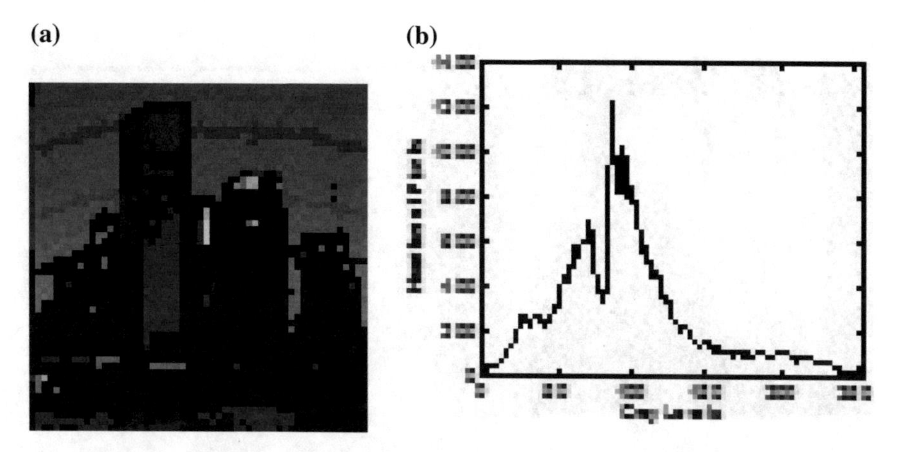

Fig. 19 **a** Original 69007 image, **b** Histogram of (**a**)

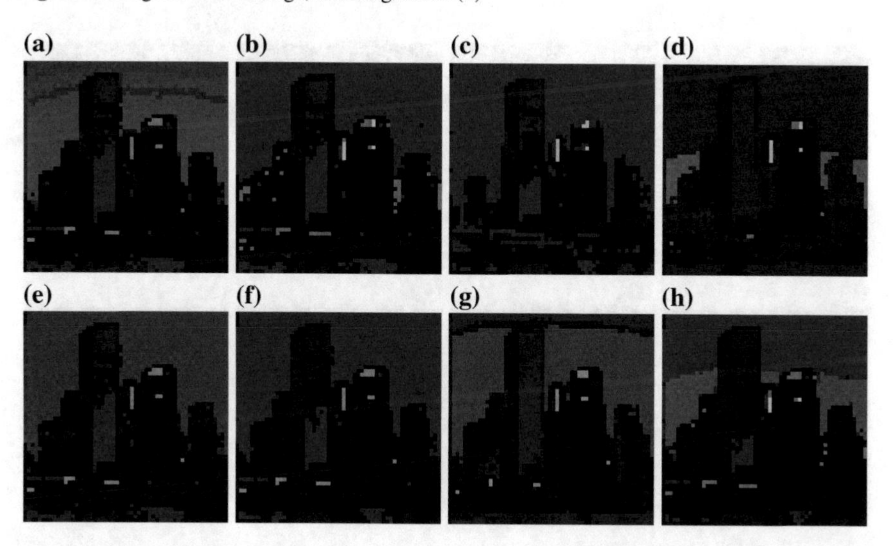

Fig. 20 Thresholded images. **a** Thresholded at level 2 with CS-BFO. **b** Thresholded at level 3 with CS-BFO. **c** Thresholded at level 4 with CS-BFO. **d** Thresholded at level 5 with CS-BFO. **e** Thresholded at level 2 with CS. **f** Thresholded at level 3 with CS. **g** Thresholded at level 4 with CS. **h** Thresholded at level 5 with CS

between the original image and the thresholded image at a particular threshold level. A higher value of FSIM closer to 1 represents a better thresholding performance [41]. This parameter is added here to strengthen our claim.

The SSIM index is calculated as:

$$SSIM(I, \tilde{I}) = \frac{(2\mu_I \mu_{\tilde{I}} + C_1)(2\sigma_{I\,\tilde{I}} + C_2)}{({\mu^2}_I + {\mu^2}_{\tilde{I}} + C_1)({\sigma^2}_I + {\sigma^2}_{\tilde{I}} + C_2)} \tag{14}$$

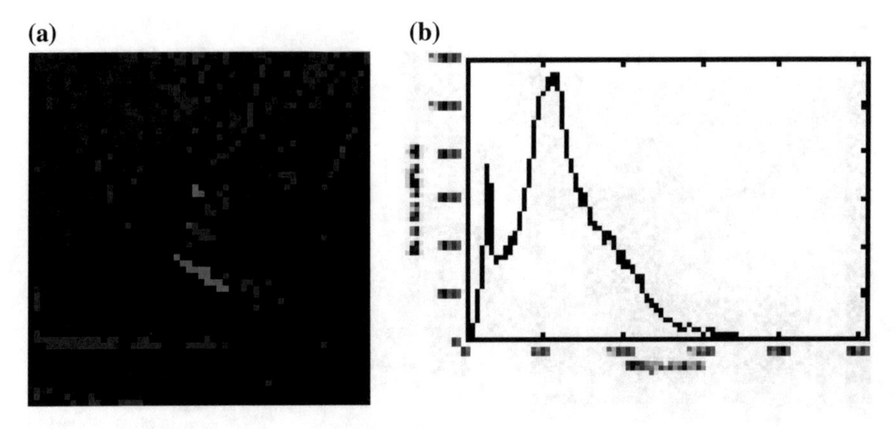

Fig. 21 **a** Original 268074 image, **b** Histogram of (**a**)

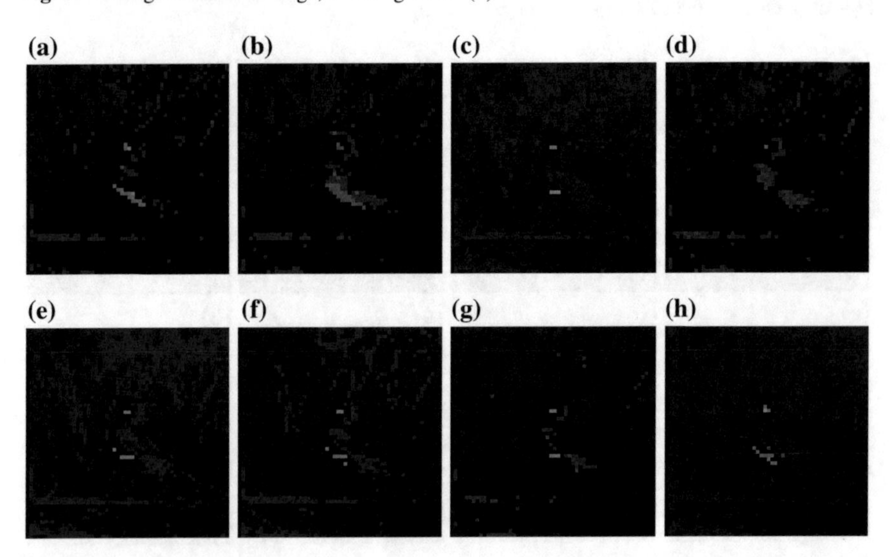

Fig. 22 Thresholded images. **a** Thresholded at level 2 with CS-BFO. **b** Thresholded at level 3 with CS-BFO. **c** Thresholded at level 4 with CS-BFO. **d** Thresholded at level 5 with CS-BFO. **e** Thresholded at level 2 with CS. **f** Thresholded at level 3 with CS. **g** Thresholded at level 4 with CS. **h** Thresholded at level 5 with CS

Here 'I' denote the original image and '$\tilde{I}$' represent the thresholded image at a particular level. Note that μ_I is the average of I, $\mu_{\tilde{I}}$ is the average of $\tilde{I}$, σ_I^2 is the variance of I, $\sigma_{\tilde{I}}^2$ is the variance of $\tilde{I}$, $\sigma_{I\tilde{I}}$ is the variance of I and $\tilde{I}$. Here C_1 and C_2 are constants and are chosen as 0.065. It is seen from Table 2 that SSIM is higher for CS-BFO method. In this sense, the proposed CS-BFO technique produces results which are visually better than other techniques. The FSIM index is calculated as:

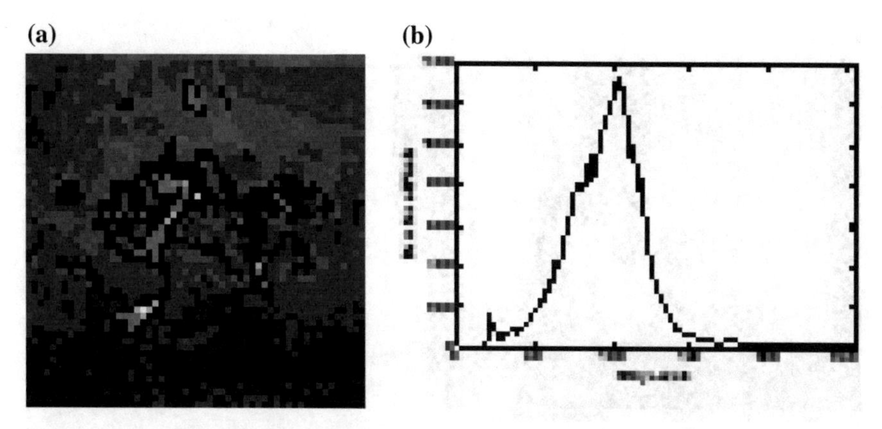

Fig. 23 a Original 289011 image, b Histogram of (a)

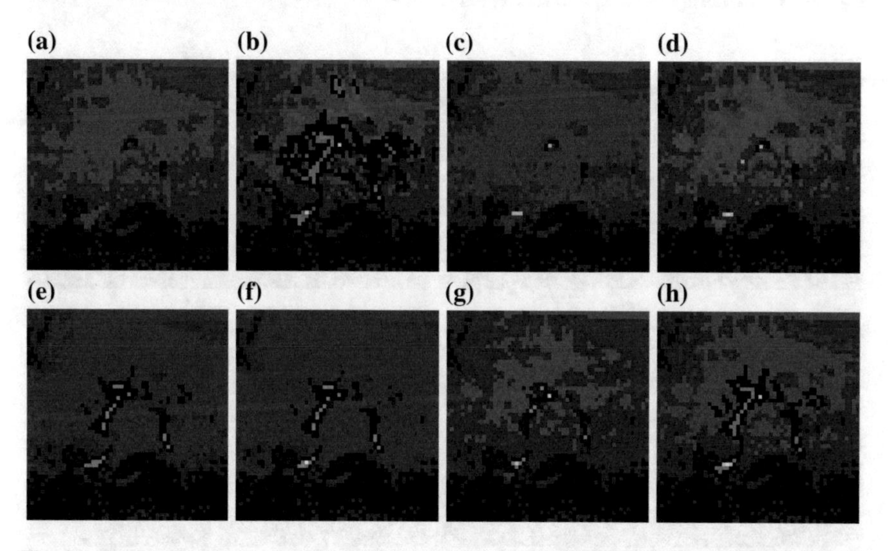

Fig. 24 Thresholded images. a Thresholded at level 2 with CS-BFO. b Thresholded at level 3 with CS-BFO. c Thresholded at level 4 with CS-BFO. d Thresholded at level 5 with CS-BFO. e Thresholded at level 2 with CS. f Thresholded at level 3 with CS. g Thresholded at level 4 with CS. h Thresholded at level 5 with CS

$$FSIM = \frac{\sum_{X \in \Omega} S_L(X) \cdot PC_m(X)}{\sum_{X \in \Omega} PC_m(X)} \tag{15}$$

where Ω is the whole image in spatial domain. Here $S_L(X)$ is the overall similarity of two images, $PC_m(X)$ is the maximum phase congruence between two images. Interestingly, FSIM is also higher for CS-BFO (as depicted in Table 2). It is worthy to claim here that improved SSIM and FSIM are obtained for CS-BFO. Hence, CS-BFO performs better than other methods. While considering SSIM and FSIM, CS is the second contestant and shows better indices than PSO and GA.

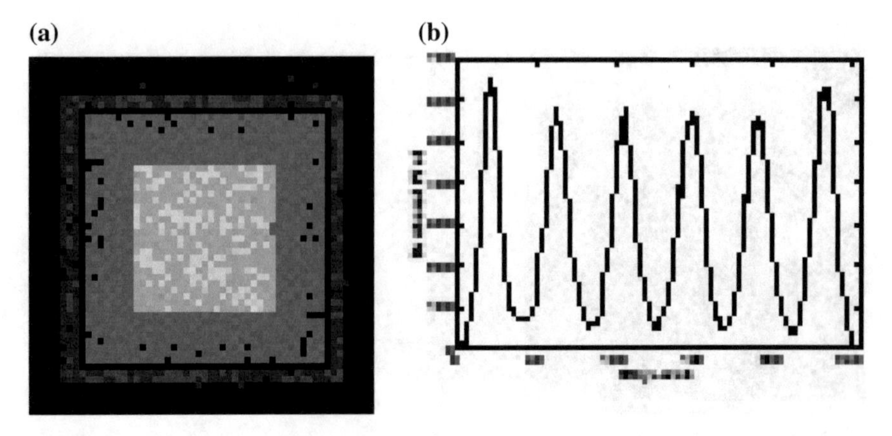

Fig. 25 **a** Original *square* image, **b** Histogram of (**a**)

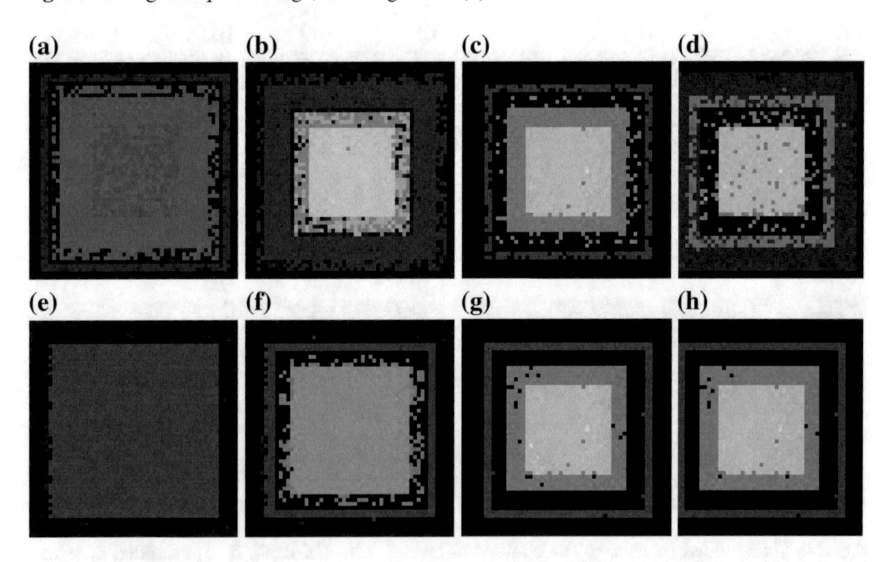

Fig. 26 Thresholded images. **a** Thresholded at level 2 with CS-BFO. **b** Thresholded at level 3 with CS-BFO. **c** Thresholded at level 4 with CS-BFO. **d** Thresholded at level 5 with CS-BFO. **e** Thresholded at level 2 with CS. **f** Thresholded at level 3 with CS. **g** Thresholded at level 4 with CS. **h** Thresholded at level 5 with CS

Contrast (CON) and Inverse Difference Moment (IDM) [28] are also used to describe the GLCM parameters. The contrast is defined in Eq. 3. Here, $C(m, n)$ represents the normalized GLCM. A high value of contrast implies better thresholded image, which is higher for CS-BFO method (as shown in Table 3).

Table 1 Best objective function values

Test images	m	Best objective function values			
		CS-BFO	CS	PSO	GA
2018	2	257.84	250.00	247.52	244.09
	3	336.41	315.25	312.50	303.47
	4	890950.00	7408.50	1005.90	862.16
	5	1175200.00	78849.00	13950.72	1030.90
5096	2	164.48	157.91	156.25	81.33
	3	210.47	192.16	188.68	84.25
	4	1375500.00	1319400.00	523670.00	434782.60
	5	1388700.00	1343200.00	1245700.00	465246.11
35049	2	266.96	260.45	243.90	118.15
	3	329.21	306.87	285.71	238.82
	4	1079644.00	943140.00	91709.46	20378.00
	5	1220600.00	973910.00	624765.71	180790.00
48017	2	230.36	222.22	220.81	114.91
	3	285.98	277.48	232.55	191.54
	4	3203100.00	2459300.00	156150.00	31609.55
	5	3266500.00	3242400.00	697934.11	407990.00
69000	2	267.52	263.15	255.49	162.88
	3	340.71	311.88	303.03	253.28
	4	1220800.00	891030.00	197667.52	145025.00
	5	1482900.00	1398200.00	817995.91	283480.00
69007	2	267.61	263.16	257.97	247.02
	3	334.77	303.03	267.98	265.72
	4	1936500.00	1188600.00	193020.00	3333.33
	5	2837700.00	2776100.00	344827.58	290860.00
118072	2	217.85	200.10	192.30	102.78
	3	260.58	223.25	217.39	106.39
	4	2858500.00	2778900.00	1758462.00	52845.73
	5	3212300.00	2864300.00	2545948.00	375869.19
268074	2	212.23	208.33	200.99	103.68
	3	268.31	232.89	212.76	208.32
	4	1469100.00	885440.00	235144.73	1971.60
	5	2361700.00	1562500.00	996160.00	23714.00
289011	2	276.40	270.27	266.86	99.75
	3	339.15	269.06	263.15	87.09
	4	2476500.00	2465591.00	1251063.40	378.73
	5	3978900.00	3123900.00	2132400.00	80901.00
Square	2	123.55	101.55	96.15	10.09
	3	152.25	118.52	88.49	80.27
	4	1025300.00	1025300.00	845220.00	798722.04
	5	1056694.00	1003200.00	905387.05	845320.00

Table 2 Optimum threshold values

Test images	m	Optimum threshold values			
		CS-BFO	CS	PSO	GA
2018	2	101, 141	59, 125	106, 155	86, 225
	3	58, 82, 164	45, 132, 225	69, 91, 171	35, 158, 244
	4	69, 95, 163, 214	30, 79, 137, 234	171, 212, 228, 242	37, 98, 156, 181
	5	46, 58, 64, 91, 161	107, 142, 161, 179, 185	44, 103, 150, 187, 233	75, 114, 150, 211, 238
5096	2	68, 85	113, 207	52, 98	52, 133
	3	38, 48, 90	135, 226, 248	37, 56, 96	99, 119, 139
	4	51, 95, 152, 187	117, 147, 176, 202	43, 92, 140, 230	12, 61, 78, 117
	5	41, 93, 140, 171, 209	67, 80, 89, 135, 210	42, 52, 94, 142, 164	61, 138, 170, 223, 240
35049	2	81, 150	116, 208	85, 157	27, 137
	3	33, 79, 153	50, 73, 112	49, 90, 140	87, 115, 208
	4	86, 130, 204, 255	61, 100, 138, 168	57, 87, 147, 222	61, 71, 195, 248
	5	43, 63, 100, 125, 216	96, 149, 180, 212, 222	27, 59, 85, 114, 189	90, 118, 178, 205, 246
48017	2	76, 133	130, 244	75, 140	41, 73
	3	54, 75, 142	35, 115, 193	50, 80, 146	79, 141, 207
	4	36, 51, 63, 150	135, 195, 227, 246	22, 62, 149, 202	47, 108, 207, 232
	5	43, 135, 182, 213, 240	16, 39, 54, 101, 230	15, 26, 40, 166, 251	33, 76, 115, 143, 223
69000	2	81, 156	50, 195	93, 164	65, 91
	3	60, 81, 159	39, 127, 229	70, 96, 167	28, 74, 160
	4	51, 122, 184, 234	110, 160, 202, 215	73, 131, 198, 221	97, 177, 200, 226
	5	73, 88, 95, 110, 156	83, 107, 118, 134, 158	38, 117, 158, 179, 213	32, 33, 44, 222, 237
69007	2	83, 157	94, 182	90, 165	65, 202
	3	26, 67, 158	56, 83, 102	71, 92, 166	159, 213, 242
	4	47, 55, 88, 169	116, 170, 208, 232	17, 57, 87, 168	87, 104, 150, 228
	5	27, 71, 104, 140, 215	116, 169, 204, 215, 231	51, 110, 161, 197, 213	70, 107, 149, 194, 218
118072	2	64, 123	116, 199	88, 114	52, 66
	3	53, 67, 121	135, 205, 230	39, 70, 131	72, 106, 161
	4	48, 65, 135, 255	40, 53, 63, 156	39, 59, 139, 200	49, 61, 105, 216
	5	66, 132, 215, 235, 255	23, 76, 97, 139, 220	53, 71, 134, 206, 222	22, 68, 111, 138, 209
268074	2	55, 147	149, 200	88, 115	34, 154
	3	22, 57, 143	47, 138, 229	41, 80, 120	56, 74, 122
	4	25, 76, 102, 150	58, 89, 166, 230	51, 80, 121, 140	103, 124, 149, 186
	5	34, 101, 136, 167, 255	16, 31, 53, 93, 195	43, 76, 107, 180, 203	7, 22, 52, 78, 118

(continued)

Table 2 (continued)

Test images	m	Optimum threshold values			
		CS-BFO	CS	PSO	GA
289011	2	80, 155	118, 225	86, 162	90, 156
	3	79, 159, 247	103, 214, 237	62, 91, 161	185, 222, 239
	4	73, 102, 180, 226	107, 116, 123, 227	76, 109, 184, 219	126, 147, 224, 250
	5	97, 144, 187, 211, 243	83, 147, 170, 195, 211	104, 155, 172, 197, 213	75, 95, 119, 153, 230
Square	2	33, 56	37, 140	45, 73	151, 243
	3	31, 67, 245	58, 73, 155	24, 43, 70	68, 80, 231
	4	64, 139, 159, 236	19, 25, 158, 203	64, 121, 158, 191	65, 76, 181, 236
	5	34, 65, 137, 148, 245	54, 159, 190, 206, 237	27, 37, 64, 140, 150	25, 138, 146, 183, 214

The IDM measures the homogeneity of the image and is calculated as:

$$IDM = \sum_{m=0}^{L-1} \sum_{n=0}^{L-1} \left[1/(1 + (m - n)^2)\right] C(m, n) \tag{16}$$

It is inversely correlated to contrast. IDM is low for inhomogeneous images and relatively higher for homogenous images. It is lower for CS-BFO technique (as displayed in Table 3), which is desirable.

Peak signal-to-noise ratio (PSNR) is used as the quality measure between the original image I and the thresholded image $\tilde{I}$ of size $M \times N$. Higher the PSNR better is the thresholded image. Table 4 displays the variation of PSNR with all the methods. It is defined as [42]

$$PSNR = 20 \log 10 \left(\frac{1}{RMSE} \right) \tag{17}$$

where RMSE is the root mean square error and is defined as

$$RMSE = \sqrt{\frac{1}{M \times N} \sum_{i=1}^{M} \sum_{j=1}^{N} (I(i, j) - \tilde{I}(i, j))^2}$$

From Table 4, it is observed that CS-BFO provides us better PSNR values as compared to other methods. In summary, the proposed CS-BFO method has shown improved results with respect to different measures but computationally expensive. Note that CS is the second contestant in this respect. However, CS shows improved speed compared to other methods. Variations of CPU time (sec) are displayed in

Table 3 Comparison of evaluation parameters for measuring thresholding performance

Test images	m	SSIM				FSIM			
		CS-BFO	CS	PSO	GA	CS-BFO	CS	PSO	GA
2018	2	0.8882	0.8401	0.8002	0.7745	0.9743	0.9674	0.8273	0.5055
	3	0.9330	0.9210	0.8812	0.8338	0.9750	0.9723	0.9693	0.7147
	4	0.9954	0.9338	0.9050	0.8627	0.9793	0.9754	0.9706	0.9635
	5	0.9795	0.9781	0.9777	0.9484	0.9890	0.9843	0.9803	0.9801
5096	2	0.9714	0.9451	0.9412	0.7893	0.9982	0.9820	0.9751	0.5817
	3	0.9949	0.9588	0.9500	0.8565	0.9974	0.9894	0.9818	0.9558
	4	0.9971	0.9594	0.9548	0.8534	0.9914	0.9855	0.9724	0.9656
	5	0.9999	0.9962	0.9619	0.9570	0.9998	0.9979	0.9875	0.9830
35049	2	0.9870	0.9750	0.9729	0.9294	0.9977	0.9786	0.9718	0.9447
	3	0.9891	0.9749	0.9721	0.8520	0.9980	0.9914	0.9908	0.9181
	4	0.9916	0.9889	0.9798	0.9604	0.9985	0.9960	0.9666	0.9558
	5	0.9901	0.9865	0.9783	0.9368	0.9973	0.9859	0.9826	0.9826
48017	2	0.9728	0.9296	0.9261	0.9191	0.9945	0.9863	0.9862	0.9776
	3	0.9745	0.9328	0.8636	0.8205	0.9955	0.9898	0.9821	0.9532
	4	0.9568	0.9532	0.9505	0.9463	0.9955	0.9943	0.9927	0.9213
	5	0.9784	0.9549	0.9509	0.9054	0.9976	0.9942	0.9941	0.9862
69000	2	0.9458	0.9082	0.8440	0.7342	0.9938	0.9879	0.9504	0.9486
	3	0.9389	0.9379	0.9088	0.7901	0.9900	0.9865	0.9846	0.9542
	4	0.9859	0.9686	0.9454	0.9388	0.9956	0.9858	0.9825	0.9685
	5	0.9619	0.9616	0.9540	0.9479	0.9962	0.9949	0.9844	0.9819
69007	2	0.9906	0.9451	0.8937	0.8579	0.9977	0.9859	0.9769	0.9702
	3	0.9412	0.9406	0.9298	0.8139	0.9857	0.9849	0.9834	0.9552
	4	0.9848	0.9663	0.9362	0.9352	0.9951	0.9921	0.9801	0.9792
	5	0.9854	0.9720	0.9533	0.9319	0.9952	0.9909	0.9834	0.9800
118072	2	0.9185	0.8908	0.8435	0.7065	0.9724	0.9662	0.9633	0.9414
	3	0.9841	0.9555	0.9359	0.9270	0.9925	0.9775	0.9762	0.9724
	4	0.9623	0.9361	0.9346	0.8726	0.9908	0.9792	0.9775	0.9651
	5	0.9868	0.9531	0.9413	0.8316	0.9838	0.9824	0.9753	0.9617
268074	2	0.9560	0.8941	0.8800	0.7896	0.9906	0.9627	0.9547	0.9399
	3	0.9660	0.9503	0.9021	0.8005	0.9926	0.9897	0.9705	0.9455
	4	0.9746	0.9704	0.9677	0.9061	0.9947	0.9931	0.9924	0.9711
	5	0.9990	0.9974	0.9870	0.9232	0.9996	0.9993	0.9956	0.9801
289011	2	0.9516	0.9459	0.9372	0.9243	0.9595	0.9551	0.9498	0.9313
	3	0.9824	0.9650	0.9410	0.9254	0.9832	0.9647	0.9529	0.9501
	4	0.9931	0.9792	0.9661	0.9644	0.9964	0.9813	0.9607	0.9603
	5	0.9993	0.9844	0.9823	0.9775	0.9997	0.9887	0.9772	0.9720
Square	2	0.8173	0.7388	0.5215	0.4206	0.9206	0.9056	0.8850	0.8798
	3	0.9117	0.8410	0.5386	0.5198	0.9329	0.9321	0.9308	0.8872
	4	0.9139	0.8976	0.8950	0.8821	0.9508	0.9472	0.9385	0.9333
	5	0.9622	0.9342	0.9181	0.8972	0.9546	0.9499	0.9486	0.9406

Table 4 Comparison of evaluation parameters for measuring thresholding performance

Test images	m	IDM				Contrast			
		CS-BFO	CS	PSO	GA	CS-BFO	CS	PSO	GA
2018	2	0.0638	0.0509	0.0501	0.0501	917.987	859.2759	526.4093	325.13
	3	0.0739	0.0726	0.0626	0.0584	775.8469	659.848	459.7879	428.8619
	4	0.0952	0.0623	0.0584	0.0494	1677.50	885.8352	761.4023	542.3399
	5	0.0768	0.0741	0.0657	0.0486	32513.0	1156.20	632.9395	599.6979
5096	2	0.0661	0.0642	0.0551	0.0490	671.113	522.2733	477.2119	366.524
	3	0.0779	0.0762	0.0499	0.0494	703.3805	700.4217	478.60	431.0529
	4	0.3333	0.0732	0.0653	0.0488	4335.0	693.8393	479.9729	434.7964
	5	0.0758	0.0713	0.0595	0.0543	584.5217	481.8288	435.2423	427.472
35049	2	0.3014	0.0640	0.0621	0.0546	663.5106	503.2428	486.6353	484.2353
	3	0.1960	0.1037	0.0973	0.0599	496.1946	427.7809	369.5752	363.9202
	4	0.0864	0.0796	0.0792	0.0610	558.592	531.883	473.3847	311.223
	5	0.2418	0.1160	0.0581	0.0573	1747.10	683.9374	646.1041	625.9706
48017	2	0.1615	0.0555	0.0553	0.0501	1295.80	706.6496	704.9634	91.3132
	3	0.0990	0.0681	0.0640	0.0541	959.2491	628.1549	626.2626	442.6567
	4	0.1479	0.1395	0.0879	0.0491	1410.90	661.40	492.6044	469.6297
	5	0.1772	0.1707	0.0888	0.0811	782.1527	675.129	581.4051	445.6095
69000	2	0.0759	0.0607	0.0578	0.0578	747.6221	747.4594	657.3651	461.0289
	3	0.0894	0.0865	0.0673	0.0629	698.1237	621.4993	454.8236	420.5664
	4	0.0777	0.0737	0.0626	0.0534	1250.40	926.9654	703.5219	663.333
	5	0.0871	0.0830	0.0594	0.0574	783.9952	757.1032	738.7755	690.8482
69007	2	0.0811	0.0645	0.0607	0.0592	839.7689	809.0918	763.0727	478.5779
	3	0.1607	0.0845	0.0710	0.0424	1863.10	775.7666	683.9459	567.2135
	4	0.2023	0.0958	0.0597	0.0504	1361.30	1106.20	949.1574	709.617
	5	0.1455	0.0932	0.0708	0.0502	1391.50	1350.20	1083.90	1019.50
118072	2	0.0690	0.0620	0.0556	0.0513	789.8287	679.7502	500.2609	471.2185
	3	0.1125	0.0791	0.0666	0.0490	826.5215	468.6492	440.3889	422.9089
	4	0.1199	0.1138	0.1036	0.0887	373.6271	354.2937	351.6347	277.7735
	5	0.1883	0.0802	0.0676	0.0517	1001.20	842.9632	505.8959	439.4412
268074	2	0.0939	0.0606	0.0420	0.0365	1216.50	897.7303	488.3368	194.6519
	3	0.1278	0.0741	0.0696	0.0562	659.1204	629.246	409.6018	400.3235
	4	0.1110	0.0612	0.0564	0.0382	1100.60	928.9911	729.4232	594.0642
	5	0.3333	0.1265	0.0901	0.0710	43350.0	970.3669	685.6124	659.8917
289011	2	0.0823	0.0758	0.0731	0.0601	440.5301	324.2832	302.3166	261.3164
	3	0.0841	0.0788	0.0645	0.0602	511.9059	406.2852	315.2156	283.0846
	4	0.0905	0.0862	0.061	0.0591	474.4105	446.9834	317.6874	294.237
	5	0.0869	0.0791	0.0676	0.0639	441.4006	421.5285	334.7603	326.6067
Square	2	0.1011	0.1005	0.0936	0.0736	269.2862	180.634	153.6706	146.818
	3	0.1175	0.1078	0.0885	0.0113	271.7993	237.1794	211.9535	146.2412
	4	0.1384	0.0861	0.0861	0.0835	289.9187	262.3659	262.2745	254.2363
	5	0.1243	0.1108	0.1039	0.0927	247.6694	220.8571	191.0518	184.7975

Table 5 Comparison of evaluation parameters for measuring thresholding performance

Test images	m	PSNR				CPU time (s)		
		CS-BFO	CS	PSO	GA	CS-BFO	CS	PSO
2018	2	14.7160	13.2521	11.5661	7.1470	3.0090	9.8773	4.2049
	3	16.1385	15.8885	15.1324	13.3272	3.4220	10.5943	4.4860
	4	20.8578	20.7641	16.2783	15.2391	3.7435	13.0737	4.5539
	5	41.9637	28.2329	21.3175	17.9136	3.8827	15.8542	5.1767
5096	2	24.9249	21.6819	21.3482	17.9107	3.0462	7.7884	4.0613
	3	32.4229	23.0494	22.1345	17.9841	3.2818	8.7753	4.3500
	4	34.8917	33.625	23.0567	22.8694	3.8286	10.3581	5.4076
	5	48.420	35.4676	23.2134	23.0032	3.8869	10.5507	5.6245
35049	2	26.2584	23.4688	21.2816	18.023	3.0960	8.1157	3.6424
	3	28.9645	26.5104	25.7788	22.6047	3.5031	10.0742	3.8158
	4	30.0235	29.3525	26.8906	26.0806	3.7044	11.9178	4.3633
	5	31.2111	30.509	30.1184	26.9561	3.8649	11.9899	4.7576
48017	2	18.9589	16.250	14.8221	10.4003	3.1511	1.6480	4.0862
	3	19.046	18.5127	18.0122	17.6767	3.5472	9.0756	4.1865
	4	23.1233	20.9672	20.5879	20.2799	3.5518	12.491	4.4676
	5	24.0408	23.1994	20.9006	20.5773	3.9654	12.7387	4.9443
69000	2	19.9516	18.4032	14.6965	13.5498	3.4985	7.8022	3.8457
	3	20.1936	20.1329	18.4869	15.7891	3.6838	10.6096	4.1081
	4	21.1112	20.8614	20.6949	20.3911	4.5515	11.8391	4.4282
	5	26.3043	22.6392	22.2657	21.6605	5.1357	15.4206	4.9353
69007	2	18.5367	17.0011	16.6326	15.0781	3.7593	7.4937	3.4841
	3	22.8099	20.7025	19.4319	18.9173	3.9878	9.8087	3.8619
	4	26.7693	23.0368	20.5901	19.6393	4.1118	11.0772	3.9145
	5	29.0884	26.9551	20.6237	20.0269	4.2310	13.3077	4.2014
118072	2	18.7568	17.4245	16.0303	13.1227	3.6736	8.8937	3.9905
	3	20.8283	19.6379	19.3115	16.0583	3.8248	12.025	4.0065
	4	22.6482	19.8089	19.7858	17.0145	3.8832	12.0978	4.5911
	5	27.2132	26.1549	21.5627	20.2257	4.9355	12.7215	5.2503
268074	2	22.7577	18.8758	18.8543	16.1552	3.5815	8.7488	4.2904
	3	25.0733	23.9331	22.1895	19.2213	4.4056	9.6852	5.2114
	4	34.8661	25.4174	24.7281	20.0167	4.9122	11.2692	5.2492
	5	40.0495	35.9001	29.0207	20.1271	5.0607	13.0562	5.4428
289011	2	22.4462	21.3519	21.3423	19.784	3.8884	8.0483	4.3764
	3	25.9696	24.6733	21.3856	19.9522	4.5956	8.4168	4.6670
	4	32.0684	27.296	23.1338	22.8942	5.6149	8.7743	4.7512
	5	41.9406	27.960	27.9035	25.946	5.6944	9.8128	5.0630
Square	2	12.114	11.5281	8.9266	7.9905	3.2944	7.9257	4.7972
	3	15.7409	13.140	9.1422	8.8001	4.2276	8.5026	5.6522
	4	16.0313	15.2728	15.1664	14.6658	5.4875	9.1629	5.8700
	5	19.2617	17.2478	16.920	14.997	5.6785	10.1280	6.1083

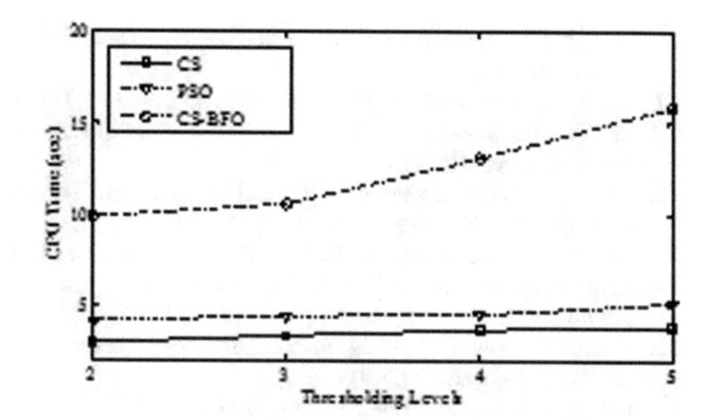

Fig. 27 Variation of CPU time for image 2018

Fig. 27. Note that CPU time for GA is not shown in Table 4. Here GA takes more time compared to all other three methods since we use OPTIM toolbox of MATLAB.

From Fig. 27, it is observed that CS-BFO takes more time as compared to CS and PSO techniques. Note that CS takes least CPU time.

6 Conclusion

The chapter presented the idea of multilevel thresholding using a novel hybrid CS-BFO algorithm based on edge magnitude information. Evolutionary computational techniques have been successfully deployed for maximizing edge magnitude of an image. Fidelity- and contrast-related measures are used to compare these evolutionary computational techniques judiciously. From the results, we observe that performance of CS-BFO is better as compared to other three techniques. In addition, the inherent constraint handling capability makes it attractive for multilevel image thresholding applications. From Table 4, it is seen that CS is faster than other methods. The reason is the fact that CS needs few parameter settings. Our proposed CS-BFO algorithm has potential and can be used for solving optimization problems, where social foraging model works.

References

1. Fu, K.S., Mui, J.K.: A survey on image segmentation. Pattern Recognit. **13**(1), 3–16 (1981)
2. Haralick, R.M.: Image segmentation survey. In: Faugeras, O.D. (ed.), Fundamentals in Computer Vision, pp. 209–224. Cambridge University Press, Cambridge (1983)
3. Haralick, R.M., Shapiro, L.G.: Image segmentation techniques. Comput. Vis. Gr. Image Process. **29**(1), 100–132 (1985)
4. Pal, N.R., Pal, S.K.: A review on image segmentation techniques. Pattern Recognit. **26**(9), 1277–1294 (1993)
5. Kapur, J.N., Sahoo, P.K., Wong, A.K.: A new method for gray-level picture thresholding using the entropy of the histogram. Comput. Vis. Gr. Image Process. **29**(3), 273–285 (1985)

6. Otsu, N.: A threshold selection method from gray level histograms. Automatica **11**(285–296), 23–27 (1975)
7. Kittler, J., Illingworth, J.: Minimum error thresholding. Pattern Recognit. **19**(1), 41–47 (1986)
8. Sahoo, P.K., Soltani, S., Wong, A.K.: A survey of thresholding techniques. Comput. Vis. Gr. Image Process. **41**(2), 233–260 (1988)
9. Zhang, Y.J.: An Overview of Image and Video Segmentation in the Last 40 years. In: Advances in Image and Video Segmentation, pp. 1–15 (2006)
10. Lee, S.U., Yoon Chung, S., Park, R.H.: A comparative performance study of several global thresholding techniques for segmentation. Comput. Vis. Gr. Image Process. **52**(2), 171–190 (1990)
11. Sankur, B., Sezgin, M.: Image thresholding techniques: a survey over categories. Pattern Recognit. **34**(2), 1573–1583 (2001)
12. Chang, C.I., Du, Y., Wang, J., Guo, S.M., Thouin, P.D.: Survey and comparative analysis of entropy and relative entropy thresholding techniques. IEE Proc. Vis. Image Signal Process. **153**(6), 837–850 (2006)
13. Yin, P.Y.: A fast scheme for optimal thresholding using genetic algorithms. Signal Process. **72**(2), 85–95 (1999)
14. Cheng, H., Chen, Y.H., Sun, Y.: A novel fuzzy entropy approach to image enhancement and thresholding. Signal Process. **75**(3), 277–301 (1999)
15. Zahara, E., Fan, S.K., Tsai, M.D.: Optimal multi-thresholding using a hybrid optimization approach. Pattern Recognit. Lett. **26**(8), 1082–1095 (2005)
16. Sathya, P.D., Kayalvizhi, R.: Optimal multilevel thresholding using Bacterial foraging algorithm. Expert Syst. Appl. **38**(12), 15549–15564 (2011)
17. Sarkar, S., Sen, N., Kundu, A., Das, S., Chaudhury, S.: A differential evolutionary multilevel segmentation of near infra-red images using Renyi entropy. In: International Conference on Frontiers of Intelligent Computing: Theory and Applications, pp. 699–706. Springer, Heidelberg (2013)
18. Sarkar, S., Patra, G.R., Das, S.: A differential evolution based approach for multilevel image segmentation using minimum cross entropy thresholding. In: Swarm, Evolutionary and Memetic Computing, pp. 51–58. Springer, Heidelberg (2011)
19. Zhang, Y., Wu, L.: Optimal multi-level thresholding based on maximum tsallis entropy via an artificial bee colony approach. Entropy **13**(4), 841–859 (2011)
20. Horng, M.H.: A multilevel image thresholding using the honey bee mating optimization. Appl. Math. Comput. **215**(9), 3302–3310 (2010)
21. Bhandari, A.K., Singh, V.K., Kumar, A., Singh, G.K.: Cuckoo search algorithm and wind driven optimization based study of satellite image segmentation for multilevel thresholding using Kapur's entropy. Expert Syst. Appl. **41**(7), 3538–3560 (2014)
22. Raja, N.S., Sukanya, S.A., Nikita, Y.: Improved PSO based multi-level thresholding for cancer infected breast thermal images using otsu. Proc. Comput. Sci. **48**, 524–529 (2015)
23. Bhandari, A.K., Kumar, A., Singh, G.K.: Modified artificial bee colony based computationally efficient multilevel thresholding for satellite image segmentation using Kapur's, Otsu and Tsallis functions. Expert Syst. Appl. **42**(3), 1573–1601 (2015)
24. Nabizadeh, N., John, N., Wright, C.: Histogram-based gravitational optimization algorithm on single MR modality for automatic brain lesion detection and segmentation. Expert Syst. Appl. **41**(17), 7820–7836 (2014)
25. Liu, Y., Mu, C., Kou, W., Liu, J.: Modified particle swarm optimization-based multilevel thresholding for image segmentation. Soft Comput. **19**(5), 1311–1327 (2014)
26. Brajevic, I., Tuba, M.: Cuckoo search and firefly algorithm applied to multilevel image thresholding. In: Cuckoo Search and Firefly Algorithm, pp. 115–139. Springer International Publishing, Heidelberg (2014)
27. Roy, S., Kumar, U., Chakraborty, D., Nag, S., Mallick, A., Dutta, S.: Comparative analysis of cuckoo search optimization-based multilevel image thresholding. In Intelligent Computing, Communication and Devices, pp. 327–342. Springer, India (2015)

28. Baraldi, A., Parmiggiani, F.: An investigation of the textural characteristics associated with gray level co-occurrence matrix statistical parameters. IEEE Trans. Geosci. Remote Sens. **33**(2), 293–304 (1995)
29. Chanda, B., Majumder, D.D.: A note on the use of the gray-level co-occurrence matrix in threshold selection. Signal Process. **15**, 149–167 (1988)
30. Gonzalez, R.C., Woods, R.E.: Digital Image Processing. Addison-Wesley Publishing, New Jersey (1992)
31. Albregtsen, F.: Statistical Texture Measures Computed from Gray Level Co-occurrence Matrices, Image Processing Laboratory. Department of Informatics, University of Oslo (1995)
32. Haralick, R.M., Shanmugam, K., Dinstein, I.H.: Textural features for image classification. IEEE Trans. Syst. Man Cybern. **6**, 610–621 (1973)
33. Mokji, M.M., Abu Bakar, S.A.: Adaptive thresholding based on co-occurrence matrix edge information. In: First Asia International Conference on Modelling and Simulation, pp. 444–450. IEEE, New York (2007)
34. Yang, X.S., Deb, S.: Cuckoo search via levy flights. In: World Congress on Nature and Biologically Inspired Computing, pp. 210–214. IEEE, New York (2009)
35. Yang, X.S., Deb, S.: Engineering optimization by cuckoo search. Int. J. Math. Model. Numer. Optim. **1**(4), 330–343 (2010)
36. Goldberg, D.E., Holland, J.H.: Genetic algorithms and machine learning. Mach. Learn. **3**(2), 95–99 (1988)
37. Kennedy, J., Eberhart, R.: Particle swarm optimization. IEEE Int. Conf. Neural Netw. **4**, 1942–1948 (1995)
38. Passino, K.M.: Biomimicry of Bacterial Foraging for Distributed Optimization and Control. IEEE Control Syst. **22**(3), 52–67 (2002)
39. The Berkeley Segmentation Dataset and Benchmark. (2007). http://www.eecs.berkeley.edu/Research/Projects/CS/vision/grouping/segbench
40. Wang, Z., Bovik, A.C., Sheikh, H.R., Simoncelli, E.P.: Image quality assessment: from error visibility to structural similarity. IEEE Trans. Image Process. **13**(4), 600–612 (2004)
41. Zhang, L., Zhang, L., Mou, X., Zhang, D.: FSIM: a feature similarity index for image quality assessment. IEEE Trans. Image Process. **20**(8), 2378–2386 (2011)
42. Akay, B.: A study on particle swarm optimization and artificial bee colony algorithms for multilevel thresholding. Appl. Soft Comput. **13**(6), 3066–3091 (2013)

REFII Model and Fuzzy Logic as a Tool for Image Classification Based on Image Example

Goran Klepac

Abstract Image segmentation as a concept has great potential for practical implementation. Image segmentation is complex concept, which can be focused on object recognition within images, or content similarity oriented concept, or some other concept which in general try to recognize similar elements within images. This chapter will introduce novel concept based on picture content similarity, which can be used as a tool for recommendation systems in situation where we operate with unknown image data sets. Simple example for that can be travel agencies web pages, where potential users are seeking for future travel destinations. Based on experience regarding previously visited locations, system calculates content similarity between mostly visited locations represented by pictures and offers new locations upon calculated preference. Generally speaking, this process can be declared as image segmentation, because segmentation is based on example picture content, and all locations similar to the chosen picture content are declared as segment. Visitor gets recommendation for his next travel destination based on previously seen locations within one web session, or more than one web session if visitor can be uniquely recognized by login and password. Method presented in this chapter should be a good solution for online systems which demands fastness and efficiency for recommendation systems. Existing methods are much precise, but demands longest computing processing. Proposed method does not demand exhaustive training to be efficient in image classification. Partially it cause with less precision in comparison with methods such as histogram-based methods, compression-based methods, algorithms based on edge detection, region growing, partial differential equation-based methods, and others, but speed up process of image classification.

Keywords REFII · Image segmentation · Content retrieval · RGB · Fuzzy logic · Fuzzy expert systems

---

G. Klepac (✉)
Raiffeisen Bank Austria, Petrinjska 59, Zagreb, Croatia
e-mail: goran@goranklepac.com

S. Bhattacharyya et al. (eds.), *Hybrid Soft Computing for Image Segmentation*, DOI 10.1007/978-3-319-47223-2_4

1 Introduction

The solution of image classification problem based on example image is interesting in situation where we try to find similar images to selected image. It is not task of object recognition which demands certain degree of accuracy regarding object recognition; it is task for finding images with similar content. Main advantage of that methodology could be finding all images which have mutual themes like night sky, forest landscape in the autumn, sea landscapes buildings.

For these purposes elements of REFII model in combination with fuzzy logic will be used. REFII model is used for time series analysis purposes, but in this case it will be used in combination with fuzzy logic as a tool for image classification. Basic idea is to decompose each pixel RGB value from the picture into time series spans, where each pixel will be represented as a time span of RGB values as well as with area under the transformed REFII time span. Fuzzy logic has a role to define similarity criteria by rules. Picture similarity is defined as a similarity ratio between correspondent pixels of selected and observed picture. Similarity on pixel level is defined as a similarity between slopes between selected and observed correspondent pixel within picture and similarity between area under curve between selected and observed correspondent pixel within picture. Similarity criteria will be defined through fuzzy logic on pixel and picture level. As an auxiliary tool for similarity evaluation fuzzy logic concept will be used. The power of presented methodology is that it does not demand indexing, tagging and previous classification for similarity recognition.

Hybridization is achieved in integration of REFII model which is time series oriented analytical tool, fuzzy logic and distance metrics commonly used for clustering and similarity calculation.

Presented methodology can have application in area of big data analytic, especially for image searching purposes, and have potential for image clustering upon defined image examples.

Integral part of the chapter is practical example for image similarity recognition and classification based on presented model. Within practical example an illustration on empirical data about how changes in similarity criteria based on fuzzy logic have influence on output results will be presented. This chapter will also give direction on future research regarding described methodology for object recognition purposes within images and for automated image tagging and indexing.

2 Background

Automatic image classification, image clustering, image recognition and all automatic image related algorithms are nontrivial tasks. First reason lies in fact that images are complex data structures with variety of information which can vary in complexity regarding used data format and additional image features [1, 2].

Processing on pixel level methodology, which is must for many methods is consuming process which demands taking care of optimization processes. In those conditions, challenging task is to find efficient and relatively robust and quick methodology for image classification by example. It is evident that we cannot expect perfect accuracy, but plausible accuracy which will show acceptable performance for solving practical problems from business practice [3–5]. Recommendation systems can be improved by inclusion of additional information regarding image sources, and that additional information can be one of the most important factors which determine some actions within recommendation system.

In a situation when we are trying to recognize specific object, Harris corner detector, scale invariant feature transform (SIFT) or similar algorithm can be used as an integral part of solution. Problem is in extension of processing time, and additional knowledge base which should be defined good enough to be operative and useful [6, 7]. Additional problem is when we are faced with new unknown case within image, then knowledge database should be extended. In situation when knowledge base should be updated very often, system efficiency drops. Additional problem in domain of image processing are matching tagged images, clustering images, indexing images [8–10]. Accuracy demands more processing time and often maintenance of those systems can be complex depending on system dynamic [11–13]. Challenge is to find solution, which will be acceptable from the perspective of accuracy and adequate processing time, adequate maintenance requirements. Significant number of solutions for image similarity, image clustering, and image object recognition are concentrated on RGB pixel values and pixel environment. Euclidean distance, K-means clustering, support vector machines, deep learning methodology with neural network usage, Bayesian networks are referent methods which are mostly used for solving described problems [14–20]. Main aim of proposed method will be solution that will be robust enough to provide results within adequate period of time, and for which there is no need for additional maintenance in condition of increasing system dynamic. Recommendation system and customer profiling/preferences can increase their performances using additional information extracted from images. Sometimes that information can be very tangible, like green landscape or sea. Sometimes information can be intangible and abstract and could not be described by exact terms, and mostly can describe appearance of some specific feelings in association with specific group of pictures regarding appearance of specific color spectra within image. That leads us to new level of image processing with intention for specific application in systems like recommendation systems or profiling, which can be a good starting point for deeper customer understanding [21–24]. Image segmentation algorithms such as histogram-based methods, compression-based methods, algorithms based on edge detection, region growing, partial differential equation-based methods, graph partitioning methods, watershed transformation, multi-scale segmentation, semi-automatic segmentation, trainable segmentation could be demanding regarding computing [19, 25–33] Usage of K-mean clustering, hierarchical clustering, spectral clustering, graph cuts are computation demanding process. It could be the problem in systems like recommendation systems which demands fast reaction based on previously unknown image. References [28–32] To develop light method which will give fast reaction based on previously unknown image the following method is developed. This method has an inspiration in REFII model, model which is developed for time series analysis purposes. Basic postulates of REFII model in combination with fuzzy logic was used for creating image classification tool.

3 REFII Model and Introduction

The main purpose of REFII model is to automate time series analysis, through a unique transformation model of time series. An advantage of this approach of time series analysis is the linkage of different methods for time series analysis, linking traditional data mining tools in time series, and constructing new algorithms for analyzing time series REFII model is not a closed system, which means that we have a finite set of methods. First, this is a model for transformation of values of time series, which prepares data used by different sets of methods based on the same model of transformation in a domain of problem space. REFII model gives a new approach in time series analysis based on a unique model of transformation, which is a base for all kind of time series analysis [34, 35].

The algorithm for time series transformation into the REFII model is done in several steps. A time series can be defined as a series of values $S(s_1, s_n)$ where S represents a time series and (s_1, s_n) represents the elements of series S.

1. Step 1 Time interpolation
   Format of an independent time series $V_i(vi_1, vi_n)$ on the interval $< 1..n >$ (days, weeks, months, quarters, years) with values of 0. It is necessary to implement the interpolation of values missing from $S(s_1, s_n)$ based on the series V_i. The result of this process is the series $S(s_1, s_n)$ with interpolated values from the $V_i(vi_1, vi_n)$ series.
2. Step 2 Time granulation
   In this step we define the degree of summarization of the time series $S(s_1, s_n)$ that is located within a basic unit of time (day, week, month...). Time series elements are summarized by using statistical functions like sum, mean or mode on the level of granular slot. That way, the granulation degree of the time series can be increased (days to weeks, weeks to months) and the result is a time series $S(s_1, s_n)$ with a greater degree of granulation. We can return to this step during the analysis process to fulfil the analysis goals, and that includes the mandatory repetition of this process in the following steps.
3. Step 3 Normalization
   The normalization procedure implies the transformation of a time series $S(s_1, s_n)$ into a time series $T(t_1, t_n)$ where each element of the array is subject to a min-max normalization procedure to the $< 0, 1 >$ interval. Time series T is made up of elements (t_1, t_n) where t_i is calculated as:

$$t_i = \frac{s_i - min(s)}{max(s) - min(s)},$$

where min(s) and max(s) are the minimum and maximum values of time series T. The time shift between basic patterns (a measure of time complexity) in a time slot on the X axis is defined as $d(t_i, t_i + 1) = a$.

4. Step 4 Transformation to REF notation

According to the formula

$$T_r = t_{i+1} - t_i, T_r > 0 \Rightarrow \text{“}R\text{”}, T_r = 0 \Rightarrow \text{“}E\text{”}, T_r < 0 \Rightarrow \text{“}F\text{”},$$

where the Y_i elements are members of the N_s series.

5. Step 5 Slope calculation based on the angle angular deflection coefficient:

$$T_r > 0 \Rightarrow R\ Coefficient = t_{i+1} - t_i$$
$$T_r < 0 \Rightarrow F\ Coefficient = t_i - t_{i+1}$$
$$T_r = 0 \Rightarrow F\ Coefficient = 0$$

6. Step 6 Calculation of the area below the curve numerical integration by the rectangle theory:

$$p = \frac{(t_i * a) + (t_i * a)}{2}$$

7. Step 7 Creating time indices
   Creating a hierarchical index tree depends on the nature of the analysis, where the element of the structured index can be located and an attribute such as the clients code.
8. Step 8 Category creation
   Creating derived attribute values based on the area below the curve and the deflection angles. It is possible to create categories by applying crisp and fuzzy logic.
9. Step 9 Connecting the REFII model's transformation tables with the relational tables that contain attributes with no time dimension.

These nine basic steps are the foundation of the algorithmic procedure underlying the REFII model whose ultimate goal is the formation of the transformation matrix. The transformation matrix is the foundation for performing future analytical procedures whose goal is time series analysis.

The REFII model combines the trends of discrete functions and the area on the time segment level and creates a basic pattern. The basic pattern is represented with three core values:

1. Growth trend code (REF)
2. Angular deflection coefficient that can be classified into categories with the aid of the classical crisp or fuzzy logic

3. The time segment area can be classified into categories with the aid of the classical crisp or fuzzy logic

Basic patterns can form complex structures of series of samples, and as such can be part of the analysis process. The basic pattern defined through the REFII model is its fundamental part. Time series transformed with REFII model can be used for dynamic Bayesian network construction.

4 Conceptual Solution Proposal

For image segmentation purposes REFII model is revised and adopted. Focus is on RGB values as elements for transformation purposes. Taking into consideration image file structure, RGB element will be observed as time spans. Focus will be on angle deflection represented through RGB values and area beneath curve also expressed through RGB values. Normalization as a step in REFII model will be skipped, based on the fact that RGB values have standard range of values. Elemental REF transformation will also be skipped due to fact that direction as a piece of information is not relevant as well for this problem domain. Instead of direction we are interesting on absolute value calculation, as information about differences between two data sources.

REFII model is inspiration for image similarity/segmentation solution. Formally we can express methodology for image data transformation as:

$$AdR_i = \frac{|AsR_i - AtR_i|}{255} * 100 \tag{1}$$

where,

AdR_i is angular difference in percentage of Red values between selected image (source image) and image for which we try to determine similarity (target image) for i-th pixel.

AsR_i is value on RGB scale for Red value for i-th pixel in selected image (source image).

AtR_i is value on RGB scale for Red value for i-th pixel image for which we try to determine similarity (target image).

$$AdG_i = \frac{|AsG_i - AtG_i|}{255} * 100 \tag{2}$$

where,

AdG_i is angular difference in percentage of Green values between selected image (source image) and image for which we try to determine similarity (target image) for i-th pixel.

AsG_i is value on RGB scale for Green value for i-th pixel in selected image (source image).

AtG_i is value on RGB scale for Green value for i-th pixel image for which we try to determine similarity (target image).

$$AdB_i = \frac{|AsB_i - AtB_i|}{255} * 100 \qquad (3)$$

where,

AdB_i is angular difference in percentage of Blue values between selected image (source image) and image for which we try to determine similarity (target image) for i-th pixel.

AsB_i is value on RGB scale for Blue value for i-th pixel in selected image (source image).

AtB_i is value on RGB scale for Blue value for i-th pixel image for which we try to determine similarity (target image).

$$Area_i = \frac{|(AsR_i * AsG_i * AsB_i) - (AtR_i * AtG_i * AtB_i)|}{16581375} * 100 \qquad (4)$$

where,

$Area_i$ is difference in volume percentage of RGB values between selected image (source image) and image for which we try to determine similarity (target image).

AsR_i is value on RGB scale for Red value for i-th pixel in selected image (source image).

AsG_i is value on RGB scale for Green value for i-th pixel in selected image (source image).

AsB_i is value on RGB scale for Blue value for i-th pixel in selected image (source image).

AtR_i is value on RGB scale for Red value for i-th pixel image for which we try to determine similarity (target image).

AtG_i is value on RGB scale for Green value for i-th pixel image for which we try to determine similarity (target image).

AtB_i is value on RGB scale for Blue value for i-th pixel image for which we try to determine similarity (target image

Next step is similarity determination, based on fuzzy logic as follows:

$$\mu_{(AsR_i,AsG_i,AsB_i)}(AS_i) = max(\mu_{(AsR_i)}, \mu_{(AsG_i)}, \mu_{(AsB_i)}) \qquad (5)$$

$$\mu_{(AS_i,Area_i)}(Similarity_i) = max(\mu_{(AS_i)}, \mu_{(Area_i)}) \qquad (6)$$

Fuzzy logic is used as similarity calculation engine on pixel level.

Calculated values on pixel level become linguistic variables which are part of rule blocks for segmentation purposes.

For image content similarity search, this approach gives a framework which can be extended on pixel area similarity search, and creating additional environmental similarity index. Basic idea is that correspondent pixel values should not vary extremely for similar images. Example for that can be forest or sea figures. Two figures with meadows or forests should have similar correspondent pixel values. Similarity means acceptable differences on RBG scale. This can be starting point for recommendation systems, or for tagging. Regarding tagging, and recommendation system fuzzy model could be strengthened with additional linguistic variables which can be used for determination regarding action within expert system.

5 Realization of Proposed Solution

Proposed solution is realized through Python program and Fuzzy expert system. Following example in Python

1. Opens two images
2. Read images using PIL (Python Imaging Library) library
3. Transfer Red, Green, Blue values for each pixel into list for each color
4. Calculate differences between Red, Green, Blue values
5. Calculate differences on pixel level for consolidated Red, Green, Blue values
6. Write calculated lists into external file.

Following code in Python shows automation of proposed algorithmic solution for difference calculation between two images on pixel level.

```
from PIL import Image
import csv
from numpy import array
img=Image.open("c:\Goran\C.jpg")
img1=Image.open("c:\Goran\D.jpg")
arr = array(img)
arr1 = array(img1)

listRs = []
listGs = []
listBs = []

listRt = []
listGt = []
listBt = []

listCubeDiff=[]

for n in arr:listRs.append(n[0][0]) #R
for n in arr:listGs.append(n[0][1]) #G
for n in arr:listBs.append(n[0][2]) #B

for n in arr1:listRt.append(n[0][0]) #R
for n in arr1:listGt.append(n[0][1]) #G
for n in arr1:listBt.append(n[0][2]) #B
```

```
listAngleDiff_R = []
listAngleDiff_G = []
listAngleDiff_B = []

listAreaDiff_R=[]
listAreaDiff_G=[]
listAreaDiff_B=[]

csv_out = open('mycsv.csv', 'wb')


mywriter = csv.writer(csv_out)

listAngleDiff_R.append('Diff_Angle_R')
listAngleDiff_G.append('Diff_Angle_G')
listAngleDiff_B.append('Diff_Angle_B')
listAreaDiff_R.append('Diff_Area_R')
listAreaDiff_G.append('Diff_Area_G')
listAreaDiff_B.append('Diff_Area_B')
listCubeDiff.append('Diff_Cube')

for i in range(len(listRs)):
listAngleDiff_R.append(abs(float(listRs[i]- listRt[i])/255 *100))
listAngleDiff_G.append(abs(float(listGs[i]- listGt[i])/255 *100))
listAngleDiff_B.append(abs(float(listBs[i]- listBt[i])/255 *100))

A_R=float((listRs[i]*255)-(listRt[i]*255))
listAreaDiff_R.append(abs(A_R/65025*100))

A_G=float((listGs[i]*255)-(listGt[i]*255))
listAreaDiff_G.append(abs(A_G/65025*100))

A_B=float((listBs[i]*255)-(listBt[i]*255))
listAreaDiff_B.append(abs(A_B/65025*100))

Cube_s=float(listRs[i]*listGs[i]*listBs[i])
Cube_t=float(listRt[i]*listGt[i]*listBt[i])
Diff_cube=abs(float(Cube_s-Cube_t))
print Diff_cube
listCubeDiff.append((Diff_cube/16581375)*100)



for row in zip(listAreaDiff_R, listAreaDiff_G,
listAreaDiff_B, listCubeDiff ):
mywriter.writerow(row)
csv_out.close()
```

Presented example taking in considers two pictures for comparison. For processing more than two pictures, existing code could be extended with following code:

```
for  figures in range(1,n):
image = Image.open("folder\\{0}.jpg".format(figures))
a = array(image)
```

Presented code was applied on following pictures (A,B,C), to prove proposed algorithm efficiency (Figs. 1, 2 and 3).

As a result we have following RGB distributions for presented pictures (Fig. 4):

Fig. 1 Figure A

Fig. 2 Figure B

Figure shows distribution of pixel differences calculated for red channels between image A and image B in range from 0 to 100. It is left leaning distribution, with value concentration in zones of low differences. Regarding this frequency distribution which represents pixel differences in area of red channels between image A and image B, hypothesis is that these two images are similar (Fig. 5).

Figure shows distribution of pixel differences calculated for green channels between image A and image B in range from 0 to 100. It is left leaning distribution, with value concentration in zones of low differences. Regarding this frequency distribution which represents pixel differences in area of green channels between image A and image B, hypothesis is that these two images are similar (Fig. 6).

Fig. 3 Figure C

Figure shows distribution of pixel differences calculated for blue channels between image A and image B in range from 0 to 100. It is left leaning distribution, with value concentration in zones of low differences. Regarding this frequency distribution which represents pixel differences in area of blue channels between image A and image B, hypothesis is that these two images are similar.

Distribution of pixel differences calculated for all channels (presented as cube area) between image A and image B in range from 0 to 100. It is left leaning distribution, with value concentration in zones of low differences. Regarding this frequency distribution which represents pixel differences for all channels (presented as cube area) between image A and image B, hypothesis is that these two images are similar (Fig. 7).

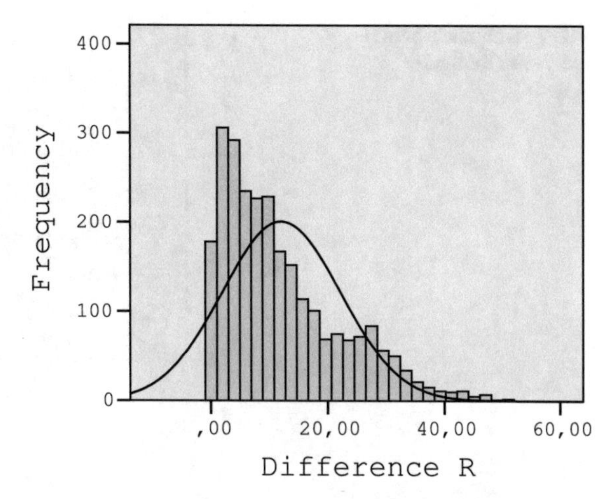

Fig. 4 Difference between red values for figures A and B

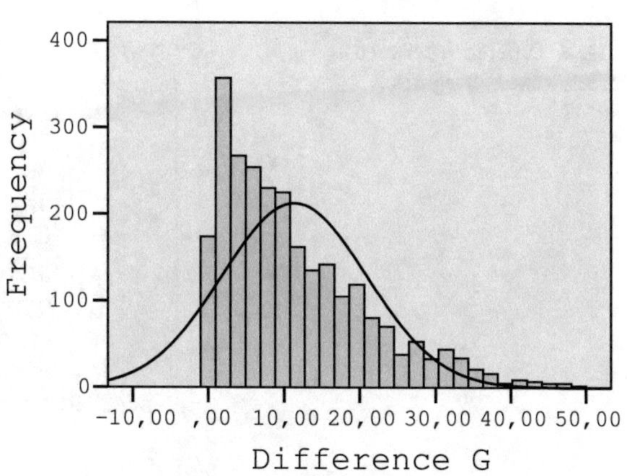

Fig. 5 Difference between green values for figures A and B

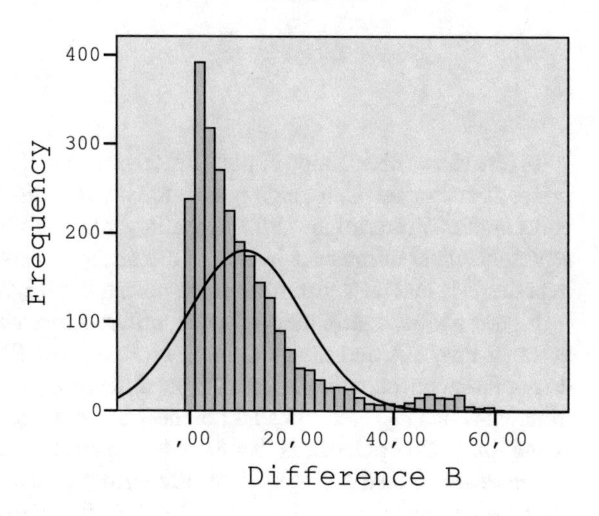

Fig. 6 Difference between blue values for figures A and B

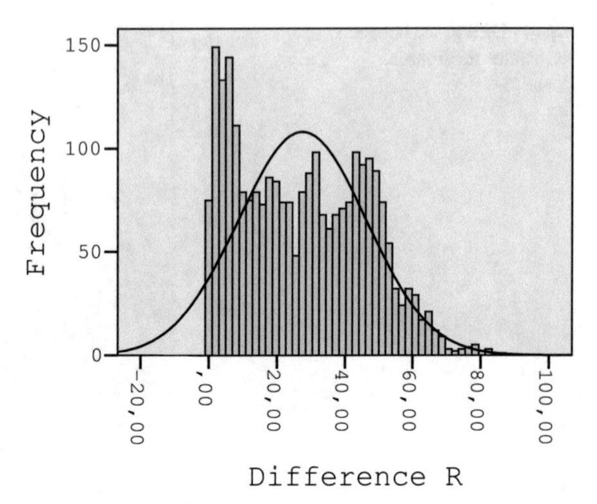

Fig. 7 Difference between red values for figures A and C

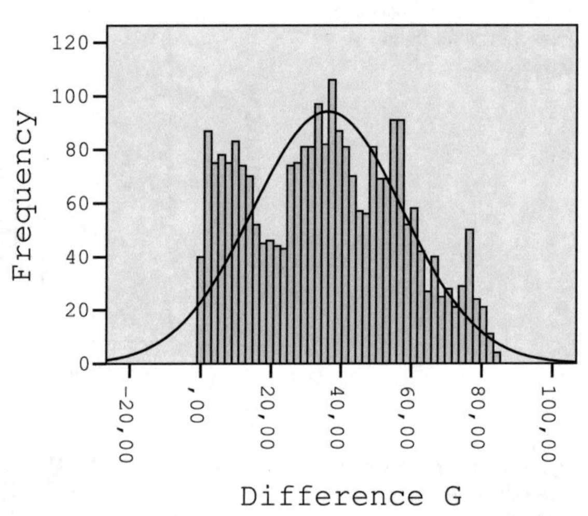

Fig. 8 Difference between green values for figures A and C

Figure shows distribution of pixel differences calculated for red channels between image A and image C in range from 0 to 100. It is distributions which do not have concentration in area of low differences. Regarding this frequency distribution which represents pixel differences in area of red channels between image A and image B hypothesis is that these two images are not similar (Fig. 8).

Figure shows distribution of pixel differences calculated for green channels between image A and image C in range from 0 to 100. It is distributions which do not have concentration in area of low differences. Regarding this frequency distribution which represents pixel differences in area of green channels between image A and image B hypothesis is that these two images are not similar (Fig. 9).

Figure shows distribution of pixel differences calculated for blue channels between image A and image C in range from 0 to 100. It is distributions which do not have

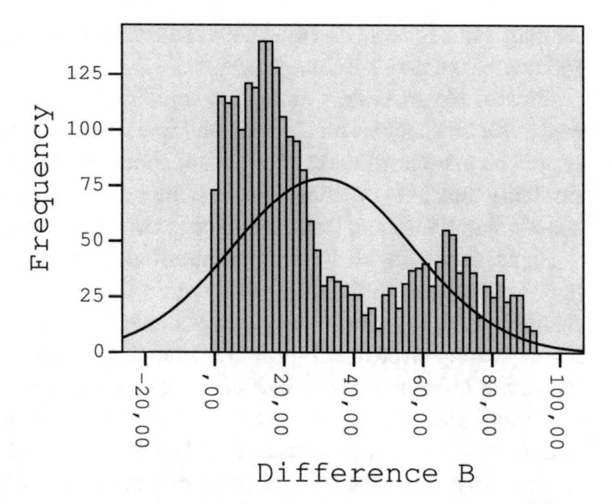

Fig. 9 Difference between blue values for figures A and C

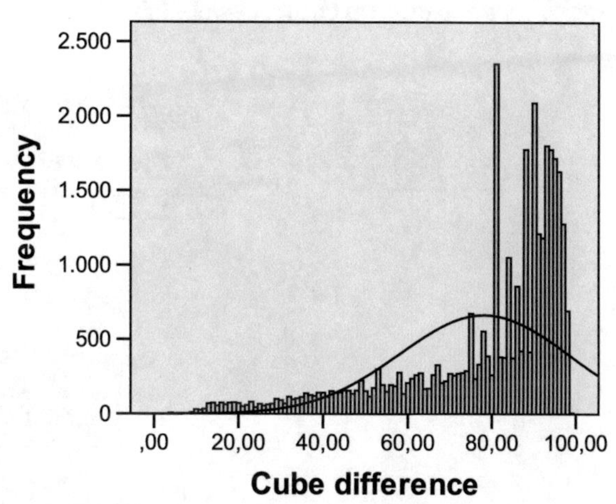

Fig. 10 Difference between cube values for figures A and C

concentration in area of low differences. Regarding this frequency distribution which represents pixel differences in area of blue channels between image A and image B, hypothesis is that these two images are not similar (Fig. 10).

Distribution of pixel differences calculated for all channels (presented as cube area) between image A and image C in range from 0 to 100. It is distributions which do not have concentration in area of low differences. Regarding this frequency distribution which represents pixel differences for all channels (presented as cube area) between image A and image C, hypothesis is that these two images are not similar.

Following figure shows fuzzy expert system for similarity evaluation between two images on pixel level. Similarity threshold can be determined in way which similarity rate is acceptable for making hypothesis about similarity between two images.

Similarity rate is defined as ratio between similar pixels between two observed images and number of pixels in image (Fig. 11).

Created expert system compares similarity for red, green and blue channel, as well as for all channels on consolidated level. Based on rule blocks, as a result expert system on pixel level makes judgement about similarity on image level. At the end, similarity threshold on image level can be expressed as a ration between similar and non-similar pixels, and that determine is two images are similar or not.

Figure shows linguistic variable definition for similarity evaluation on red channel. In presented case study, thresholds within linguistic variables are the same for all channels. As it is visible from given example, similarity redefinition is very flexible, and depending on problem space area regarding image subject, borders for similarity evaluation in linguistic variables can be quickly redefined (Fig. 12).

Integral part of fuzzy expert system is rule definition, which determines output from fuzzy system. In this case, fuzzy expert system makes decision about similarity on pixel level. Following table shows rule definition within fuzzy expert system for similarity evaluation on channel level (Table 1).

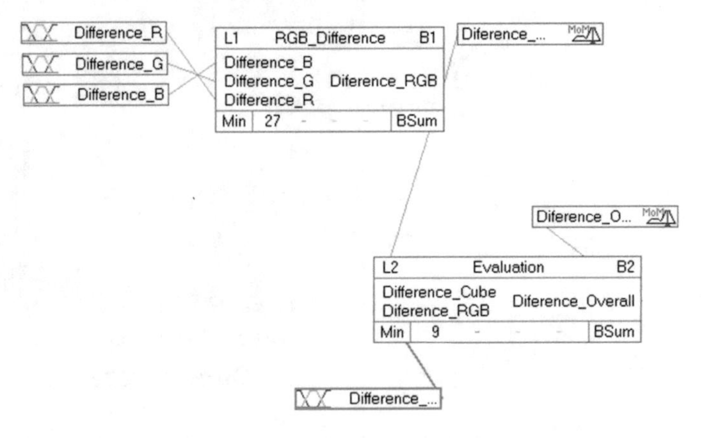

Fig. 11 Fuzzy expert system for evaluation similarity on pixel level

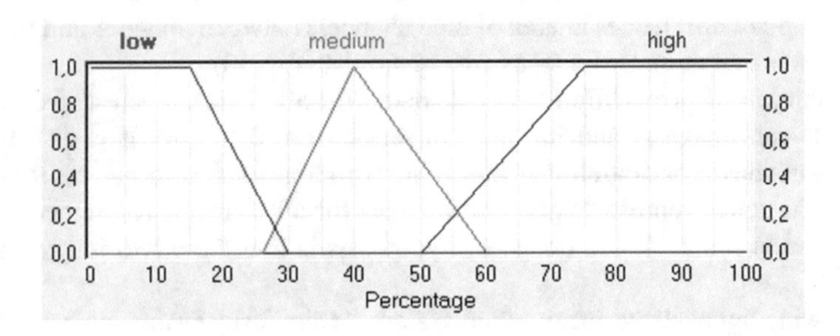

Fig. 12 Linguistic variable definition

Table 1 Rule block definition for similarity on channel level

Difference B	Difference G	Difference R	Calculated difference RGB
Low	Low	Low	Low
Low	Low	Medium	Low
Low	Low	High	High
Low	Medium	Low	Medium
Low	Medium	Medium	Medium
Low	Medium	High	High
Low	High	Low	High
Low	High	Medium	High
Low	High	High	High
Medium	Low	Low	Low
Medium	Low	Medium	Medium
Medium	Low	High	High
Medium	Medium	Low	Medium
Medium	Medium	Medium	Medium
Medium	Medium	High	High
Medium	High	Low	High
Medium	High	Medium	High
Medium	High	High	High
High	Low	Low	High
High	Low	Medium	High
High	Low	High	High
High	Medium	Low	High
High	Medium	Medium	High
High	Medium	High	High
High	High	Low	High
High	High	Medium	High
High	High	High	High

Table 2 shows rule definition within fuzzy expert system for similarity evaluation on pixel level. It is final definition from fuzzy expert system which determines overall similarity between two pixels from observed images.

As a result, final instance in fuzzy expert system makes evaluation about pixel similarity between two images. Flexibility of the system is visible through fact that border definition in linguistic variables can be easily changed, and similarity threshold can be changed on image level (Table 3). This mechanism gives opportunity for customization of image similarity evaluation. This functionality can be usable in situation where we are dealing with different expected images subjects. Images for which we are evaluating similarity can be most similar to one another than in some other situation. That depends on problem space area.

Table 2 Rule block definition for overall similarity on pixel level

Difference RGB	Difference cube	Overall difference
Low	High	High
Medium	Low	Medium
Medium	Medium	Medium
Medium	High	High
High	Low	High
High	Medium	High
High	High	High

Table 3 Test results

Categories	Number of images in category	Correct selection percentage
Horse	91	83
Roses	83	79
Elephants	32	85
Landscape	72	78

Images A and B are similar and image C is not similar with A and B. Presented algorithm should recognize similarity/dissimilarity between images. As we are dealing with content similarity search, without intention for object recognition, focus will be on RGB values for content similarity.

Additional test was performed on image samples from: www.vision.caltech.edu image_datasets/caltech256.

Basic idea was to perform test on image dataset which contains images of horses, roses, elephants and landscapes in way that model should select appropriate image from bunch of selected images based on previously selected random image from the one of mentioned categories. Fully created dataset contained 278 images. One of the images was randomly selected during each iteration and model had task to recognize most similar image from dataset. Criteria for correct selection were selection of image from the same category. Test results are given in the following table:

As it is visible from the table, obtained results show good performance for single picture selection. Instead of single picture selection, it is possible to adopt algorithm to unite all similar images to centroid image which represent some specific category.

6 Model Application

Image content retrieval is concentrated on content as a subject of examination, not on specific object. There are numerous advantages and disadvantages regarding that concept, and proposed model. Advantage is quick analysis with instant results.

Disadvantages are generalization and imprecision. However this concept could be efficient and useful in situation where we would like to find similar content between figures in way where we would like to generalize content. Example for that is travel agency, where this concept can be used for their web pages as an integral part of recommendation system. Anonymous visitor can try to find its destination, by clicking images of some destinations. Hypothesis is that he has some idea about his destitution in way that is it sea, mountains or desert. Each clique gives to system new information regarding preferences and system suggest destinations regarding previous visits, by image content retrieval using proposed model. Image content retrieval should not be only mechanism for recommendation; it can include other methods like Bayesian networks, logistic regression or neural networks, and all that methods could be based on click streams and similar data sources. Proposed model could be used for tagging purposes in way not only to put obvious tags like: sea, desert, mountains. Tagging could be oriented on other term definition like: green environment, blue environment, and gray buildings.

In situation where visitor identify himself by login and password usage, it is much easier situation for recommendation. Based on his past activities it is possible to develop recommendation system. Proposed model which recognize image content retrieval, can be additional tool for recommendation.

Proposed model can be useful in recommendation systems like retail oriented web shop, or tour operator sites where is a great dynamic of new items or locations publishing in portfolio of company. In such conditions it is not appropriate to tag each item image separately. Proposed model gave instant solution for recommendation systems based on image content, even we are talking about new images. Proposed method should have power to recognize most similar item/items from the images as well as image category. That can be solution for recommendation systems on web based on image content. Proposed method is not such precise in comparison with existing models/methodology for image segmentation based on K-Nearest neighborhood, Bayes classifier, support vector machines, graph cuts, but it is faster in data processing, and provides instant quick solution. It is important when we are talking about online systems in situation where we do not have privilege to classify newcomer images frequently. It is light solution from perspective of data processing which can be used for online shops, which has frequent image inputs of new items. Existing model can be improved in way to concentrate on localized shapes within images, or to make some rules regarding prioritization in recommendation system taking in consideration visitors preferences upon his previous activities.

7 Future Research

Presented model as it was previously emphasized is relatively simple and efficient for quick similarity analysis. Situations which demands a quick response regarding similarity between two or many images is application area for proposed model. This concept gives general evaluation regarding similarity. Precision could be improved by introducing neighborhood pixel similarity index. That is the existing model could be extended with additional evaluation. Basic idea is that diversity in RGB scale is not very high in near pixel neighborhood if we are talking about content. It is especially evident in situation between images with similar content like seashore, meadow, and other environments. User can vary radius of pixel neighborhood as parameter like in self-organizing maps and learning vector quantization 1, 2 and 3 algorithms. In this way presented algorithm could give on accuracy. We have to have in mind that presented method are not object recognition oriented, and primary aim of proposed method is content recognition. It means that content could be declared in many different ways with focus on scenery, ambient, associations and other categories. That can be base for tagging. Tags can be used not only for selling purposes, but also for therapeutic purposes. For example users in bad mood can be faced with peaceful and encouraging image content and it can be combined with selling activities regarding their preferences. For that, recommendation is usage of traditional predictive modeling along with proposed model. Understanding the models and client behavior, gaining insight into their profiles by using proposed methods enabled a very different, defensive approach to be taken towards the market segment at strengthening customer loyalty.

8 Conclusion

Proposed method is ambitious to be useful in situation where we do not have trained models based on supervised learning. This approach has some advantages and some disadvantages. Advantage is that each new case (image) could be categorized, tagged, classified without additional learning process relatively quickly. Disadvantage of proposed method is that it is hard to expect high accuracy and reliability from that method. Reason for that also lies in fact that images for which we would like to find similar images are not precisely defined, and we are not seeking specific object or precisely defined clusters or image content. Algorithm seeks for similarity in way of comparison differences on pixel level. This method can be improved by introducing neighborhood pixel similarity index within model as additional element. Consumption is that diversity in RGB scale is not very high in near pixel neighborhood if we are talking about content for correspondent image. Presented model has uses in variety of situations like in traveling agencies as a part of recommendation system, web shops as a part of recommendation system as well and for solving similar problems. Generally speaking appropriate classification of images is done by their dominant characteristics on RGB scale. Precise classification could be done by considering many different aspects, as some other methods do. The main purpose is to introduce simple and efficient method which shows plausible results, and which can be applied quickly and efficiency in conditions where we do not have room for supervised model usage.

References

1. Bezdek, J., Krisnapuram, R., Pal, N.: Fuzzy Models and Algorithms for Pattern Recognition and Image Processing (The Handbooks of Fuzzy Sets). Springer, New York (2005)
2. Burger, W.: Digital Image Processing: An Algorithmic Introduction Using Java. Springer, New York (2008)
3. Canti, M.: Image Analysis, Classification and Change Detection in Remote Sensing: With Algorithms for ENVI/IDL and Python. CRC Press, New York (2014)
4. Celebi, E.: Partitional Clustering Algorithms. Springer, Berlin (2014)
5. Doermann, D.: Handbook of Document Image Processing and Recognition. Springer, New York (2014)
6. Goshtasby, A.: Image Registration: Principles, Tools and Methods (Advances in Computer Vision and Pattern Recognition). Springer, Berlin (2012)
7. Kainmueller, D.: Deformable Meshes for Medical Image Segmentation. Springer Fachmedien Wiesbaden, Germany (2015)
8. Leondes, C.: Image Processing and Pattern Recognition, Volume 5 (Neural Network Systems Techniques and Applications). Academic Press, San Diego (1997)
9. Nagamalai, D., Renault, E., Dhanuskodi, M.: Advances in Digital Image Processing and Information Technology. Springer, Berlin (2011)
10. Nixon, M.: Feature Extraction & Image Processing for Computer Vision. Academic Press, San Diego (2012)
11. Shih, F.: Image Processing and Pattern Recognition: Fundamentals and Techniques. Wiley-IEEE Press, New York (2010)
12. Toennies, K.: Guide to Medical Image Analysis: Methods and Algorithms Advances in Computer Vision and Pattern Recognition. Springer, Berlin (2012)
13. Zhou, K.: Medical Image Recognition, Segmentation and Parsing: Machine Learning and Multiple Object Approaches. Academic Press, London (2015)
14. Castelman, K.: Digital Image Processing. Prentice Hall, Upper Saddle River (1995)
15. De, S., Bhattacharyya, S., Chakraborty, S.: Multilevel and color image segmentation by NSGA II based OptiMUSIG activation function. In: Bhattacharyya, S., Banerjee, P., Majumdar, D., Dutta, P. (eds.) Handbook of Research on Advanced Hybrid Intelligent Techniques and Applications, pp. 321–348. IGI Global, Hershey (2016)
16. Gogniat, G., Milojevic, D., Morawiec, A.: Erdogan: Algorithm-Architecture Matching for Signal and Image Processing. Springer International Publishing, Netherlands (2011)
17. Koprowski, R.: Image Analysis for Ophthalmological Diagnosis. Springer International Publishing, Switzerland (2016)
18. Klette, R.: Concise Computer Vision. Springer, London (2014)
19. Parker, J.: Algorithms for Image Processing and Computer Vision. Wiley, New York (2010)
20. Szeliski, R.: Computer Vision: Algorithms and Applications (Texts in Computer Science). Springer, Berlin (2011)
21. Criminisi, A.: Decision Forests for Computer Vision and Medical Image Analysis (Advances in Computer Vision and Pattern Recognition). Springer, New York (2013)
22. Garcia, G., Suarez, O.: Learning Image Processing with OpenCV. Packt Publishing - ebooks, (2015)
23. Gonzales, R.: Digital Image Processing. Pearson Education, New Delhi (2014)
24. Hasanzadeh, M., Kasaei, S.: Fuzzy Image Segmentation Using Membership Connectedness. EURASIP J. Adv. Signal Process. (2008). doi:10.1155/2008/417293
25. Amar, M., Ben Ayed, I.: Variational and Level Set Methods in Image Segmentation. Springer, Berlin (2011)
26. Ameer, M.: Image Segmentation: A Fuzzy Clustering Framework. VDM Verlag Dr. Mller, Germany (2010)
27. Batra, D., Kowdle, A., Parikh, D., Luo, J., Chen, T.: Interactive Co-segmentation of Objects in Image Collections. Springer, New York (2011)

28. Brock, K.: Image Processing in Radiation Therapy (Imaging in Medical Diagnosis and Therapy). CRC Press, Boca Raton (2013)
29. Han, X.: Medical Image Segmentation Using Level Set Method and Digital Topology: Concepts and New Developments. VDM Verlag Dr. Mller, Germany (2008)
30. Russ, J.: The Image Processing Handbook, 6th edn. CRC Press, Boca Raton (2011)
31. Salem, M.: Medical Image Segmentation: Multiresolution-Based Algorithms. VDM Verlag Dr. Mller, Germany (2011)
32. Quan, L.: Image-Based Modeling. Springer, Berlin (2010)
33. Yoo, T.: Insight into Images: Principles and Practice for Segmentation, Registration, and Image Analysis. CRC Press, Boca Raton (2004)
34. Klepac, G.: Discovering behavioural patterns within customer population by using temporal data subsets. In: Bhattacharyya, S., Banerjee, P., Majumdar, D., Dutta, P. (eds.) Handbook of Research on Advanced Hybrid Intelligent Techniques and Applications, pp. 321–348. IGI Global, Hershey (2016)
35. Klepac, G., Kopal, R., Mrsic, L.: REFII model as a base for data mining techniques hybridization with purpose of time series pattern recognition. In: Bhattacharyya, S., Dutta, P., Chakraborty, S. (eds.) Hybrid Soft Computing Approaches, pp. 237–270. Springer India, Berlin (2015)

Microscopic Image Segmentation Using Hybrid Technique for Dengue Prediction

Pramit Ghosh, Ratnadeep Dey, Kaushiki Roy, Debotosh Bhattacharjee and Mita Nashipuri

Abstract An application of hybrid soft computing technique for early detection and treatment of a most common mosquito-borne viral disease Dengue, is discussed thoroughly in this chapter. The global pictures of dengue endemics are shown clearly. The structure of dengue virus and the infection procedure of the virus are also discussed. A detailed analysis of dengue illness, diagnosis methods, and treatments has been done to conclude that platelet counting is needful for early diagnosis of Dengue illness and for monitoring the health status of the patients. The main challenge in developing an automated platelet counting system for efficient, easy, and fast detection of dengue infection as well as treatment, is in the segmentation of platelets from microscopic images of a blood smear. This chapter shows how the challenges can be overcome. Color-based segmentation and k-means clustering cannot provide desired outputs in all possible situations. A hybrid soft computing technique efficiently segments platelet and overcomes the shortcomings of the other two segmentation techniques. This technique is the combination of fuzzy c-means technique and adaptive network-based fuzzy interference system (ANFIS). We have applied three different segmentation techniques namely color-based segmentation, k-means, and the hybrid soft computing technique on poor intensity images. However, only the hybrid soft computing technique detects the platelets correctly.

Keywords ANFIS · Average filter · Color segmentation · Dengue · Fuzzy c-means · L*a*b color space · Platelet counting · K-means clustering

P. Ghosh (✉)
RCC Institute of Information Technology, Kolkata 700015, India
e-mail: pramitghosh2002@yahoo.co.in

R. Dey · K. Roy · D. Bhattacharjee · M. Nashipuri
Jadavpur University, Kolkata 700032, India
e-mail: ratnadipdey@gmail.com

K. Roy
e-mail: kaushiki.cse@gmail.com

D. Bhattacharjee
e-mail: debotoshb@hotmail.com

M. Nashipuri
e-mail: mitanasipuri@gmail.com

S. Bhattacharyya et al. (eds.), *Hybrid Soft Computing for Image Segmentation*, DOI 10.1007/978-3-319-47223-2_5

1 Introduction

Dengue fever is a vector-borne viral disease, caused by the mosquito species named Aedes Aegypti. The dengue virus (DENV) is RNA based. It spreads all over the world, especially in tropical regions. The name Dengue comes from a Swahili word which means Bone breaking fever. The dengue fever was first reported in China during the Jin Dynasty (265–420 AD). In 2012, the dengue fever was the most common mosquito-borne viral disease in the world [1]. As per World Health Organization, each year 50–100 million dengue infected cases are reported all over the world. India is also in the danger of dengue. In India, dengue season is from the month of July to the month of December. During this time span a large number of dengue infected cases are reported.

Symptoms of dengue are similar to other viral diseases like influenza, measles, chikungunya, malaria, Human Immunodeficiency virus (HIV) seroconversion illness, yellow fever, etc. [2]. So, detection of dengue is a very challenging task. Many diagnostic procedures are available to detect the presence of dengue virus in the human body. The detailed description of these diagnosis methods has been presented in Sect. 4. As an early diagnosis of dengue infection, doctors prescribe platelet count. Patients suffering from dengue-like symptoms and belonging from dengue endemic areas are instructed for platelet count. There are no specific medicines available to kill dengue viruses [3]. Patients who survive dengue are immensely due to the medical care of physicians and nurses. Physicians apply medicines that help the human body to fight against dengue viruses. In addition to that, regular monitoring is done to keep track of the health status of the patients. For monitoring, platelet count plays an important role. The status of dengue infection can be identified by the platelet count of the patient. The count goes low according to the severity of illness, and it becomes normal when the patient recovers from the infection. So, for early detection of dengue infection and for monitoring the health of a patient, platelet count is very crucial.

We have made an automated system that can count platelets from microscopic images of a blood smear. Our primary goal was to count platelets. To accomplish this, the foremost task is to segment platelets from other blood cells and blood plasma.

Many researchers have worked on blood cell image segmentation. The watershed transformation was done [4, 5] to segment RBCs. RBC counting system was developed in Android [6]. In this work, a RBC counting app was made using immersion simulation-based self-organizing transform, a modified version of watershed transformation. Pearl P. Guan [7] proposed a blood cell segmentation technique using Hough Transform and Fuzzy curve tracing algorithm which performed better than a watershed algorithm. Hough transform and morphological operations were used to segment RBCs from HSV converted image [8]. The Histogram thresholding was used here to compute threshold values. In another approach S. Kareem et al. [9] used Annular Ring Ration transform to segment RBCs. M. Mohammed et al. [10] proposed a WBC segmentation method. At first, the input RGB image was converted to gray scale by applying Gram–Schmidt orthogonalization technique. Then the grayscale image was converted to a binary image using grayscale thresholding followed by minimum filtering. The Otsus thresholding method was used to calculate the threshold value and morphological operations were used to remove other smaller objects. Binary images, thus generated, were used as a mask on the original image to segment WBC nucleus. In another proposed technique of WBC segmentation [11], the WBC region was first segmented by color, gradient, and brightness. Then for better segmentation regularity detection method followed by Hough transform was used. Many hybrid techniques were applied for better segmentation of blood cells. S. Mohapatra et al. [12] segmented leucocytes using a hybrid approach in which Rough Sets and Fuzzy sets were combined. The nucleus and cytoplasm of leukocytes were extracted from the background by using rough-fuzzy c-means algorithm. The input RGB image was converted to L*a*b color space and the images were clustered among four clusters. The clusters have contained a nucleus, cytoplasm, RBCs, and background separately. This technique could not work well for the images containing overlapped WBCs. A combination of neural network and the genetic algorithm was applied [13] for feature extraction in case of blood cell nucleus segmentation. This hybrid soft computing-based feature extractor produced optimized classification results. In another approach, the WBC nucleus was segmented using a combination of Fuzzy c-means and SVM [14]. The input images were clustered using Fuzzy c-means and then classified using SVM. In the case of platelet segmentation, we first apply color-based segmentation which is discussed thoroughly in Sect. 10.2. This segmentation technique has some shortcomings. To overcome those limitations we have applied k-means clustering technique to segment platelets. But k-means did not produce the desired results. So, we have applied a hybrid soft computing technique. It is a hybrid cascade technique where the clustering has been done using fuzzy c-means clustering technique and the classification has been done using adaptive network-based fuzzy interference system (ANFIS). This segmentation technique is efficient to segment platelets from the microscopic image of a blood smear.

This chapter is organized as follows: Sect. 2 shows the global scenario of dengue infection; Sect. 3 describes dengue virus and its effect in human body; Sect. 4 discusses about dengue virus detection and treatment procedure of dengue infected patients; the proposed hybrid technique is analyzed in Sect. 5, and finally, Sect. 6 concludes the chapter.

2 Dengue Scenario All over the World

Dengue fever is tropical and subtropical area disease. Countries located in those regions are in threat of Dengue fever. Dengue fever is mainly detected in the regions like Asia–pacific, Western Pacific, European region, South–East Asia, Eastern Mediterranean region, and Africa region. Figure 1 shows the severity of dengue outbreak in the tropical regions.

Within last five decades, dengue became the most dangerous viral disease in the world. The number of dengue incidences increased rapidly. Every year 50–100

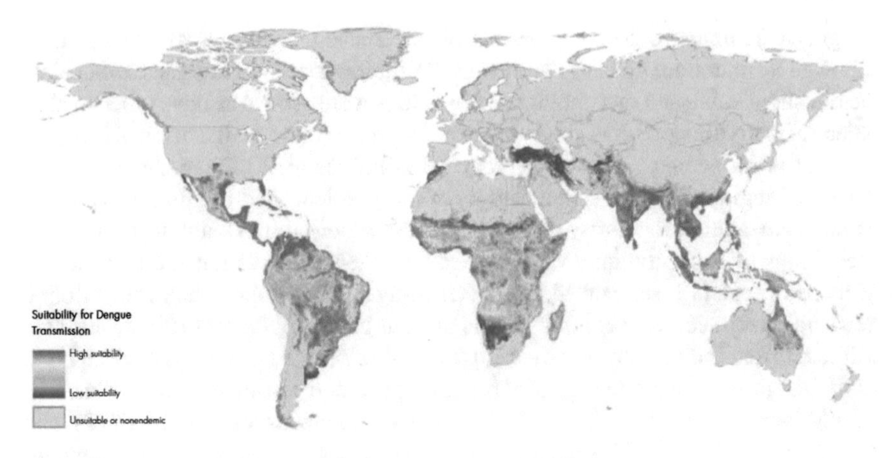

Fig. 1 Distribution of Dengue outbreak all over the world [15]

million dengue cases are reported by WHO all over the world [16]. In some mostly affected countries, the burden of dengue is as much as or higher than the diseases like HIV, malaria, hepatitis, etc. [17]. In recent years, dengue spread more than 150 countries. In Asia–pacific region 1800 million people are at risk of Dengue. In the Western Pacific region, dengue-affected cases have been reported in more than 30 countries. In 2013–2014, Dengue serotype 3 was reported in island nations like Fiji. The interesting information is that this serotype was absent in those countries for 30 years. Malaysia and Singapore are also in threat of Dengue. After 55 years, dengue incidences happened in Europe. In 2010, it was first reported in France and Croatia. In 2012, almost 2200 cases were reported in Portugal [18]. In 2015, Dengue cases were reported in China [19]. In India, earlier dengue endemic states were Andhra Pradesh, Goa, Gujarat, Haryana, Karnataka, Kerala, Madhya Pradesh, Maharashtra, Punjab, Rajasthan, Tamil Nadu, Uttar Pradesh, West Bengal, Chandigarh, Delhi, and Pondicherry. But now it has spread to the rest of the states. The numbers of dengue incidences are increasing rapidly, but the mortality rate due to dengue is decreasing. In 2007, the mortality rate was 1.2 % and in 2013, it dropped to 0.25 %. Compared to the other countries in South–East Asia the mortality rate due to dengue is very less in India [20].

3 The Dengue Virus and Its Effect in Human Body

In the previous section, it is cleared that dengue disease is spread all over the world, and it becomes a threat to all countries. In this section, we discuss the virus structure and its effects in human body.

The Dengue viruses belong to the genus Flavivirus in the family Flaviviridae. Dengue is caused mainly by four closely related viruses named DEN-1, DEN-2,

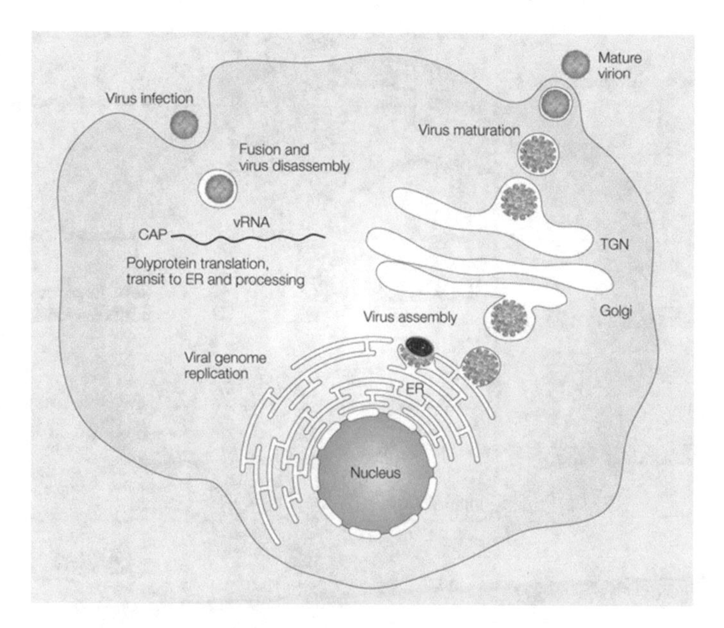

Fig. 2 Dengue virus infection in human cell

DEN-3, and DEN-4 [21]. The dengue virus is mainly transmitted by a mosquito named Aedes Aegypti. When an infected Aedes Aegypti female mosquito bites a person then the dengue virus spreads from the salivary gland of the mosquito to the human blood [22].

The dengue viruses first attack two types of skin cells (i) Keratinocytes cells [23] and (ii) Langerhans cells [24]. Figure 2 shows how the viruses infect a human cell. The virus enters in a human cell by a process named endosome. The virus releases its RNA genome into the cytoplasm. The RNA genome then starts replicating by using the human cell replication process. The newly formed dengue viruses travel through the Golgi apparatus complex, and they become mature forming infections. When they become fully mature, they leave the cell and transmit to other cells to infect them [25]. When the Langerhans cells are infected, they go to the lymph system, and the immune system of the human body becomes active. The human immune system produces two types of cells B-cells and T-cells to kill the viruses. The B-cells produce two antibodies IgM and IgG in blood stream whereas the T-cells kill the dengue infected cells. A dengue-affected person feels fever when the immune system fights against these viruses. However, the immune system lasts only for two to three months. In that period, the human body is protected from all four types of dengue virus serotypes (DEN-1, DEN-2, DEN-3, and DEN-4). After that time, any dengue virus can affect the human body. When the dengue viruses further affect a human body, the previous antibodies help the viruses to spread throughout the body [26].

Dengue fever has three phases. Figure 3 shows a graphical display of changes in the human body during three phases of Dengue fever. The symptoms of dengue

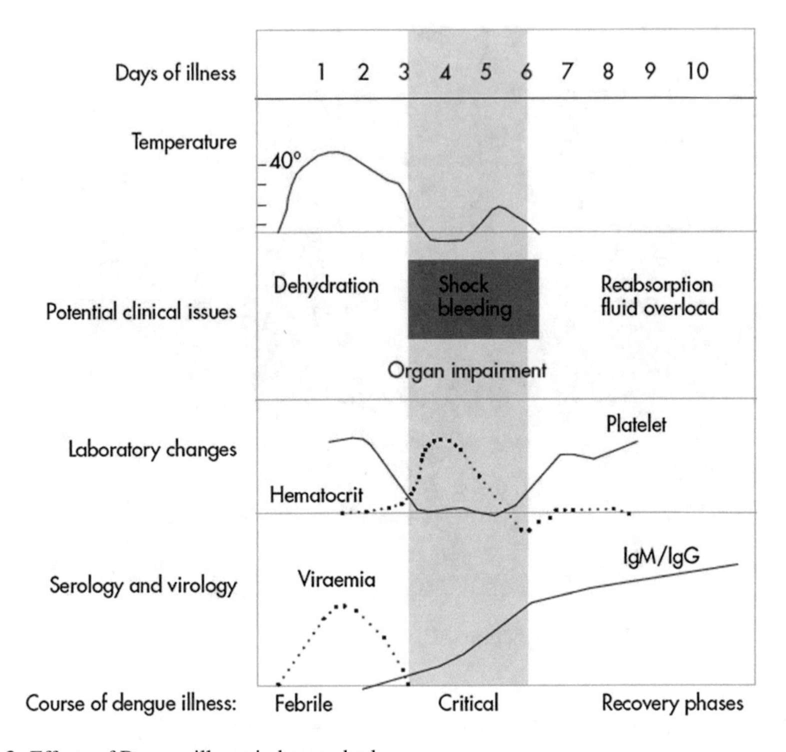

Fig. 3 Effects of Dengue illness in human body

fever usually are noticed after an incubation period of 4–13 days after the bite of the infected mosquito. After the incubation period, the illness is followed by three main phases: febrile, critical, and the recovery phase [27, 28].

Febrile Phase

The duration of febrile phase is mainly for 2–7 days. During these period patients commonly, suffer from following problems [29]

- Facial flushing
- Skin erythema
- Myalgia
- Arthralgia
- Generalized body ache
- Photophobia
- Rubelliform Exanthema
- Retro-orbital eye pain
- Headache.

In addition to the above symptoms some patients also suffers from a sore throat, conjunctivitis, nausea, vomiting, etc. Mucosal Bleeding and petechiae may be seen for mild hemorrhagic nature of dengue [30, 31]. In this phase, the working capability of patients become very low [32].

A positive tourniquet test is often a confirmatory test in this phase to detect dengue [30, 33]. The tourniquet test determines the hemorrhagic tendency of a patient. In this period thrombocytopenia (low platelet count) is followed by a progressive decrease in total WBC count [26]. Thus, from the above we conclude that low platelet count and low WBC count are the main confirmatory tests of this period.

Critical Phase

The critical phase is usually on days 3–7 [27] of illness. Patients feel lethargic but remain mentally alert. Mostly dengue patients die in critical phase [31]. In critical phase patients suffer

- Persistent Vomiting
- Abdominal Pain
- Postural Hypotension
- Bleeding
- Leukopenia [30] - WBC count becomes less than 5000 cells/mm^3
- Thrombocytopenia [30] - Platelet count becomes less than 100 000 cells/mm^3
- High Haematocrit [34, 35]
- Plasma leakage
- Shock - Body temperature drops to 37.5–38 °C or less. Low blood pressure.
- Liver enlargement.
- Organ failure.

The vomiting and abdominal pain leads to plasma leakage in the patients body which is further accompanied by leucopenia followed by thrombocytopenia. High hematocrit is the result of plasma leakage. The plasma leakage leads to shock in patients. Mortality for dengue is due to this shock.

Recovery Phase

If the patient survives the 24–48 h of the critical phase, then a slow recovery of the patient takes place in the following 48–72 h. In the recovery phase, the hematocrit stabilizes, the WBC count rises, the platelet count also increases but its recovery is slower than that of white blood cell count [34, 35].

Dengue Warning Signs

According to WHO, Dengue in classified [36] into two types (i) Dengue Fever (DF) and (ii) Dengue Hemorrhagic Fever (DHF). The Dengue Hemorrhagic Fever is further classified into two types (i) Mild (grade I and II) and (ii) Severe (grade III and IV).

There are some warning signs of dengue infection. Depending on these signs the severity of infection can be identified. The warning signs are

- Abdominal pain or tenderness
- Vomiting

- Clinical fluid accumulation
- Mucosal bleeding
- Lethargy, restlessness
- Liver enlargement >2 cm
- Increase in Hematocrit accompanied by rapid decrease in platelet count.

4 Dengue Diagnosis and Treatment

In this chapter, we will discuss the diagnosis method and treatment of dengue disease. The dengue disease symptoms are very similar to many other diseases [37–45]. Therefore, the diagnosis of dengue is a very challenging task. The treatment is also very critical, as there is no vaccine and medicine available to kill dengue virus.

4.1 Dengue Diagnosis

The diagnosis [46] of dengue illness is carried out in two ways (i) Isolation of Dengue viruses and (ii) Serological Tests. Table 1 shows different types of dengue illness with their symptoms in the human body and necessary laboratory diagnosis for each of

Table 1 Different types of Dengue illness with symptoms and laboratory diagnosis

DF/DHF	Grade	Symptoms	Laboratory findings
DF	–	Fever with (any two or more) - • Headache • Retro-orbital pain • Myalgia • Arthralgia	• Leucopenia Low WBC count • Thrombocytopenia Low platelet count
DHF	I	Symptoms of DF with positive tourniquet test and plasma leakage	• Platelet count lower than 1,00,000/cu mm • Haematocrit rise 20 % or more
DHF	II	All symptoms of DHF I with • bleeding from skin and organs • abdominal pain	• Platelet count lower than 1,00,000/cu mm • Haematocrit rise 20 % or more
DHF	III	All symptoms of DHF II with circulating failure (Pulse rate $\leq$ 20; Hypotension with cold calm skin and restlessness)	• Platelet count lower than 1,00,000/cu mm • Haematocrit rise 20 % or more
DHF	IV	Heavy shock with undetectable blood pressure. 20 % rise in Haematocrit	• Platelet count lower than 1,00,000/cu mm • Haematocrit rise 20 % or more

them. In the table DF means Dengue Fever and DHF means Dengue Hemorrhagic Fever.

From the above table, it can be seen clearly that the laboratory diagnosis for all types of dengue illness is very similar. In practice, when a person suffers from periodic fever, vomiting, rash in skin, then the physicians suggests platelet count. If the count goes low, then there is a chance for dengue infection. When the platelet count decreases rapidly then chances of further dengue increase. From Fig. 3 it is clear that the platelet count goes very low in febrile and critical phase. So, platelet count is very important to detect dengue, and, it helps to keep track of the health status of dengue infected persons.

In pathological labs, platelet is counted manually using Hemocytometer chamber. Venous blood is diluted in 1 % ammonium oxalate stored in 4 °C. Then Hemocytometer chamber is filled with the diluted blood. Then platelets are counted by placing the chamber under the microscope in 40x objective [47]. This process gives accurate results. But the problem is that this process is very tough to carry out. It needs efficient lab technicians, and, it is a very time-consuming process. Though this process is a theoretically correct process, now pathologists are not using this process due to its high complexity. Pathologists count platelets from blood slides. In this process blood smear is drawn on a glass slide. Then it is stained using Leishmans stain and is placed under the microscope. Platelets are then counted using 100x objective of the microscope.

4.2 Dengue Treatment

The Dengue patient can be differentiated in three category group A, group B, and group C. Each of these patients needs a different type of treatment [2, 48–52] and monitoring procedure. The treatment strategies are discussed below.

Group A

This group of patients suffers mild Dengue infection. The dengue warning signs (mentioned in Sect. 3) are absent in these type of patients.

Treatment

Doctors prescribe

- Adequate bed rest.
- Plenty of fluids Coconut water, rice water, fruit juice, soup, etc. Cold drinks with high isotonic level must be avoided.
- Paracetamol for patients with high fever 10 mg/kg/dose is recommended. 3–4 times for children in a day and not more than 3 g/day for adults.

Monitoring

- Decreasing WBC
- Health status presence of any warning signs

Group - B

This group of patients has warning signs of dengue, but they do not suffer shock. The hematocrit level of the patient increases rapidly.

Treatment

Intravenous fluid therapy with Ringer Lactate or 0.9 % saline or Hartmann's solution is applied. At first, the therapy is applied for 1–2 h with dose 5–7 ml/kg/h. Then, it is reduced to 3–5 ml/kg/h for 2–4 h and after that the dose gradually decreases to 2–3 ml/kg/h. If the hematocrit level remains same or increases slightly then intravenous fluid therapy must be continued for dose 2–3 ml/kg/h for 2–4 h. If there is a further increase in hematocrit level, the dose of intravenous fluid therapy must be 5–10 ml/kg/h for 1–2 h. This intravenous fluid therapy should be continued for 24–48 h. The application of intravenous fluid becomes nominal when the urine rate is 0.5 ml/kg/h.

Monitoring

- Perfusion
- Urine output
- Haematocrit
- Blood glucose
- Renal profile
- Liver Profile
- Coagulation profile.

Group - C

This group of patients suffers from severe dengue infection. Shock is present in those patients and mortality is high for them. These patients suffer severe hemorrhages with organ failure, and they require emergency treatment.

Treatment

- **Compensated Shock** - At first, the hematocrit level needs to be checked. If it becomes very high, then the intravenous fluid therapy with isotonic crystalloid solution needs to be carried out. The dose is 5–10 ml/kg/h for 1 h in case of adults and for children, the dose is 10–20 ml/kg/h for 1 h. After applying the first dose, the health status of the patient needs to be monitored. The parameters of the health status include vital signs, hematocrit, urine output, capillary refill time. If the condition of the adult improves then the dose of intravenous fluid therapy is gradually decreased to 5–7 ml/kg/h for 1–2 h; then 3–5 ml/kg/h for 2–4 h and finally 2–3 ml/kg/h for 24–48 h. For children, this doze is first decreased to 10 ml/kg/h for

1–2 h, then 7 ml/kg/h for 2 h, then 5 ml/kg/h for 4 h and finally to 3 ml/kg/h for 24–48 h. When the patient becomes capable of taking fluid orally, then the intravenous fluid therapy becomes nominal. The fluid therapy must not be applied for more than 48 h.

If the hematocrit level for adult patients increase (>50 %) further, then the fluid therapy must be applied with dose 10–20 ml/kg/h for 1 h followed by 7–10 ml/kg/h for 1–2 h. If a similar condition arises in children, then the dose of fluid therapy must be 10–20 ml/kg/h for 1 h, then 10 ml/kg/h for the next hour and is reduced finally to 7 ml/kg/h for another 1 h.

- **Profound Shock** - Patients are applied intravenous fluid resuscitation with crystalloid or colloid solution at 20 ml/kg for 15–20 min to bring them out of shock. For adults, this is followed by crystalloid/colloid infusion with dose 10 ml/kg/h for 1 h and then it is reduced to 5–7 ml/kg/h for the next 1–2 h then 3–5 ml/kg/h for 2–4 h and finally to 2–3 ml/kg/h for 24–48 h. For children, the dose starts from 10 ml/kg/h and is reduced to 7.5 ml/kg/h, which is further reduced to 5 ml/kg/h for 4 h and finally to 3 ml/kg/h for 24–48 h.

5 Platelet Segmentation

The previous discussions in Sect. 4 provide enough evidence to state that Thrombocytopenia is one of the important diagnosis measures to identify early stage dengue fever and severe dengue infection, and it is needed to monitor the health status of dengue infected patients. We want to build an automated system for platelet counting which can be used for efficient early stage diagnosis and treatment of dengue. In this system, platelet segmentation from microscopic blood smear image is the most challenging task. Previously, we have applied color-based segmentation method for segmentation of platelets. However, the algorithms have some shortcomings. To eliminate them we applied K-means clustering. However, this method cannot erase the shortcomings of the color-based segmentation technique completely. Therefore, we finally applied hybrid soft computing technique, which segments platelets almost perfectly.

5.1 Image Acquisition

We have used a light microscope, OLYMPUS CX 21I and a digital camera fitted with the microscope to capture images from Leishman's stained blood slides at 100X objective. The 100x objective refers to 1000 times magnification of objects. Platelets are very small in size, and hence, high magnification is required to view them. The objective turret lens is set at 100x magnification, and the eye-piece lens is fixed at 10x magnification. So the overall magnification is $100 \times 10 = 1000$.

5.2 *Platelet Segmentation Using Color-Based Segmentation*

We have proposed a segmentation method [53] to segment platelets from the microscopic image of a blood smear. In this work, we have applied color-based segmentation and mathematical morphological operations to segment platelets. Since platelets have no specific shape and size, we discard shape-and size-based segmentation. Platelets are a small bluish cell, and this feature distinguishes them from other blood cells. In our work, at first, we have converted the input images from RGB to l*a*b color space. The l*a*b color space has three layers luminosity layer L and two chromaticities layer a* and layer b*. The luminosity layer or L represents lightness value of the image. The a* layer represents red-green color, and the b* layer represents blue-yellow layers. We have extracted b* layers from a* layers to get platelets. But this image consists of some parts of overlapping RBCs and WBC nucleus. These parts have been removed using morphological operations. However, this algorithm is completely dependent on the intensity of the image. For the low-intensity image, the algorithm does not provide appropriate results.

5.3 *Platelet Segmentation Using K-Means Clustering Technique*

K-means clustering [54] aims at partitioning n observations into k different clusters in such a way where observations in same clusters are highly similar whereas objects in different clusters are highly dissimilar. The main idea is to define k different centers, one for each cluster. The algorithm [54] for k means clustering is as follows:

Let n = $\{n_1, n_2, n_3, \ldots, n_x\}$ be the set of x different data points and c = $\{c_1, c_2, c_3, \ldots, c_x\}$ be the centers of k different clusters.

Step 1 We have to select randomly k different cluster centers. These centers should be chosen in such a way that they are highly far away from each other to ensure that objects in different clusters are highly dissimilar.

Step 2 The distance between each data points and cluster centers needs to be calculated.

Step 3 Each of the data points should be assigned to that cluster center with which its distance is minimum.

Step 4 After assigning each of the data points to one of the clusters, the cluster centers need to be recalculated using the formula $c_i = 1/x(\sum x_i)$ where c_i is the ith cluster center and $\sum x_i$ is the sum of all data points in the ith cluster and x is the total number of observations in that cluster.

Step 5 Recalculate the distance between each data points and the newly obtained cluster centers.

Step 6 If no data points are reassigned then the process stops, otherwise steps 3–5 needs to be repeated.

This algorithm works well when the data sets are distinct and are well separated from each other. However, the main shortcomings of this algorithm are that the initial centroids are chosen randomly from the input data, so in each run of the same data the number of iterations and the total elapsed time changes.

When k-means algorithm is applied on microscopic images, then cluster contains some RBC pixels as well as platelets. Nevertheless, our notion was to get a cluster consisting of only platelets for better segmentation. However, this is not possible in k-means. The main reason behind this is that the intensity difference between platelets and RBCs are not much. As already discussed k means algorithm gives a good result only when the data points are distinct. Since this is not the scenario here, RBCs and platelets are in the same cluster. Therefore, for further improvement we used fuzzy c-means clustering technique.

5.4 Platelet Segmentation Using Hybrid Soft Computing Technique

Figure 4 shows the block diagram of our proposed hybrid soft computing segmentation technique. The input image is first preprocessed and then converted to L*a*b color space. A soft computing clustering technique named Fuzzy c-means is then applied. In the next step, a hybrid soft computing classification algorithm ANFIS is used to classify the correct cluster. In the last step, some morphological operations were done to appropriate segment platelets.

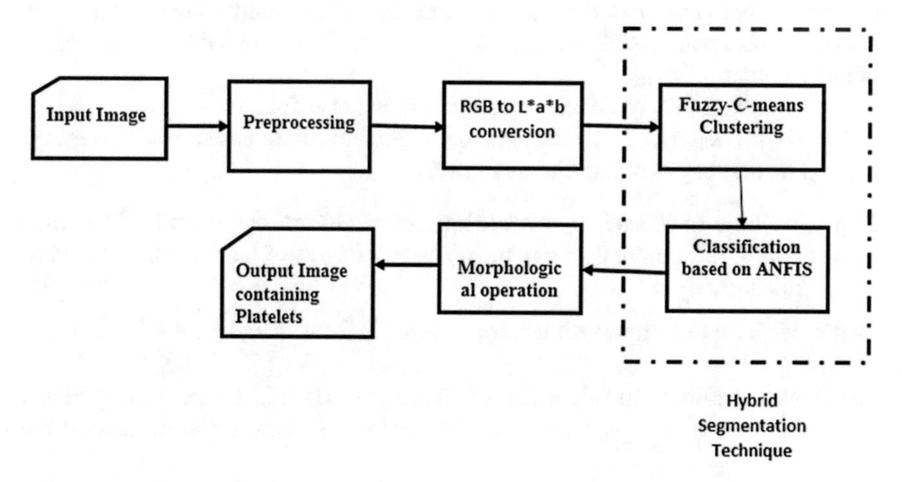

Fig. 4 Block Diagram of Hybrid soft computing technique for platelet segmentation

5.4.1 Image Preprocessing

This is the first phase of any image-processing task. This step is required to remove any unwanted noise from the image and to enhance the image for further processing. In microscopic images of blood smear, some noise may be present due to bad staining, the presence of dust particle, etc. Therefore, it is necessary to preprocess the image for further processing. In our work we have used an average filter [55] of size 3×3 because higher window size causes severe smoothing.

5.4.2 Color Conversion

In L*a*b [56] color space, the L layer represents lightness and a* and b* layers represent chromaticity. The red and green colors fall in a* layer and the blue, and yellow colors fall in b* layers. For better visual segmentation, the RGB image was converted to L*a*b color space.

5.4.3 Fuzzy C-Means Clustering

Fuzzy c-means [57] algorithm removes some of the shortcomings present in the k-means algorithm. Fuzzy c-means algorithm gives better results than k-means when the data sets are highly overlapping. Fuzzy c-means clustering algorithm assign membership to each data points corresponding to each cluster center according to the cluster center and data points distance. Membership of data points are more if the data is more nearer to the cluster center. Unlike K-means algorithm, in fuzzy c-means, a data point could belong to more than one cluster but with a certain degree of membership.

The algorithm [57] of Fuzzy c means clustering is as follows:

Let n $= \{n_1, n_2, n_3, \ldots, n_x\}$ be the set of x different data points and c $= \{c_1, c_2, c_3, \ldots, c_x\}$ be the centers of k different clusters.

Step 1 We have to choose appropriate number of clusters and initialize $U = [u_{ij}]$ matrix, U^0 randomly. Here u_{ij} represents degree of membership of data n_i in cluster c_j.

Step 2 We have to compute centroids c_j using the formula. $c_j = \frac{\sum_{j=1}^{N} u_{ij}{}^m X_j}{\sum\limits_{j=1}^{N} u_{ij}{}^m}$

Step 3 New membership values u_{ij} is to be computed using the following formula $U_{ij} = \frac{1}{\sum\limits_{k=1}^{C} \frac{\|x_i - C_j\|}{\|x_i - C_k\|}^{\frac{2}{m-1}}}$. Here c = total number of clusters and m denotes the degree of fuzziness and its value is necessarily greater than 1, typically 2.

Step 4 Update the membership matrix $U^{k+1} \leftarrow U^k$.

Step 5 Steps 2–4 needs to be repeated until change of membership values is very less, $U^{k+1} - U^k < e$ where e is a very small value typically 0.01.

This clustering algorithm works well for our purpose. A cluster can be found which contains only platelets with very small amount of overlapping RBCs. The shortcomings of k-means clustering algorithm can be avoided by this hybrid clustering techniques.

5.4.4 Soft Computing Technique Based on ANFIS for Cluster Identification

After obtaining the different clusters, we are left with the challenging task of identifying the correct cluster among the others. For this, we have used a hybrid learning algorithm, known as ANFIS [58], where neural network and fuzzy logic are combined. The main idea behind combining different soft computing techniques is to develop a hybrid technique that exploits the advantages of different soft computing techniques and overcomes their individual limitations. Table 2 demonstrates some of the advantages and shortcomings of different soft computing techniques.

Table 2 Advantages and Drawbacks of soft computing techniques

Soft computing technique	Advantages	Disadvantages
Artificial neural networks	Requires less formal statistical learning, possess ability of implicitly detecting complex nonlinear relationships between dependent and independent variables etc	Greater computational resources are required for neural network modeling and it is prone to over fitting
Support vector machines	Produces very accurate classifiers, less overfitting and it is robust to noise, does not get trapped in local minima	SVM is a binary classifier, so to do a multi-class classification pairwise classification is to be done that is testing one class against all others. Also it takes longer learning and training time and so it runs slow
Evolutionary computing	Works well even for problems where no good method is available like noisy problems, problems with discontinuities, nonlinear constraints, etc. Also parallel implementation is easier with evolutionary computing	Does not guarantee optimal solutions in a finite amount of time
Fuzzy logic	Easily understandable, possess ability to model nonlinear dependencies; representation of expert knowledge is easier	Estimation of membership function is difficult. Unlike expert systems, fuzzy logic does not possess the ability of learning and adapting after solving a problem. Another limitation of fuzzy logic is it is unable to solve those problems where no one knows the solution. Experts must exist to create the rule set of fuzzy logic system

As evident from Table 2, each of the existing soft computing techniques has some limitations. Therefore, to overcome these limitations different techniques need to be combined. As previously stated, we have used ANFIS for correct cluster identification. ANFIS stands for artificial neural network fuzzy inference system. It uses a hybrid learning algorithm to train the parameters of a Sugeno-Type fuzzy inference system.

Sugeno Fuzzy Inference Model

Let x and y be two inputs of fuzzy inference system and z be its output. A first-order sugeno fuzzy model [58] has the following rules:

Rule 1 if (x is A_1) and (y is B_1) then ($f_1 = p_1x + q_1y + r_1$)
Rule 2 if (x is A_2) and (y is B_2) then ($f_2 = p_2x + q_2y + r_2$)

Here x and y represents the input, A_i and B_i are the fuzzy sets. f_i are the outputs. The design parameters are p_i, q_i, and r_i which are learned during the training process.

ANFIS Architecture

Figure 5 represents the ANFIS architecture [58] which implements the two rules mentioned above.

In Fig. 5, the circles indicate a fixed node whereas the squares indicate adaptive nodes. In the first layer all the nodes are adaptive nodes. The output of this layer is the membership grade of the fuzzy set (A_1, A_2, B_1, B_2) given by Eqs. 1 and 2.

$$O_i{}^1 = \mu_{A_i}(x) \qquad i = 1, 2 \tag{1}$$

$$O_i{}^1 = \mu_{B_{i-2}}(y) \qquad i = 3, 4 \tag{2}$$

Here $\mu_{A_i}(x)$ and $\mu_{B_{i-2}}(y)$ can be any fuzzy membership function like bell-shaped membership function represented by Eq. 3.

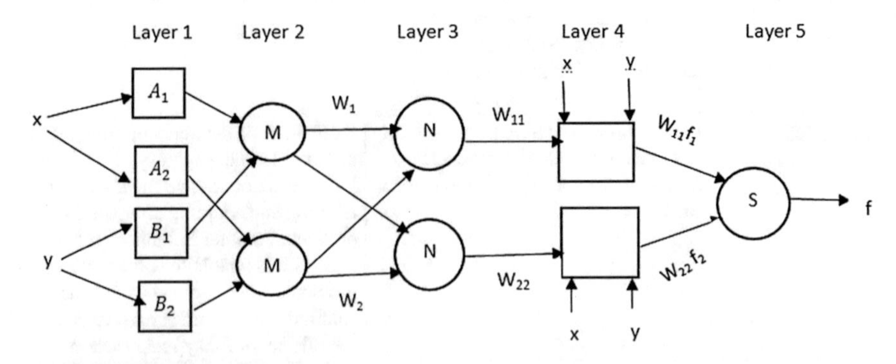

Fig. 5 ANFIS architecture

$$\mu_{A_i}(x) = \frac{1}{1 + \frac{x - c_i}{a_i}^{2b_i}} \tag{3}$$

Here a_i, b_i, c_i are the parameter set.

The second layer consists of all the fixed nodes. These are marked with M to signify that they act like simple multipliers. The outputs of this layer are represented by Eq. 4.

$$O_i^{\,2} = W_i = \mu_{A_i}(x)\mu_{B_{i-2}}(y) \qquad i = 1, 2 \tag{4}$$

The third layer labeled with N is also fixed nodes. The output of this layer is represented by Eq. 5.

$$O_i^{\,3} = W_{ii} = \frac{w_i}{w_1 + w_2} \qquad i = 1, 2 \tag{5}$$

The fourth layer comprises of all adaptive nodes. The output of all nodes of this layer is represented by Eq. 6.

$$O_i^{\,4} = W_{ii} * f_i = W_{ii}(p_i x + q_i y + r_i) \qquad i = 1, 2 \tag{6}$$

As previously stated p_i, q_i, r_i are known as design parameters which are learned during the training process. The fifth layer has only one fixed node labeled by S. This node sums all incoming signals. The outcomes of this layer is given by Eq. 7.

$$O_i^{\,4} = \sum_{i=1}^{2} W_{ii} f_i \tag{7}$$

So out of the four layers only the first and fourth layers are adaptive layers. The three modifiable parameters of the first layers are namely a_i, b_i, c_i. The three modifiable parameters of the fourth layer are namely p_i, q_i, r_i, corresponding to the first-order polynomial.

ANFIS Learning Algorithm

The algorithm [59] models the training data set by combining least squares and backpropagation gradient descent methods. In the forward pass, ANFIS uses least square method to tune the parameters of layer four namely p_i, q, r_i. In the backward pass the errors are propagated backwards to update the premise parameters namely a_i, b_i, c_i by gradient descent method. The main goal of the learning algorithm is to tune the modifiable parameters in such a way that the ANFIS result matches the training data. In our work, we have used five features to train the ANFIS system namely clusters center, mean, standard deviation, kurtosis, skewness. We have used a total of 30 images for training the system. The training matrix in our case is of dimension 6×30 where the columns correspond to the number of samples, the first five rows correspond to the five features used in training, and the last row corresponds to the class label either 1 or 0. Now when an input RGB image is given to this ANFIS trained system, the three clusters obtained from it using fuzzy c means technique serves as the test data. Now when these three clusters are given as input to the system, then only the desirable cluster is labeled as 1. The undesirable clusters are labeled as 0. After obtaining the desirable cluster further morphological operations are performed on it to remove the few overlapping RBCs.

5.4.5 Result and Discussion

In this section, we discuss the result obtained by our system and analyze them in detail. At first, the detail result of color-based segmentation is discussed. Then, the results obtained by k-means clustering are shown and finally the results of hybrid soft computing technique are discussed.

Figure 6a shows a good intensity microscopic image of a blood smear. Good intensity means that the image is clear and high contrast image. Color-based segmentation discussed in Sect. 5.2 is applied to the image. All the platelets have been identified efficiently in this case. The segmented image is shown in Fig. 6b. We apply this segmentation method on a low-intensity image shown in Fig. 7a and b shows the corresponding segmented image. The low-intensity images are low contrast images.However, in this segmented image, not all platelets have been identified. In Figs. 6a and 7a, platelets are marked by circles. This is the drawback of the color-based segmentation technique. It produces a very good result when it is applied to a good intensity image. But the result becomes erroneous when the intensity of input image is very low.

To remove the shortcomings of color-based segmentation we have applied the k-means algorithm in it. For our purpose, we have taken three clusters. Figures 8, 9 and 10 represent objects in first, second and third clusters, respectively. The main

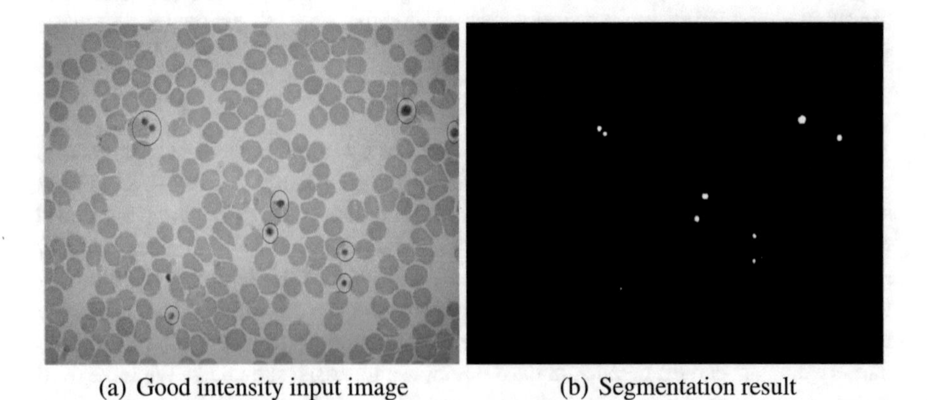

(a) Good intensity input image (b) Segmentation result

Fig. 6 Color-based segmentation applied on good intensity image

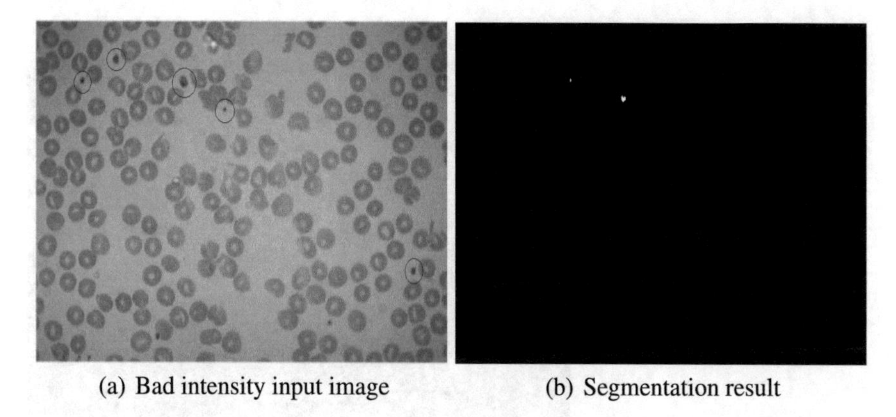

(a) Bad intensity input image (b) Segmentation result

Fig. 7 Color-based segmentation applied on low-intensity image

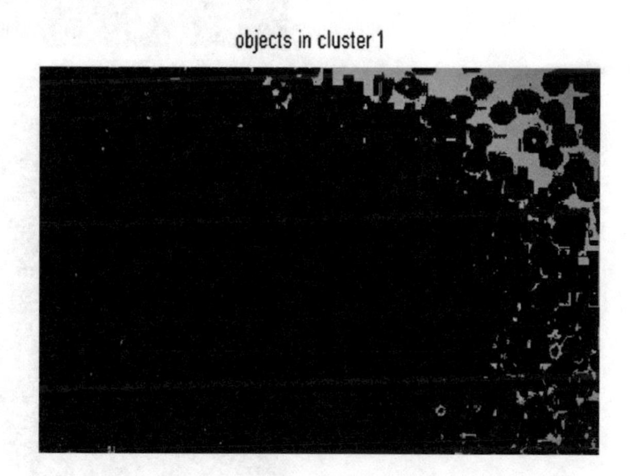

Fig. 8 Cluster 1 containing some overlapping RBCs

goal of our work was to separate the platelets from the other blood cells and the background.

As illustrated in Fig. 9, cluster 2 contains some RBC pixels as well as platelets. However, our notion was to get a cluster consisting of only platelets for better segmentation. However, this is not possible in the algorithm namely k-means. Therefore, for further improvement, we have used fuzzy c-means clustering technique.

We applied fuzzy c means technique on the low-intensity image, Fig. 7a. Figures 11, 12 and 13 represents objects in first, second, and third clusters, respectively.

As evident from Fig. 13, cluster 3 contains mostly the bluish colored pixels these are the platelets with few overlapping RBCs. So far, in our work, this clustering technique is better than the k-means because here platelet segmentation becomes easier. Since the intensity difference between overlapping RBCs and platelets is not significant, some parts of those are present in cluster 3 along with the platelets. After obtaining the three clusters, the main challenge was to identify the correct cluster, comprising majorly of platelets. We apply ANFIS learning algorithm to identify

Fig. 9 Cluster 2 containing the RBC pixels as well as platelets

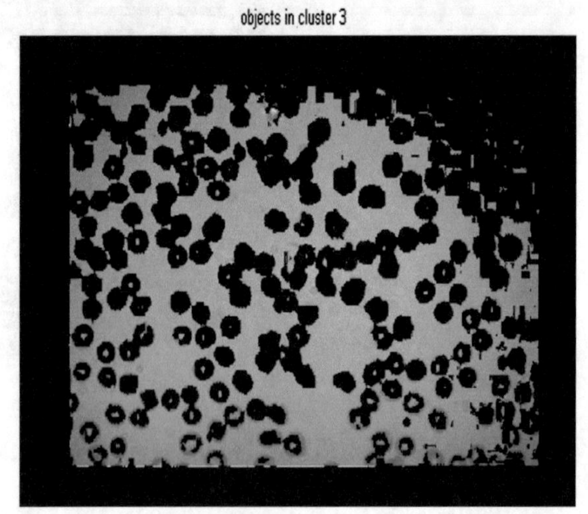

Fig. 10 Cluster 3 containing the background pixels

correct clusters. During classification phase for each of the 30 training images, their red channel information was extracted. Then the mean, standard deviation, kurtosis, skewness of all these pixels were calculated and were used as features during training. Our desired clusters are the ones in which the red components are minimum as because it is composed primarily of bluish colored platelets. The system was trained in such a way that the desired clusters were labeled as 1. In the test phase, only those clusters were labeled 1 and marked as the correct cluster whose mean, standard deviation were minimum, and having both kurtosis, skewness maximum. Figure 14 shows the desirable cluster obtained using ANFIS trained system. Thus, our problem of identifying the correct cluster from the three clusters as shown in Figs. 11, 12, and 13 gets solved. However, the correctly identified cluster contains some parts of

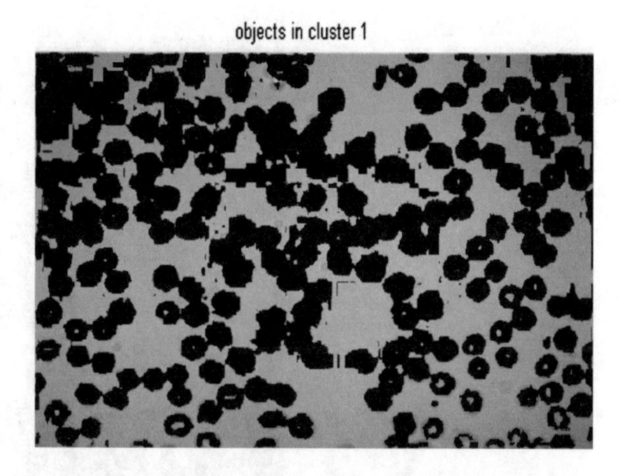

Fig. 11 Cluster 1 containing the background pixels

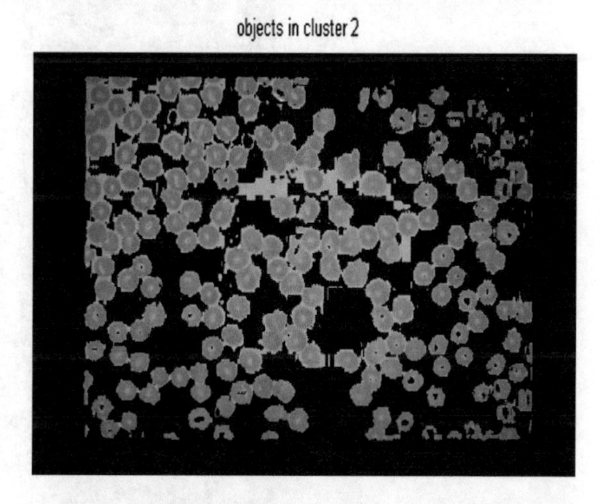

Fig. 12 Cluster 2 containing the RBC pixels

Fig. 13 Cluster 3 containing platelets and some overlapping RBCs

Fig. 14 The correctly identified cluster

Fig. 15 Final output image

overlapping RBCs with platelets. Further binarization and morphological operations have been applied to Fig. 14 to obtain the final image having only platelets. Figure 15 depicts the final output image. It is a binary image where the small white dots on the black background are the platelets. All platelets, marked in input image Fig. 7a, are identified properly. Figure 7a is a bad intensity image. As stated earlier, proper results were not obtained when normal segmentation was applied to it. This is clear from Fig. 7b where only two platelets were found after segmentation. However, using a hybrid technique comprising of fuzzy c-means and ANFIS all the platelets were detected correctly, as shown in Fig. 15.

We have applied this hybrid technique and normal color-based technique on 100 images. Then, we compute platelet count of each image using both two techniques. We find that hybrid technique provides a very efficient result on a manual count. Our nearby pathology lab provides the manual count. Table 3 shows the comparison of

Table 3 Different imaging modalities and their sensitivity and specificity approved by the FDA for early breast cancer detection

Sl no.	Manual count	Hybrid approach	% error for hybrid approach	Color-based segmentation	% error for color-based segmentation
1	8	7	12.5	8	0
2	11	11	0	10	9.09
3	8	7	12.5	9	12.5
4	11	11	0	7	36.36
5	9	12	33.3	9	0
6	11	11	0	11	0
7	8	4	50	8	0
8	11	8	27.3	7	36.36
9	6	6	0	4	33.3
10	7	6	14.28	6	14.3
11	4	4	0	5	25
12	10	10	0	9	10
13	10	9	10	6	40
14	7	8	14.28	10	42.9
15	9	9	0	5	44.4
16	7	6	14.28	5	28.6
17	8	8	0	3	62.5
18	9	8	11.11	6	25
19	4	4	0	4	0
20	6	6	0	1	83.3
21	7	7	0	4	42.9
22	6	6	0	2	66.67
23	9	9	0	2	77.78
24	10	11	10	3	70
25	8	8	0	7	12.5
26	8	8	0	5	37.5
27	15	15	0	10	50
28	14	14	0	10	28.6
29	13	15	15.4	10	23.1
30	9	10	11.11	6	33.3

platelet count for two segmentation processes normal color-based segmentation and Hybrid approach with a manual count. In Table 3, we include platelet counts for 30 images. Percentage error rate for each technique is also calculated. From the table, it is clearly shown that hybrid segmentation technique performs better on normal color-based segmentation.

We have plotted the platelet count is given by Hybrid segmentation technique, color-based segmentation and manual platelet count in a graph shown in Fig. 16. In this graph, y-axis corresponds to platelet count and the x-axis shows a serial number of images. The blue line represents manual platelet count; the pink line represents platelet count obtained from hybrid soft computing technique, and yellow line represents platelet count provided by normal color-based segmentation. In this graph, it is easily seen that the blue line and the pink line overlapped many of time. In a graph shown in Fig. 17, the percentage error rate for two segmentation techniques is plotted. The x-axis shows the serial number of images and the y-axis corresponds to percentage error rate. The blue line represents percentage error rate for hybrid segmentation technique, and pink line represents percentage error rate for normal color-based segmentation technique. The blue line goes very close to x-axis whereas pink line goes very far from the x-axis. The average accuracy of the hybrid segmentation technique is 97.7 % with respect to manual count of the platelet. In other hand, the average accuracy of color-based segmentation is 74 % when applying on the same images.

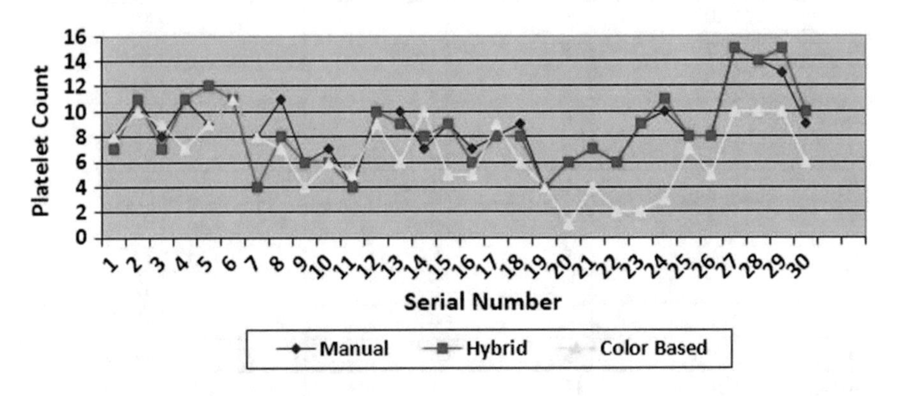

Fig. 16 Platelet count comparison graph

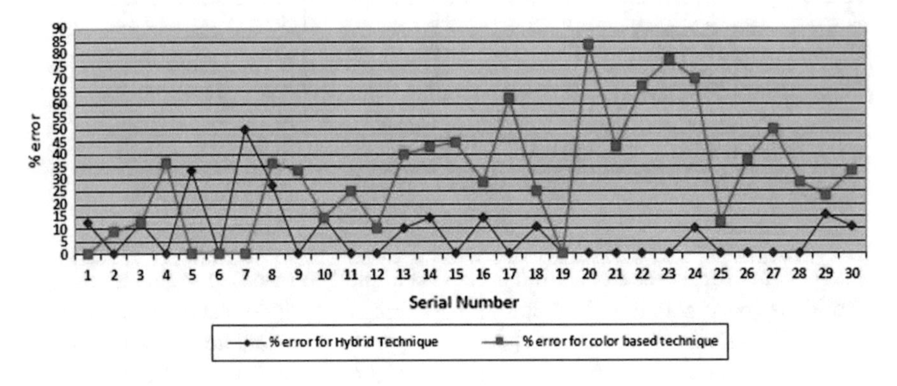

Fig. 17 Error percentage comparison graph

According to above discussion, it is clear that Platelet segmentation can be performed well than normal color-based segmentation and k-means clustering technique. Normal color-based segmentation cannot segment platelets from the bad intensity images and k-means cannot cluster platelets. Only hybrid soft computing technique segment platelets efficiently. This technique does not depend on the intensity of images. The table and graphs are evidence that hybrid technique efficient in platelet segmentation from microscopic images of a blood smear.

6 Conclusion

This chapter proposes a hybrid soft computing technique a combination of Fuzzy c-means and ANFIS. In this chapter, we have discussed the problems related to normal segmentation techniques. To overcome these shortcomings, the need for soft computing techniques arises. But each of these techniques has some limitations which have been discussed in brief. So, to obtain better segmentation results we have used a combination of more than one soft computing technique. This hybrid soft computing technique is shown to be efficient in segmenting platelets from microscopic blood smear images.

In this chapter, we have also discussed that platelet count is a very important technique for early diagnosis of dengue. Also, it is important to monitor the health status of dengue infected patients for further treatment. Our main goal was to develop an automated platelet counting system for dengue diagnosis and treatment. The main challenge in developing this system is to segment platelets appropriately. Our proposed hybrid soft computing technique helped us to achieve our purpose and obtain almost accurate results.

Acknowledgments Authors of this chapter are paying their thanks to the Department of Bio-Technology, Govt. of India for sanctioning and funding the project (Letter No - BT/PR8456/MED/29/739/2013).

References

1. Global strategy for dengue prevention and control 2012–2020; World Health Organization (2012). ISBN 978 92 4 150403 4
2. Handbook for clinical management of dengue; World Health Organization (2012). ISBN 978 92 4 150471 3
3. Global report on dengue, WHO. www.who.int/mediacentre/factsheets/fs117/en/ (Updated February 2015). Accessed 4th Feb 2016
4. Wang, W., Song, H., Zhao, Q.: A modified watersheds image segmentation algorithm for blood cell. In: International Conference on Communications, Circuits, and Systems Proceedings, vol. 1, Guilin (2006)

5. Sharif, J.M., Miswan, M.F., Ngadi, M.A., Salam, S.H., bin Abdul Jamil, M.M.: Red blood cell segmentation using masking and watershed algorithm: a preliminary study. In: International Conference on Biomedical Engineering (ICoBE), Penang, pp. 258–262, 27–28 Feb 2012
6. Karunakar, Y., Dr Kuwadekar, A.: An unparagoned application for red blood cell counting using marker controlled watershed algorithm for android mobile. In: Fifth International Conference on Next Generation Mobile Applications and Services, Cardiff, pp. 100–104 (2011)
7. Guan, P.P., Yan, H.: Blood cell image segmentation based on the hough transform and fuzzy curve tracing. In: Proceedings of the 2011 International Conference on Machine Learning and Cybernetics, Guilin, 10–13 July 2011
8. Venkatalakshmi, B., Thilagavathi, K.: Automatic red blood cell counting using hough transform. In: Proceedings of the 2013 IEEE Conference on Information and Communication Technologies (ICT 2013)
9. Kareem, S., Morling, R.C.S., Kale, I.: A novel method to count the red blood cells in thin blood films. In: IEEE International Symposium on Circuits and Systems (ISCAS) (2011)
10. Mohamed, M.M.A., Far, B.: A fast technique for white blood cells nuclei automatic segmentation based on Gram–Schmidt orthogonalization. In: IEEE 24th International Conference on Tools with Artificial Intelligence, Athens, pp. 947–952 (2012)
11. Liao, Q., Deng, Y.: An accurate segmentation method for white blood cell images. In: Proceedings of the IEEE International Symposium on Biomedical Imaging (2002)
12. Mohapatra, S., Patra, D., Kumar, K.: Unsupervised leukocyte image segmentation using rough fuzzy clustering. In: International Scholarly Research Network ISRN Artificial Intelligence, vol. 2012, Article ID 923946
13. Rovithakis, G.A., Maniadakis, M., Zervakis, M.: A hybrid neural network/genetic algorithm approach to optimizing feature extraction for signal classification. In: IEEE Transactions on Systems, Man, and Cybernetics Part B: Cybernetics, pp. 695–703, vol. 34, no. 1, Feb 2004
14. Laosai, J., Chamnongthai, K.: Acute leukemia classification by using SVM and K-means clustering. In: Proceedings of the International Electrical Engineering Congress (2014)
15. Simmons, C.P., Farrar, J.J., van VinhChau, N., Wills, B.: Dengue. N. Engl. J. Med. **366**, 1423–1432 (2012)
16. WHO.: Dengue and Severe Dengue (Fact Sheet No. 117, Revised January 2012). World Health Organization, Geneva (2012)
17. Shepard, D.S., et al.: Cost-effectiveness of apaediatric dengue vaccine. Vaccine **22**, 1275–1280 (2004)
18. Dengue: prevention and control, World Health Organisation, 21 Nov 2014
19. Update on the Dengue situation in the Western Pacific Region; WHO, 17th Nov 2015
20. Dengue Cases and Deaths in the Country since 2009; National Vector Borne Disease Control Programme, Directorate General of Health Services, Ministry of Health & Family Welfare, Govt. of India. http://nvbdcp.gov.in/den-cd.html. (Accessed 3rd Dec 2015)
21. Dengue Viruses. http://www.nature.com/scitable/topicpage/dengue-viruses-22400925
22. Dengue Transmission. http://www.nature.com/scitable/topicpage/dengue-transmission-22399758
23. Eckert, R.L., Rorke, E.A.: Molecular biology of keratinocyte differentiation. Env. Health Perspect. **80**, 109–116 (1989)
24. Chomiczewska, D., Trznadel-Budko, E., Kaczorowska, A., Rotsztejn, H.: The role of Langerhans cells in the skin immune system. Pol Merkur Lekarski **26**(153), 173–177 (2009)
25. Dengue Viruses. http://www.nature.com/scitable/topicpage/dengue-viruses-22400925
26. Host Response to the dengue virus. http://www.nature.com/scitable/topicpage/host-response-to-the-dengue-virus-22402106
27. Dengue guidelines for diagnosis, treatment, prevention and control, A joint publication of the World Health Organization (WHO) and the Special Programme for Research and Training in Tropical Diseases (TDR) (2009)
28. http://www.who.int/denguecontrol/faq/en/index2.html
29. Rigau-Prez, J.G., et al.: Dengue and dengue haemorrhagic fever. Lancet **352**, 971977 (1998)

30. Kalayanarooj, S., et al.: Early clinical and laboratory indicators of acute dengue illness. J. Infect. Dis. **176**, 313–321 (1997)
31. Balmaseda, A., et al.: Assessment of the World Health Organization scheme for classification of dengueseverity in Nicaragua. Am. J. Trop. Med. Hyg. **73**, 1059–1062 (2005)
32. Lum, L.C.S., et al.: Quality of life of dengue patients. Am. J. Trop. Med. Hyg. **78**(6), 862–867 (2008)
33. Cao, X.T., et al.: Evaluation of the World Health Organization standard tourniquet test in the diagnosis ofdengue infection in Vietnam. Trop. Med. Int. Health **7**, 125–132 (2002)
34. Srikiatkhachorn, A., et al.: Natural history of plasma leakage in dengue hemorrhagic fever: a serial ultrasonic study. Pediatric Infect. Dis. J. **26**(4), 283–290 (2007)
35. Nimmannitya, S., et al.: Dengue and chikungunya virus infection in man in Thailand, 196264. Observations onhospitalized patients with haemorrhagic fever. Am. J. Trop. Med. Hyg. **18**(6), 954–971 (1969)
36. Clinical Practice Guidelines on Management of DengueInfection in Adults (Revised 2nd Edition). Ministry of Health Malaysia, Academy of Medicine Malaysia (2010)
37. Morens, D.M.: Dengue outbreak investigation group. Dengue in Puerto Rico: public health response to characterize and control an epidemic of multiple serotypes. Am. J. Trop. Med. Hyg. **1986**(35), 197–211 (1977)
38. Wilder-Smith, A., Earnes, A., Paton, N.I.: Use of simple laboratory features to distinguish the early stage of severe acute respiratory syndrome from dengue fever. Clin. Infect. Dis. **39**(12), 1818–1823 (2004)
39. Kularatne, S.A., et al.: Concurrent outbreaks of Chikungunya and Dengue fever in Kandy, Sri Lanka, 2006–2007: a comparative analysis of clinical and laboratory features. Postgrad. Med. J. **2009**(85), 342–346 (1005)
40. Yamamoto, K., et al.: Chikungunya fever from Malaysia. Intern. Med. **49**(5), 501–505 (2010)
41. Cabie, A., et al.: Dengue or acute retroviral syndrome? Presse Med. **29**(21), 1173–1174 (2000)
42. Martinez, E.: Diagnstico diferencial. In: Dengue, pp. 189–195. Rio de Janeiro, Fiocruz (2005)
43. Dietz, V.J., et al.: Diagnosis of measles by clinical case definition in dengue endemic areas: implications for measles surveillance and control. Bull. World Health Organ. **70**(6), 745–750 (1992)
44. Flannery, B., et al.: Referral pattern of leptospirosis cases during a large urban epidemic of dengue. Am. J. Med. Trop. Hyg. **65**(5), 657–663 (2001)
45. Mcfarlane, M.E.C., Plummer, J.M., Leake, P.A., Powell, L., Chand, V., Chung, S., Tulloch, K.: Dengue fever mimicking acute appendicitis: a case report. Int. J. Surg. Case Rep. **4**(11), 1032–1034 (2013)
46. Guideline for clinical management of Dengue Fever, Dengue Haemorrhagic Fever and Dengue Shock Syndrome, Directorate of National Vector Borne Diseases Control Programme, Government of India
47. Bain, B.J., Bates, I., Laffan, M.A., Lewis, S.M.: Dacie and Lewis Practical Haematology. Elsevier Churchill Livingstone, Edinburgh (2012)
48. Dung, N.M., Day, N.P., Tam, D.T.: Fluid replacement in dengue shock syndrome: a randomized, double-blind comparison of four intravenous fluid regimens. Clin. Infect. Dis. **29**, 787–794 (1999)
49. Ngo, N.T., Cao, X.T., Kneen, R.: Acute management of dengue shock syndrome: a randomized double-blind comparison of 4 intravenous fluid regimens in the first hour. Clin. Infect. Dis. **32**, 204–213 (2001)
50. Wills, B.A., et al.: Comparison of three fluid solutions for resuscitation in dengue shock syndrome. N. Engl. J. Med. **353**, 877–889 (2005)
51. Hung, N.T., et al.: Volume replacement in infants with dengue hemorrhagic fever/dengue shock syndrome. Am. J. Trop. Med. Hyg. **74**, 684–691 (2006)
52. Wills, B.A.: Management of dengue. In: Halstead, S.B. (ed.) Dengue, pp. 193–217. Imperial College Press, London (2008)
53. Dey, R., Roy, K., Bhattacharjee, D., Nasipuri, M., Ghosh, P.: An automated system for segmenting platelets from microscopic images of blood cells. In: 2015 International Symposium on Advanced Computing and Communication (ISACC), India (2015)

54. Zalizam, T., Mudaa, T., Salamb, R.A.: Blood cell image segmentation using hybrid K-means and median-cut algorithms. In: 2011 IEEE International Conference on Control System, Computing and Engineering, Penang, pp. 237–243
55. Jain, R., Kasturi, R., Schunck, B.G.: Machine Vision, Chap. 4 Image Filtering (1995)
56. Kaur, A., Kranthi, B.V.: Comparison between YCbCr color space and CIELab color space for skin color segmentation. Int. J. Appl. Inf. Syst. (IJAIS) **3**(4) (2012). (ISSN: 2249-0868 Foundation of Computer Science FCS, New York, USA, July 2012)
57. Wang, W., Zhang, Y., Li, Y., Zhang, X.: The global fuzzy c-means clustering algorithm. In: Proceedings of the 6th World Congress on Intelligent Control and Automation, Dalian, China, 21–23 June, 2006
58. Garg, V.K., Dr. Bansal, R.K.: Soft computing technique based on ANFIS for the early detection of sleep disorders. In: International Conference on Advances in Computer Engineering and Applications (ICACEA 2015), IMS Engineering College, Ghaziabad, India
59. AbdulRazzaq, M., Ariffin, A.K., El-Shafie, A., Abdullah, S., Sajuri, Z.: Prediction of fatigue crack growth rate using rule-based systems. In: 4th International Conference on Modeling, Simulation and Applied Optimization (ICMSAO) (2011)

Extraction of Knowledge Rules for the Retrieval of Mesoscale Oceanic Structures in Ocean Satellite Images

Eva Vidal-Fernández, Jesús M. Almendros-Jiménez, José A. Piedra and Manuel Cantón

Abstract The processing of ocean satellite images has as goal the detection of phenomena related with ocean dynamics. In this context, Mesoscale Oceanic Structures (MOS) play an essential role. In this chapter we will present the tool developed in our group in order to extract knowledge rules for the retrieval of MOS in ocean satellite images. We will describe the implementation of the tool: the *workflow* associated with the tool, the *user interface*, the *class structure*, and the *database* of the tool. Additionally, the experimental results obtained with the tool in terms of fuzzy knowledge rules as well as labeled structures with these rules are shown. These results have been obtained with the tool analyzing chlorophyll and temperature images of the Canary Islands and North West African coast captured by the SeaWiFS and MODIS-Aqua sensors.

Keywords Remote sensing · Satellite images · Mesoscale oceanic structures · Image processing · Tools · SeaWiFS · MODIS-Aqua

1 Introduction

With the increasing in the last decades of environmental problems like the global climate change [1–3] and the ocean primary production changes [4], the development of tools focused on intelligent search of objects and regions in satellite images stored in large databases of spatial and environmental agencies, should be a major goal. In this context, the global oceanic circulation plays an essential role in the global climate, because the ocean covers more than 70 % of the surface of our planet. Most of the oceanic circulation is mesoscale (i.e., scale from 50 to 500 km and 10–100 days), and the energy of Mesoscale Oceanic Structures (MOS) is at least one order of magnitude greater than the general circulation. Thus, MOS are crucial for ocean dynamic study and global change analysis [5, 6].

E. Vidal-Fernández · J.M. Almendros-Jiménez (✉) · J.A. Piedra · M. Cantón
Department of Informatics, University of Almería, 04120 Almería, Spain
e-mail: jalmen@ual.es

S. Bhattacharyya et al. (eds.), *Hybrid Soft Computing for Image Segmentation*, DOI 10.1007/978-3-319-47223-2_6

The main types of MOS are upwellings, upwelling filaments, cold/warm eddies, and wakes [7–11]. A coastal *upwelling* can be defined as the interaction between the bottom and surface water. upwellings are mainly due to oceanic currents which are, at the same time, influenced by wind dynamics. They occur when dense cold water at the bottom of the ocean rises to the surface near the coast, transporting nutrients, and enabling the development and proliferation of phytoplankton [12]. These upwellings occur regularly (during the whole year with varying intensity) along the North West African coast (see Fig. 1) and others like the Peruvian, Californian and South African coasts, among others, where the wind conditions are suitable [1, 3, 13–16]. The analysis and prediction of upwellings, for which satellite images are a powerful tool, are very important to commercial fishing. There are some mesoscale cross-shore structures along the upwelling front called *upwelling filaments* which are tongue-shaped cold upwellings. They are important nutrient carriers from coast to the open ocean, often found near capes [17] (e.g., off Cape Ghir or Cape Blanc) (see Fig. 1). *Eddies* are highly morphological and contextual variable structures [8, 11]. They may appear rounded near islands or in the open ocean. In *cool eddies*, cold nutrient-rich water rises to the surface [18], while *warm eddies* (see Fig. 1) drag water with organic material to the ocean floor and keep in the warm surface. eddy water differs from the surrounding water in salinity and temperature, and can also travel long distances for long periods of time without mixing with the surrounding

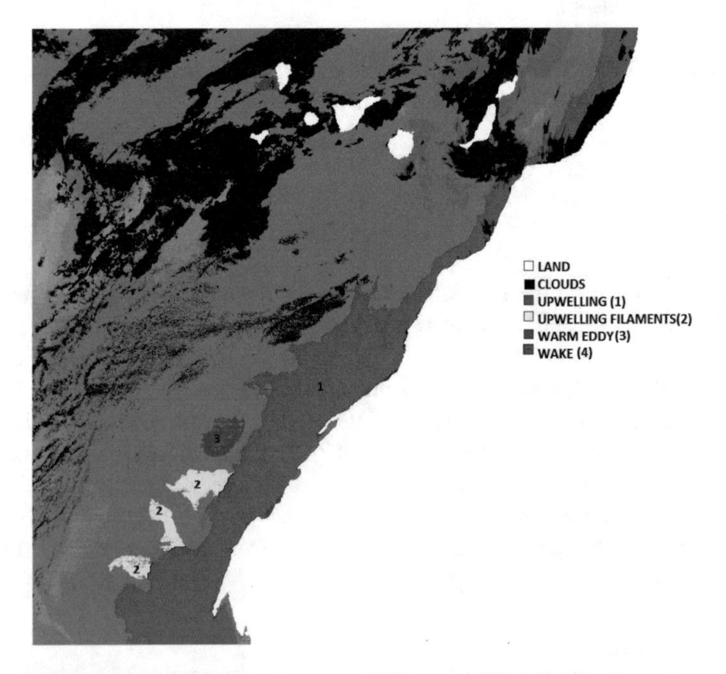

Fig. 1 MODIS-Aqua-Sea surface temperature (SST) scene (2003-03-24). Ocean structure map where the *upwelling* has been colored in *pink* for visualization (*1*), *upwelling filaments* (Cap Blanc) in *yellow* (2), *warm* in *blue* (3), and *wake* (west of La Palma) in *red* (*4*)

water [12, 19]. The cool (cyclonic) eddies are created by calm wind conditions, while warm (anticyclonic) eddies are formed under stronger wind conditions. *Wakes* are warm oceanic structures associated with islands [12, 20], and can be created by obstacle formed by islands. This interaction causes very thin tongues of hot, nutrient poor water (as compared to the surrounding water) to form close to the southwest islands (see Fig. 1). These structures are shown in Fig. 1.

The images used in our analysis correspond to a wide temporal range (from 1997 to 2013, and for each month), with clouds partially hiding structures, and as consequence MOS are very different from one time to another. Even when region recognition is automatic (by image segmentation), there are some factors that difficult MOS detection. Some factors are the presence of noise and clouds, as well as regions with holes, and regions which cannot be distinguished from the surrounding water (due to their size or their color). To overcome some of these handicaps, the initial color scale used in the image processing stage can be modified. To overcome the handicap of clouds and noise, which split a region into small regions, regions can be merged.

1.1 Contributions of the Chapter

In this chapter we will describe the implementation of a tool for the *extraction of fuzzy knowledge rules*. Such knowledge rules are required to retrieve regions from the tool database in order to be *labeled*. knowledge rules can thus considered as queries against the database. The *experimental results* obtained with the tool in terms of knowledge rules as well as labeled structures with these rules are shown. These results have been obtained with the tool analyzing chlorophyll and temperature images of the Canary Islands and North West African coast captured by SeaWiFS (Sea-viewing Field-of-view Sensor) and MODIS-Aqua (Moderate Resolution Imaging Spectroradiometer, on board the Aqua satellite) sensors. The tool will be described as follows:

1. **Workflow for Labeling MOS**: The workflow we present in this chapter describes the steps required to label a MOS. It provides a number of steps to be followed by the oceanographer in order to label a MOS in reasonable time. Following these steps a novice user of the tool would be able to label MOS. Nevertheless, the needed time depends on the image characteristics. For instance, whether the image includes more than one MOS, whether MOS are divided in pieces due to clouds, etc. The workflow shows a complete case distinction, enabling to get good results even for novice users.
2. **Extraction of knowledge rules**: A method for extraction of knowledge rules is presented. The core of the workflow for labeling MOS is the extraction of rules. These knowledge rules are described in terms of fuzzy concepts, and the method for extraction is a trial and error procedure which refines knowledge rules in each step from an initial test.

3. **Experimental Results**: The dataset used to test our tool consists on MOS described by 64 image descriptors. The experimental results provide knowledge rules which make possible to retrieve and label 365 MOS. The knowledge rules are built from fuzzy features (15 from the initial 64 features). In some cases, some regions have been merged to get a compact and well-contoured MOS. In particular, the tool enables to manually merge regions, using the so-called ROI (Region of Interest). Experimental results are described by number and type of MOS, whether or not ROI is used, as well as the knowledge rules for each case.

1.2 Related Work

In our approach primary objects (output of the segmentation phase) evolve toward objects of interest (*regions*), with the help of the knowledge of an expert (oceanographer) [21] and specific techniques of segmentation for each type of object [22]. It is crucial in the context of MOS, which are characterized by high morphological and contextual variability [23–25]. Some examples of techniques and tools for region retrieval from images can be found in [26–29].

Our research group has a wide experience in the development of methods and techniques for the analysis of ocean satellite images, mainly focused on sensors capturing the area of study (i.e., Canary Islands and North West African coast) (see [28–32], for the most recent contributions). In particular, the same dataset (and thus of the same sensor) was used in two previous works [28, 29]. The tool we present here has permitted the automatization of the process of region labeling and merging, and the automatization of knowledge rule extraction. This is crucial in a high morphological and contextual variability context, in which a trial and error method has to be used to detect MOS. In [29], we had the same problems of MOS identification: presence of noise and clouds, as well as regions with holes, and regions which cannot be distinguished from the surrounding water. We already there modified the initial color scale to get well-contoured and compact regions from segmentation. However, we found there that it can be still not enough, and manual merging is required. The tool we present here is able to automatically merge regions from manual selection of the ROI by the expert. Thus manual (although assisted by the tool) merging can be done now. One the other hand, the definition of knowledge rules is hard when it is manually done. The tool enables to try knowledge rules playing from the user interface with several choices of fuzzy descriptors and range of values, and more interesting to compare with previously labeled regions and previously defined knowledge rules, which reduces significantly the time required to define new knowledge rules. Thus, the development of the tool has permitted to reduce drastically the time for labeling and extraction of knowledge rules, and also reuse, detection of patterns, etc. The tool (i.e., workflow, procedure of knowledge rule extraction and image comparison) is the main contribution of this work, and in order to validate the tool, we will show some examples of images in which the manual merging is required, and report experimental results of manual merging. The number and type of knowledge rules, involving fuzzy descriptors and values, is also of great interest for oceanographers interested in the analyzed area.

1.3 Structure of the Chapter

The structure of the chapter is as follows. Section 2 will describe the dataset. Section 3 will show the workflow of the tool. Section 4 will present the knowledge rules and labeling results obtained from the analyzed images. Section 5 will describe the implementation of the tool. Finally, Sect. 6 will conclude and present future work.

2 Dataset

We have analyzed 212 satellite images of the Canary Islands and North West African coast captured by SeaWiFS sensor on board the Orbview-2 satellite (1.1 km resolution) and MODIS-Aqua sensor on board the Aqua satellite (1.1 km resolution), obtained from the Ocean Color Web of NASA [33]. Many studies have reveal that SeaWiFS and MODIS-Aqua-OC sensor chlorophyll images and MODIS-Aqua-SST temperature sensor images are a suitable choice for the study of MOS [34–37], primary production [38], and global change [39]. In [40] the authors analyze data from chlorophyll images of a 10-year period, comparing them with in situ collected data. They also compare results with different sensors, finding highly correlated coefficients. A total of 212 images have been analyzed. The number of processed images was larger, but most of them were discarded by the tool due to several reasons: too many clouds, out of the study area, faulty images, etc. Of them, 92 images are chlorophyll images acquired from the SeaWiFS sensor, from 1997 to 2004, and covering all the seasons (9-1997; 14-1998; 25-1999; 14-2000; 7-2001; 5-2002; 7-2003, and 11-2004). These images are derived from *MLAC (Merged Local Area Coverage product)* products. The MLAC products consolidate all of the LAC from different receiving stations available for the same orbit in geographic regions with multiple HRPT (*High Resolution Picture Transmissions*) stations. Overlapping scenes are evaluated to acquire a single best image without duplication. The MLAC product was chosen because it generates high-quality L2 level images, including radiometric calibration and geometric correction. The other 120 images are from the MODIS-Aqua sensor, for 2003–2013 and divided into 61 chlorophyll and 59 SST images. These images are derived from L2 level LAC (*Local Area Coverage*) products.

In the case of MODIS sensor, we worked with two products: those about both chlorophyll (OC) and temperature (SST), and thus providing two images, and those about only chlorophyll or only temperature, from which a single image is obtained. This is the reason why the number of chlorophyll images is larger than the number of temperature images for products L2 Level LAC of MODIS. We analyzed in terms of years: 2003-2 OC and 2 SST; 2004-5 OC and 4 SST; 2008-19 OC and 19 SST;

2009-12 OC and 13 SST; 2010-10 OC and 16 SST; 2011-4 OC and 1 SST; 2012-6 OC and 3 SST; and 2013-3 OC and 1 SST, covering all the seasons.

Therefore, a total of 212 satellite images of the Canary Islands and the Northwest African coast have been analyzed, where a large number of MOS can be found. A total of 365 Mesoscale Oceanic Structures were detected of which 284 MOS correspond to upwellings or parts of upwellings (when clouds keep parts upwellings from being merged), 44 upwelling filaments, 10 cool eddies, four warm eddies and 23 wakes.

3 Workflow of the Tool

Figure 2 shows the steps (from 1 to 20) to be followed by an user of tool. The first step consists of the selection of (1) color table (chlorophyll or temperature) and (2) images to process. The color scale is represented by a table in which each entry is an assignment of a color range of Chl-a (i.e., chlorophyll concentration of the sea surface) or SST (i.e., Sea Surface temperature) to a given color. In such a way that each value of the HDF (Hierarchical Data Format) matrix is converted into a colored pixel. HDF files include several groups of data. One of the groups contains geophysical data with matrices of Chl-a, SST and L2-flags. In order to convert HDF values into colored pixels, for each pixel of the Chl-a (or SST) matrix, the flag 32 is used to know whether the value is correct or wrong. In the case the value is correct, the flag 2 is used to know whether it belongs to land. In case it is correct and not a land value, the value of the Chl-a (or SST) matrix reports the value of the color table.

The choice of a suitable clustering of colors (i.e., color table) for a given image, enables better recognition of regions: well-contoured and compact regions and regions obtained from several small regions of almost identical color. In this step, starting from the initial color table (see Table 1 for chlorophyll), colors are grouped (although not necessarily). For instance, in Table 2 the color clustering "Red to Yellow" is shown. Such a color clustering means that colors from yellow and beyond are grouped and identified to "Red". Similar tables for temperature images can be also considered.

Figure 3 shows the processing result of an image of SeaWiFS (i.e., chlorophyll) using the original table, and the color clustering "Red to Yellow". The "Red to Yellow" color clustering permits the recognition of well-contoured and compact regions without holes, facilitating MOS detection and labeling. In this image, a big upwelling from Canary Islands to Cape Blanc and a cold eddy in the south of Gran Canaria island are detected and showed in red.

After steps (1) and (2) (selection of color tables and images) processing is carried out (3). This automatic task involves several subprocesses: decompression and extraction of the HDF file content: the matrix of chlorophyll data (Chl-a) or temperature data (SST), transformation to a color image according to the color table, and segmentation (combining smoothing filter with thresholding techniques and edge detection). Additionally (spectral, morphologic and contextual) descriptors of regions are computed.

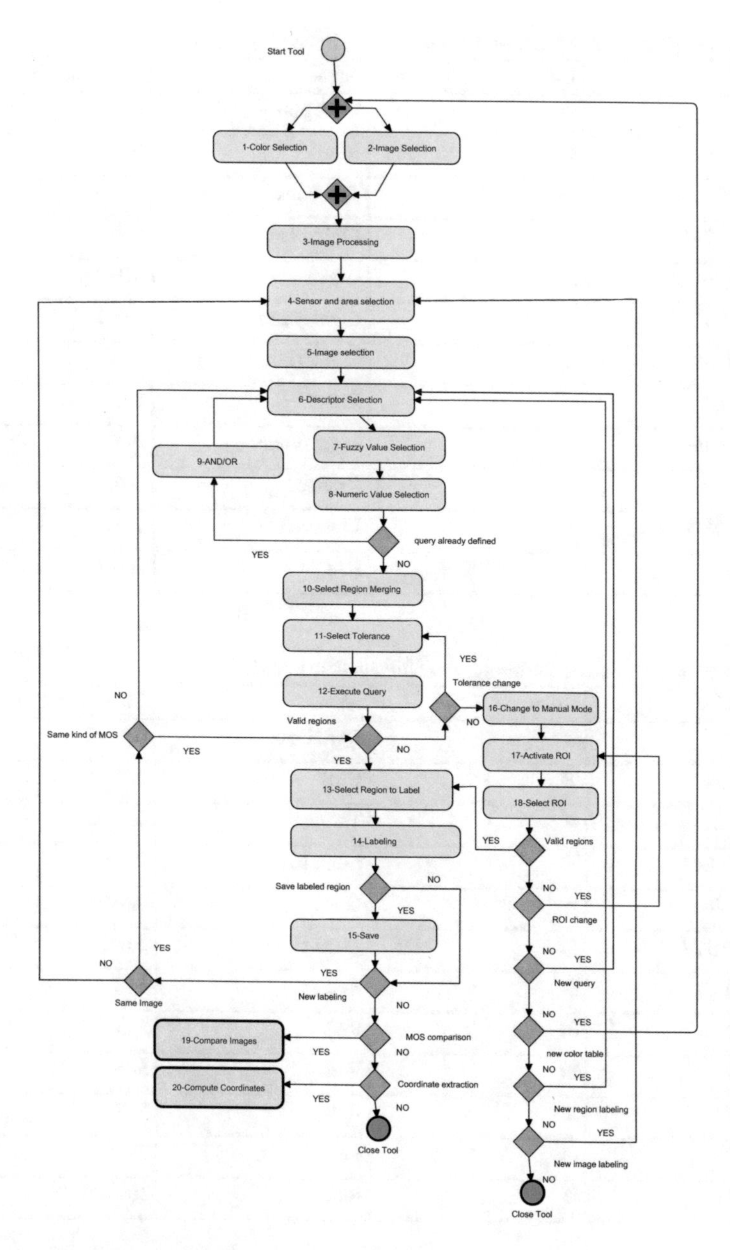

Fig. 2 Workflow of the tool

Table 1 Original SeaWiFS and MODIS sensor OC image color table

RGB Color	Color	Name	Chl-a (mg/m^3)
1;1;223	■	Dark blue	0.03–0.06
46;46;254	■	Blue	0.06–0.1
88;250;244	■	Turquoise	0.1–0.2
129;247;243	■	Light blue	0.2–0.3
46;254;46	■	Green	0.3–0.6
129;247;129	■	Light green	0.6–1
244;250;88	■	Light Yellow	1–2
255;255;0	■	Yellow	2–3
254;154;46	■	Light orange	3–5
255;128;0	■	Orange	5–7
254;46;46	■	Light red	7–10
255;0;0	■	Red	10–100

Table 2 Example of color clustering: *from light yellow to red*

RGB color	Color	Name	Chl-a (mg/m^3)
1;1;223	■	Dark Blue	0.03–0.06
46;46;254	■	Blue	0.06–0.1
88;250;244	■	Turquoise	0.1–0.2
129;247;243	■	Light blue	0.2–0.3
46;254;46	■	Green	0.3–0.6
129;247;129	■	Light Green	0.6–1
255;0;0	■	Red	1–2
255;0;0	■	Red	2–3
255;0;0	■	Red	3–5
255;0;0	■	Red	5–7
255;0;0	■	Red	7–10
255;0;0	■	Red	10–100

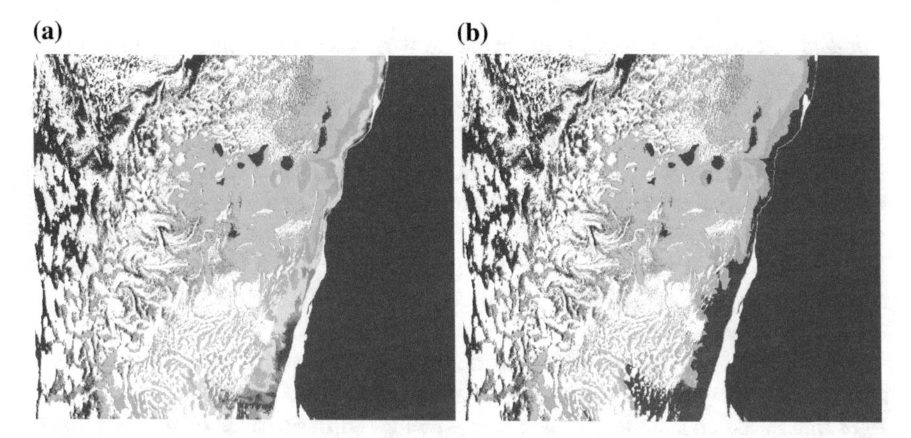

Fig. 3 SeaWiFS scene (2004-07-09). **a** Result of processing with original color table. **b** Result of processing with *Yellow* to *Red* color table

MOS labeling is the next step. It also involves two steps: sensor selection (i.e., SeaWiFS, MODIS-Aqua-OC or MODIS-Aqua-SST) and earth area selection (i.e., Canary Islands) (4) and image selection (5). In previous tasks (i.e. (1) (2) and (3)) more than one image can be processed, and now selection of earth area and sensor filters from the batch of processed images.

3.1 Extraction of Knowledge Rules

After the selection of the image to be labeled, the oceanographer has to define the knowledge rule for the retrieval of the MOS. The extraction of knowledge rules involves the selection of fuzzy spectral, morphologic and contextual features (6) (i.e., size, island distance, continent distance, temperature and chlorophyll concentration), the fuzzy value (7) (i.e., near, medium and far for distances, small, medium and big for size, very warm, warm, temperate, cold and very cold for temperature, and very low, low, medium, high and very high for concentration) and the range of the fuzzy numeric value (8) from zero to one (for instance, ≥ 0.9 and < 0.5). The tool works with fuzzy membership functions, for spectral, morphologic and contextual features, given by triangular functions, in such a way that a membership degree is assigned to each fuzzy value. The goal of the oceanographer is to select, for instance, those regions whose membership degree to a given fuzzy value of size, for instance, big, is greater or equal than a given value, for instance, 0.9; in other words, the oceanographer wants to retrieve a big region. Analogously, the oceanographer can select regions whose membership to the fuzzy value warm is greater or equal than 0.7; in other words, the oceanographer wants a warm region. The oceanographer can add several

Boolean conditions (9) of this type, in such a way that a query can be considered as a knowledge rule, with ANDs and ORs.

The method for extraction is a trial and error procedure which refines knowledge rules in each step from an initial test. The initial test and refinements take into account the particular MOS characteristics (upwelling, wake, eddy) as well as the image characteristics (temperature, chlorophyll), as well as the time of capture (cold or warm water, presence of clouds, etc.). The best strategy is to select limit values of fuzzy ranges (for instance, high concentration of chlorophyll in case of upwelling) and to refine the rule adding new fuzzy descriptors (size or distance), until a complete MOS is retrieved.

When the knowledge rule has been defined, the oceanographer can select automatic region merging (10) according to a given tolerance level (11). The merging algorithm based on a technique of region growing, unifies regions of similar properties, whenever they are separated by a given distance, called tolerance, measured by pixels. The merging algorithm has been designed in order to merge regions according to fuzzy knowledge rules. The algorithm takes as input a set of regions and returns a merged region. Two conditions are simposed for region merging: regions satisfy the same fuzzy knowledge rule and regions are closed. The algorithm uses a ROI (region of interest), which is a rectangle enclosing the region. In order to merge regions, the regions should be closed and thus the ROIs of the regions should intersect. The algorithm builds a new ROI each time two regions are merged, and the algorithm iterates over the set of regions until a unique ROI has been built. In case a ROI of a region does not intersect with the computed ROI the region is discarded. The algorithm has a parameter called tolerance allowing a weaker intersection of ROIs. By default, tolerance is 0, but it can be customized from 0 to 20.

Once the knowledge rule is executed (12), a list of regions are retrieved (some of them are merged regions whenever the automatic merging mode has been selected). The list includes simple and merged regions. Now, the oceanographer has to answer the following question: Are the regions valid? In the positive case, the oceanographer selects the region (simple or merged) to be labeled (13), and the oceanographer proceeds to label (15). In Fig. 4a an example of labeled region (a pink upwelling) is shown. The tool permits to save the labeled image (15), to label another MOS of the same image, to label a new image, and to compare labeled regions of other images (19) or the same image. Additionally, the coordinate matrix of the processed image can be computed and visualized (20).

In the negative case (i.e., regions are not valid), the oceanographer can go back and change tolerance. When tolerance change does not give a better result, the oceanographer can select manual mode. Manual mode (16) permits to activate and use a ROI (i.e., region of interest) (17), which is a rectangle, making possible to select more than one region of the image. With manual mode, regions can be forced to be merged and unified. The ROI selection (i.e., the regions) is manual, but merging is automatic. In Fig. 4b a cold eddy is shown in pink color. This image is the same as the image of Fig. 4a, but the eddy of Fig. 4b in pink color is bigger than the eddy of Fig. 4a in red color.

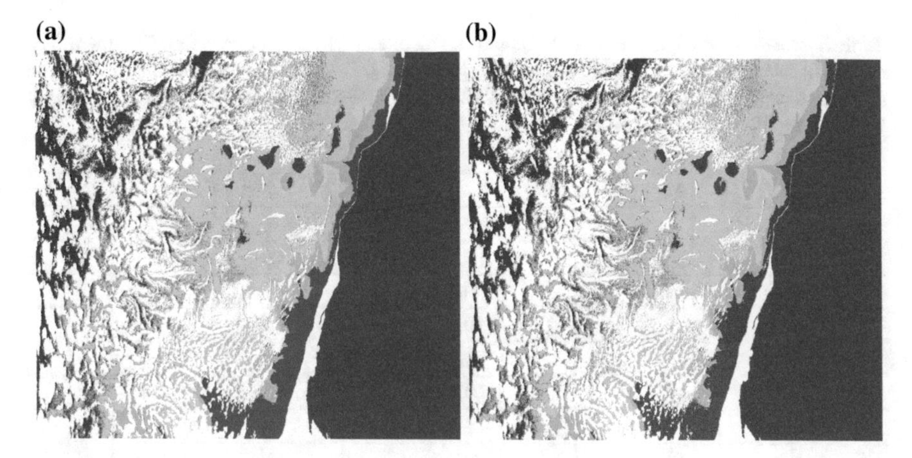

Fig. 4 SeaWiFS scene (2004-07-09). **a** Upwelling labeled in *pink color*. **b** Cold eddy labeled in *pink color*. Manual mode

When manual mode does not improve the result, the oceanographer has to go back to knowledge rule definition (redefining or modifying the knowledge rule), or to image processing (selecting a new color table).

4 Experimental Results

In this section, we report the results obtained with our tool in terms of knowledge rules and labeled structures for SeaWiFS and MODIS-Aqua images. On one hand, we show knowledge rules for the identification of MOS in these images. One the other hand, type and number of structures for each class of image are given.

4.1 Knowledge Rules

31 optimal and valid knowledge rules (12 chlorophyll rules and 19 temperature rules) have been obtained from 212 processed images and 365 labeled MOS. Table 3 shows 10 basic knowledge rules for chlorophyll (SeaWiFS and MODIS-Aqua-OC). They can be used alone, or in combination with other, in total 12, shown in Table 4. The chlorophyll images have been processed using the color clustering "Red to Yellow," the best for MOS detection and labeling for chlorophyll images.

These knowledge rules have been applied to 153 chlorophyll images, labeling all the MOS of the images. In some cases, manual mode is used. Next, we specify which rules are used in each case, and the needed for ROI in each case. The most

Table 3 Basic knowledge rules of chlorophyll (AND rules). **Ch**: chlorophyll; **CD**: Continent Distance; **ID**: Island Distance; **VL**: Very Low (Fuzzy_value_1); **L**: Low (Fuzzy_value_2); **M**: Medium (Fuzzy_value_3); **H**: High (Fuzzy_value_4); **VH**: Very High (Fuzzy_value_5); **N**: near; **M**: medium; **F**: far; **S**: small; **M**: medium; **B**: big

Rule	**Ch**					**CD**			**ID**			**Size**		
	VL	**L**	**M**	**H**	**VH**	**N**	**M**	**F**	**N**	**M**	**F**	**S**	**M**	**B**
C1					≥0.25	≥0.5								
C2			≥0.5			≥0.5								
C3			≥0.25			≥0.5								
C4					≥0.25	≤0.5								
C5			≥0.5			≤0.5								
C6			≥0.25			≤0.5								
C7					≥0.25									
C8			≥0.25											
C9	≥0.1													
C10	≥0.1								≥0.5			≥0.4		

Table 4 Knowledge rules

Knowledge rules		Chlorophyll Rules (**CR**)		Temperature Rules (**TR**)	
CR		**TR**		**TR**	
RC1	C1	RT1	T1	RT10	T8
RC2	C1 OR C2	RT2	T2	RT11	T9
RC3	C1 OR C3	RT3	T3	RT12	T10
RC4	C4	RT4	T4	RT13	T11
RC5	C4 OR C5	RT5	T5	RT14	T12
RC6	C4 OR C6	RT6	T1 OR T5	RT15	T13
RC7	C5	RT7	T1 OR T4	RT16	T14
RC8	C6	RT8	T2 OR T6	RT17	T15
RC9	C7	RT9	T2 OR T7	RT18	T16
RC10	C7 OR C8	RT19	T17	RT19	T17
RC11	C9				
RC12	C10				

used descriptors are, obviously, those for chlorophyll concentration, near continent distance as well as near island distance, according to MOS definitions.

For temperature images, the original color table is used because a color clustering suitable for all the cases cannot be found. It is due since there are scenes from several times of year, and temperature considerably varies from one time to another. We took this decision to have a criterion for comparison. Testing images individually, a color clustering suitable for surrounding water temperature, makes possible more compact and well-contoured regions (without merging), enabling recognition and

Table 5 Basic knowledge rules for temperature (AND rules). **T**: temperature; **CD**: Continent Distance; **ID**: Island Distance; **VC**: Very Cold (Fuzzy_value_1); **C**: Cold (Fuzzy_value_2); **M**: Medium (Fuzzy_value_3); **W**: Warm (Fuzzy_value_4); **VH**: Very Hot (Fuzzy_value_5); **N**: near; **M**: medium; **F**: far; **S**: small; **M**: medium; **B**: big

Rule	**T**					**CD**			**ID**			**Size**		
	VC	**C**	**M**	**W**	**VH**	**N**	**M**	**F**	**N**	**M**	**F**	**S**	**M**	**B**
T1	≥0.1					≥0.5								
T2		≥0.1				≥0.5								
T3		≥0.3				≥0.5								
T4		≥0.6				≥0.5								
T5		≥0.8				≥0.5								
T6			≥0.8			≥0.5								
T7			≥0.6			≥0.5								
T8			≥0.3			≥0.5								
T9		≥0.1				≤0.5								
T10			≥0.6											
T11			≥0.8			≤0.5								
T12		≥0.1												
T13			≥0.5											
T14				≥0.1										
T15			≥0.8											
T16			≥0.1											
T17		≥0.1							≥0.5			≥0.4		

labeling. Table 5 shows 17 basic knowledge rules for temperature images (MODIS-Aqua-SST), and Table 4 shows the knowledge rules used for MOS labeling.

Again, temperature, near continent distance and near island distance are the most used descriptors. temperature images have a higher chromatic variability, and thus a large number of knowledge rules are required for MOS labeling. Automatic detection of MOS is prioritized, but some cases require manual processing.

4.2 *Labeling*

A total of 31 knowledge rules are able to label 365 MOS. Labeled MOS can be simple or merged. In Table 6 the number of simple and merged labeled regions are shown, indicating in each case the knowledge rules, as well as the origin (SeaWiFS, MODIS-Aqua-OC and MODIS-Aqua-SST, respectively). A total of 111 MOS for MODIS-Aqua-SST have been labeled (86 merged and 25 simple regions), and 158 MOS for SeaWiFS (131 merged and 27 simple regions). Finally, 96 MOS for MODIS-Aqua-OC have been labeled (88 merged and 8 simple regions).

Table 6 Result. **MR**: Merged Region; **SR**: Simple Region

Result								
SeaWiFS	**MR**	**SR**	**MODIS-OC**	**MR**	**SR**	**MODIS-SST**	**MR**	**SR**
RC1	84	5	RC1	26	3	RT1	2	0
RC2	28	0	RC2	19	0	RT2	19	0
RC3	5	0	RC3	29	2	RT3	17	0
RC4	5	7	RC4	4	0	RT4	5	0
RC5	2	1	RC5	1	0	RT5	10	1
RC6	0	1	RC6	2	0	RT6	5	0
RC7	1	0	RC7	0	0	RT7	2	0
RC8	5	1	RC8	6	1	RT8	18	0
RC9	0	4	RC9	0	1	RT9	2	0
RC10	1	3	RC10	0	0	RT10	2	0
RC11	0	2	RC11	0	1	RT11	0	5
RC12	0	3	RC12	1	0	RT12	0	1
						RT13	1	0
						RT14	0	1
						RT15	0	1
						RT16	2	4
						RT17	1	0
						RT18	0	10
						RT19	0	2

Table 7 Labeling result of SeaWiFS. **Yes**: ROI; **No**: No ROI

Rule	Upwelling		Upwelling Filament		Coldeddy		Warmeddy		Wake	
	No	**Yes**	**No**	**Yes**	**No**	**Yes**	**No**	**Yes**	**No**	**Yes**
RC1	88	1								
RC2	28	0								
RC3	5	0								
RC4			12	0						
RC5			2	1						
RC6			1	0						
RC7			1	0						
RC8			0	6						
RC9					0	4				
RC10					0	4				
RC11							0	2		
RC12									0	3

Table 8 Labeling result of MODIS-OC. **Yes**: ROI; **No**: No ROI

Rule	Upwelling		Upwelling Filament		Cold eddy		Warm eddy		Wake	
	No	**Yes**	**No**	**Yes**	**No**	**Yes**	**No**	**Yes**	**No**	**Yes**
RC1	29	0								
RC2	19	0								
RC3	30	1								
RC4			4	0						
RC5			1	0						
RC6			1	1						
RC7			0	0						
RC8			2	5						
RC9					0	1				
RC10					0	0				
RC11							0	1		
RC12									0	1

Some knowledge rules work fine for some specific structures, where automatic merging is enough (without ROI). In some other cases, manual merging is required. Tables 7, 8 and 9 specify the success of the rules in each type of MOS as well as whether the manual mode and ROI is required. From 158 labeled MOS in SeaWiFS images, manual mode (ROI) is used 27 times. Rules RC1 to RC3 facilitate upwelling detection (122), where only one time manual mode is required. Rules RC4 to RC8 are used for upwelling filaments (23). Manual mode is required 7 times in this case. Cold eddies (8 by rules RC9 and RC10), warm eddies (2 by rule RC11) and wakes (3 by rule RC12) require manual mode. From 96 labeled MOS in MODIS-Aqua-OC images, manual mode (ROI) is used 10 times. Rules RC1 to RC3 facilitate upwelling detection (79), where only one time manual mode is required. Rules RC4 to RC8 are used for upwelling filaments (14). Manual mode is required 6 times in this case. Cold eddies (1 by rules RC9 and RC10), warm eddies (1 by rule RC11) and wakes (1 by rule RC12) require manual mode. From 111 labeled MOS in MODIS-Aqua-SST images, manual mode (ROI) is used 28 times. Rules RT1 to RT10 facilitate upwelling detection (83), where only one time manual mode is required. Rules RT11 to RT13 are used for upwelling filaments (7). Manual mode is required 6 times in this case. Cold eddies (1 by rules RT9 and RT14), warm eddies (1 by rule RT15) and wakes (11 by rules RT16 to RT19) require manual mode.

In summary, in most of the cases upwelling are easily detected in automatic mode (using a suitable tolerance level). Upwelling filaments, eddies and wakes are smaller in size, and more imprecise structures with respect to spectral and contextual properties, and thus in most of the cases they require manual mode to delimitate the area in which the region is found.

Table 9 Labeling result of MODIS-. **Yes**: ROI; **No**: No ROI

Rule	Upwelling		Upwelling Filament		Cold eddy		Warmeddy		Wake	
	No	**Yes**	**No**	**Yes**	**No**	**Yes**	**No**	**Yes**	**No**	**Yes**
RT1	2	0								
RT2	19	0								
RT3	17	0								
RT4	5	0								
RT5	11	0								
RT6	5	0								
RT7	2	0								
RT8	18	0								
RT9	1	1								
RT10	2	0								
RT11			0	5						
RT12			0	1						
RT13			1	0						
RT14					0	1				
RT15							0	1		
RT16									0	6
RT17									0	1
RT18									0	10
RT19									0	2

5 Implementation

In this section, we describe the implementation of the tool. First, the main elements of the user interface are described, and next the class and data modeling of the tool are shown. The tool has been implemented in Java, making use of the OpenCV C library for image processing. The Java code includes HDF processing with the HDF5 Java library, as well as the user interface implemented in Java Swing, and the business logic. Data access to SQL with JDBC from Java permits image and region data storing.

5.1 User Interface

The user interface has been designed to facilitate image processing and visualization as well as region merging and labeling. The selection of a color table is carried out from the user interface, using a text file containing the table items (see Fig. 5). To swap from a color clustering to another, color table files are stored in disc, and can

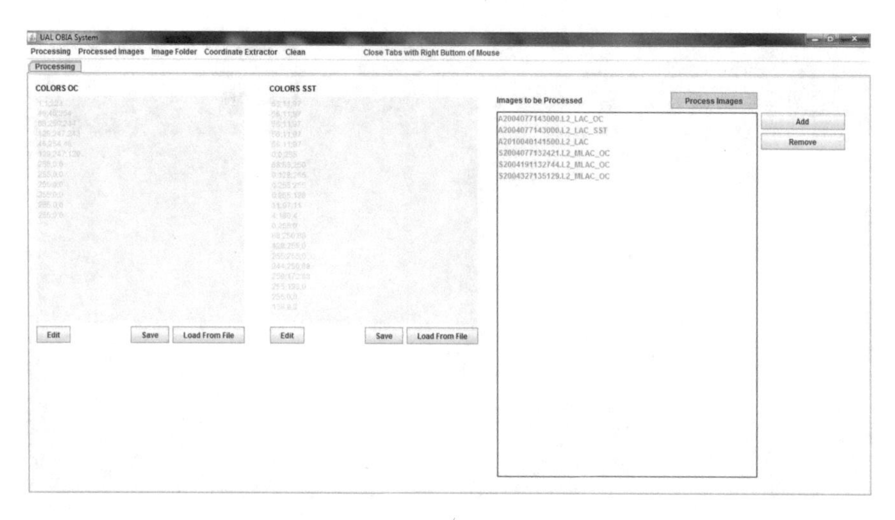

Fig. 5 Processing window

be selected any time. From this window, the oceanographer can select images to be processed, and can process images with the selected color table.

When segmentation is optimal (i.e., compact and well-contoured regions and complete MOS), the oceanographer can select the image to be labeled (see Fig. 6). From this window, images from a given sensor and earth area are shown. Selecting an image, the oceanographer can proceed to merge and label regions of the image.

Figure 7 shows the window from which knowledge rules are defined and executed. With the aim the user interface is equipped with a menu for selecting the fuzzy

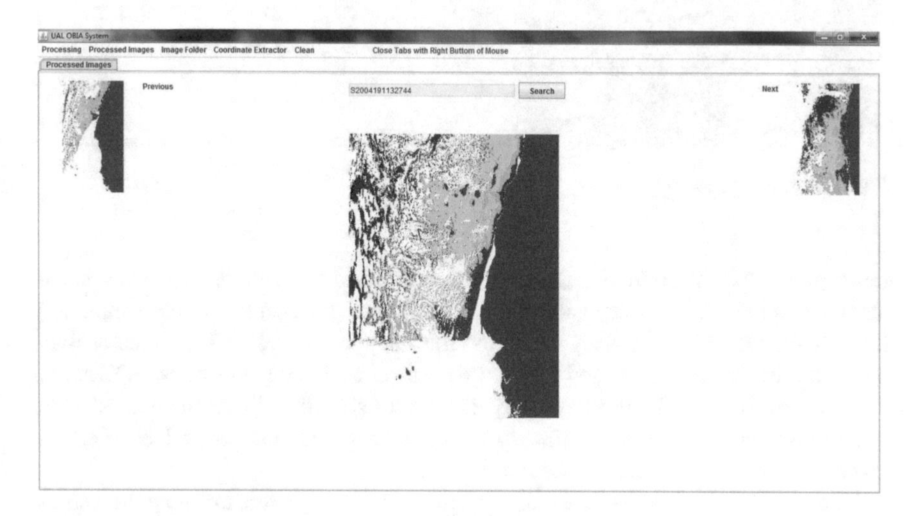

Fig. 6 Image selection window

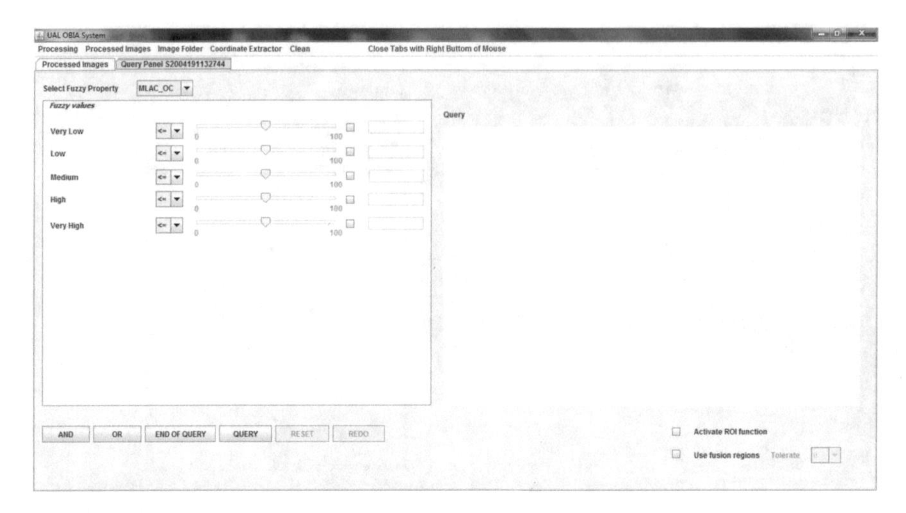

Fig. 7 Knowledge rule window

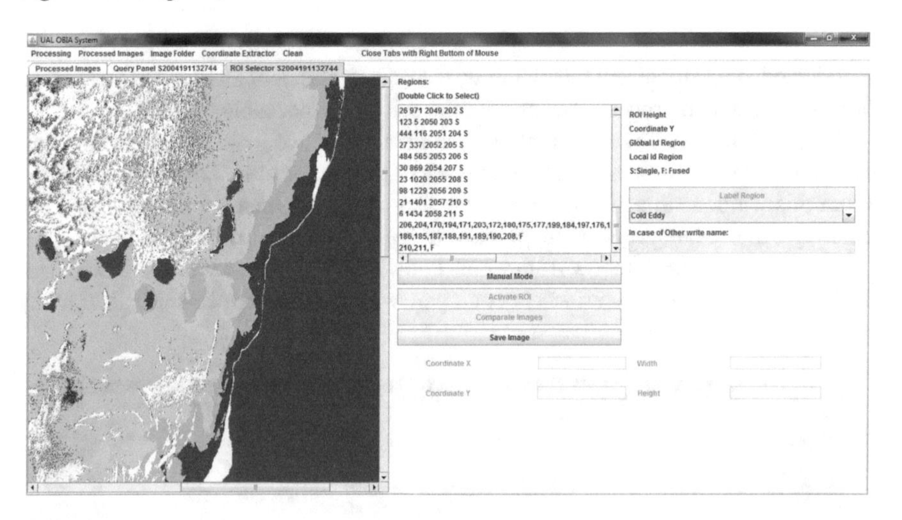

Fig. 8 Result window

descriptor (i.e., chlorophyll concentration, temperature, continent distance, island distance and size), the fuzzy value (i.e., in each case, the corresponding value: very high, high, medium, low, very low), and the range of the value (i.e., greater than, less than, equal, etc.,), to facilitate the definition of knowledge rules. ANDs and ORs connectives can be used as many as the user specifies. From this window, the oceanographer can select automatic merging of regions, and can select tolerance level. He/she can also activate automatic ROI.

The window used to list the results is shown in Fig. 8. In this example the knowledge rule $Fuzzy\_value\_5 \geq 0.25$ *OR* $Fuzzy\_value\_3 \geq 0.25$ has been used, where the

Fuzzy_value attribute correspond to fuzzy values of chlorophyll (1-very low, 2-low, 3-medium, 4-high, 5-very high) and temperature (1-very cold, 2-cold, 3-temperate, 4-warm, 5-very warm). In this example regions of very high (i.e., *Fuzzy_value_5*) or medium (i.e., *Fuzzy_value_3*) chlorophyll concentration have to be found, typical of upwellings and cold eddies. The window shows the image, and a list of regions retrieved by the knowledge rule. For each listed region, the tool shows the height of the ROI (i.e., height of the Bounding box of the region), the Y coordinate in the image, the identifier in the database, and S/F, which means simple (S) or fused (F) (i.e., merged) region. In the case of merged regions, the tool shows the identifier of each single region taking part of the merged region. By double-click on any element of the list, the tool emphasizes in pink color the region to be easily localized. The knowledge rule returns an empty list if it is not suitable for the retrieval of regions. Otherwise, the oceanographer can select and label a region from the list. Additionally, the oceanographer can select manual mode. In manual mode the manual ROI (i.e., a rectangle) is activated, and the selected regions can be grouped. When regions are grouped, the oceanographer can also label. The tool automatically unifies the regions, and they are labeled by the oceanographer (button "Label Region"). For instance, for labeling the cold eddy in the south of Gran Canaria, a ROI is selected with regions of the area.

The tool also permits, from this window, to compare with previously labeled regions. The tool retrieves images with regions of similar properties, to assist in the identification of structures in the current image (see Fig. 9). In Fig. 9 the tool shows, on the left-hand side, the current image with the selected region in pink color (i.e., a cold eddy), and on the right-hand side, an image (a SeaWiFS scene from 2004-03-17) in which a wake has been previously labeled (with similar properties to the eddy).

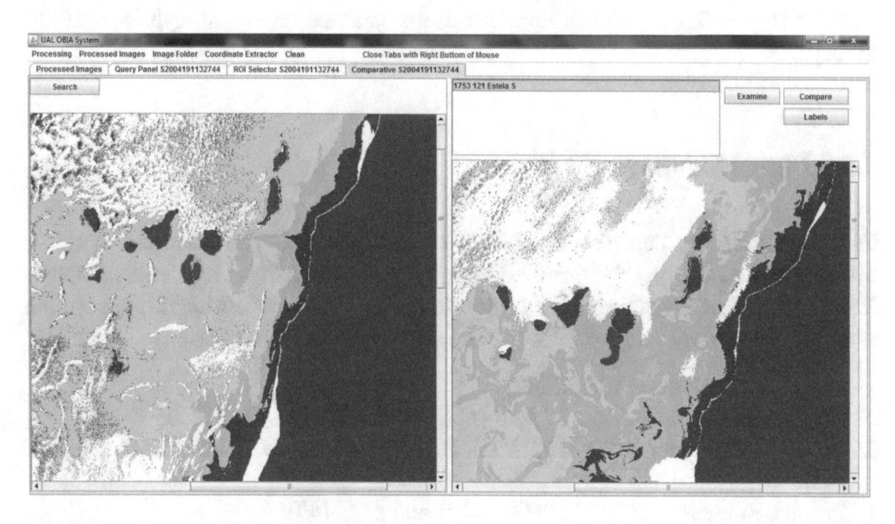

Fig. 9 image comparison window

The tool facilitates to modify the knowledge rules any time, with the goal to refine them by selecting other fuzzy descriptors, by changing value range, by deselecting automatic merging mode, and by increasing/decreasing tolerance level.

5.2 Class Modeling

In this section, we describe the class modeling of the tool. With the aim to have a friendly, robust and extensible tool, UML (Unified Modeling Language) has been used for modeling. Class modeling is shown in Fig. 10. A modular and extensible structure has been modeled. The model consists of three main elements: the user interface, the business logic, and the data logic. The user interface has been implemented with frames and panels of the Java Swing library (they are stereotyped in Fig. 10).

The main frame is a menu from which the tasks of the workflow (see Fig. 2 of Sect. 3) can be executed. The oceanographer can go back any time to previous tasks. Several images and regions to label can be selected at the same time. The main panels are *PanelProcessing*, *PanelFolder*, *PanelSelector*, *PanelQuery* and *PanelImages*. *PanelProcessing* is in charge of color clustering, image selection for processing, and processing. *PanelFolder* shows the folder (original, gray scale, mask, etc.,) for image visualization. *PanelSelector* permits the selection of the image to be labeled. *PanelQuery* facilitates the edition and execution of knowledge rules. Finally, *PanelImages* is used for labeling and manual merging of regions, as well as for comparison of images. The business logic is implemented by the class *interface*, which is in charge of the connection between the application and the database, which is handled by the class *Database*. *Database* class implements the database access using the Java JDBC library.

5.3 Data Modeling

In this section, we describe the design of the database. The database model is described by an entity–relationship diagram and shown in Fig. 11. The main table is *Region* (on the right-hand side) and stores region data. On the left-hand side a similar table is used to store merged regions. The table stores region descriptors including:

1. Hu moments: *Hu1*, ..., *Hu7* [41],
2. Maitra moments: *MM1*, ..., *MM6* [42],
3. Tensorial moments: *M00*, *M01*, *M02*, *M03*, *M10*, *M11*, *M12*, *M20*, *M21* and *M30* [43],
4. Zernike moments: *MZ1*, ..., *MZ6*, *MP1* and *MP2* [44],
5. *Perimeter*, *Area*, *Circularity*, *Eccentricity*, *MajorAxis*, *MinorAxis* [45], *CentroidX*, *CentroidY*, *Cirscumpcription* and *Orientation* descriptors,

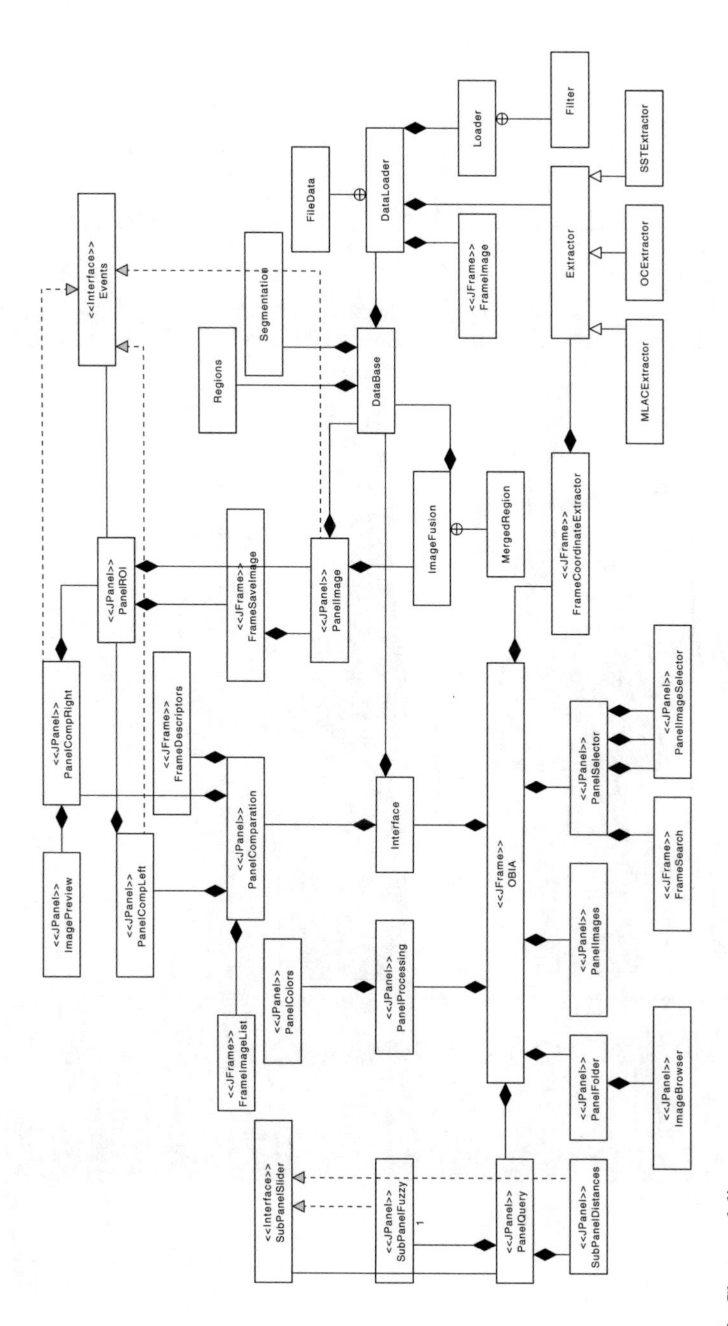

Fig. 10 Class modeling

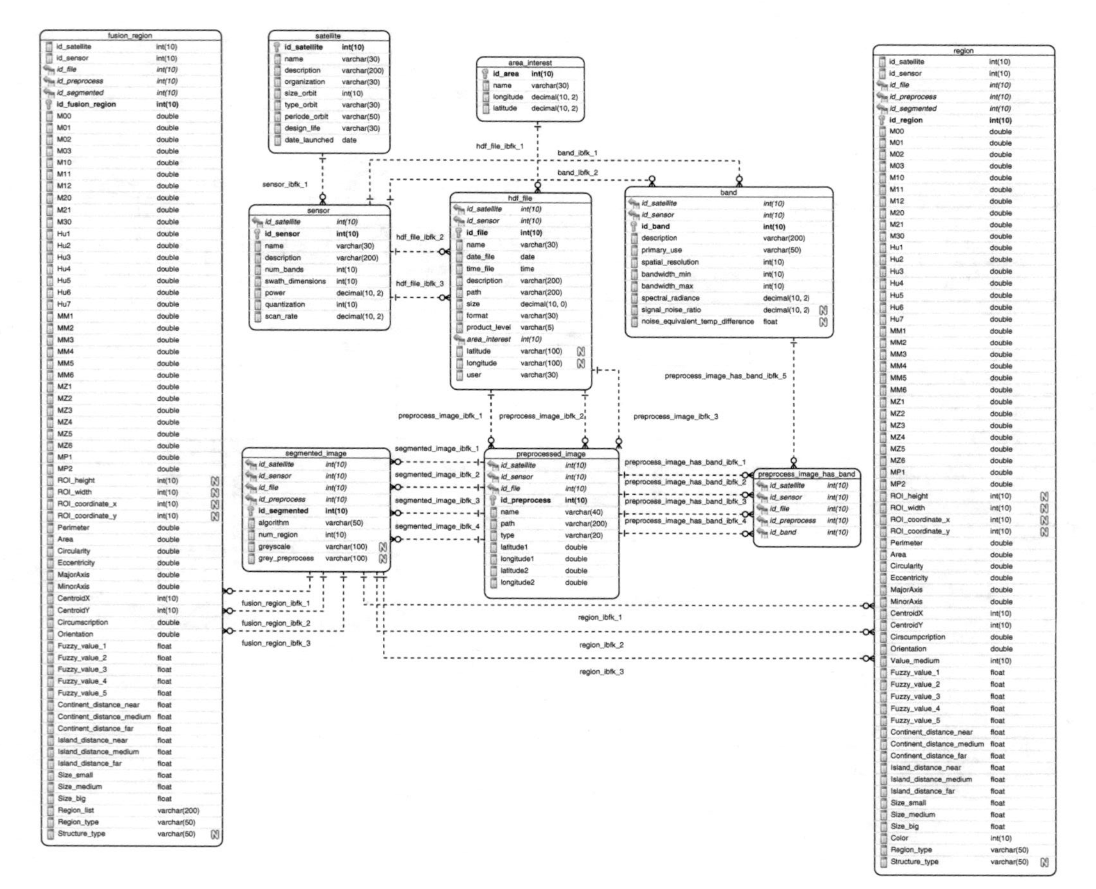

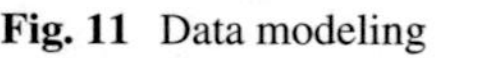

Fig. 11 Data modeling

6. ROI descriptors (i.e., *ROI_height*, *ROI_width*, *ROI_coordinate_X*, *ROI_coordinate_Y*),
7. And the fuzzy values of size, island distance, chlorophyll concentration and temperature.

The data model also uses tables for storing satellite data (table *Satellite*), sensor data (table *Sensor*), as well as areas of interest (table *Area_interest*) and sensor bands (table *Band*). HDF files data are stored in the table *hdf_file*. Finally, *segmented_image*, *preprocessed_image* and *preprocessed_image_has_band* tables store data about segmented and preprocessed images.

6 Conclusions and Future Work

In this chapter we have described the implementation of a tool for the identification and labeling of MOS. We have described the workflow of the tool, the user interface, the class diagram, and the database. As part of the workflow we have described how to extract knowledge rules for the labeling of MOS. We have presented the results obtained with the tool from the analysis of images of the Canary Islands and North West African coast captured by the SeaWiFS and MODIS-AQUA sensors. A total of 31 knowledge rules (12 for images and 19 for temperature images) have been obtained, which are required to label 365 MOS. The tool is available for download in http://indalog.ual.es/obia.

The extraction of the knowledge rules took several months of trial and error. The goal was to label each image and a larger number of MOS in each image. The goal was also to have a reduced number of knowledge rules. In our opinion, oceanographers would find useful to have a small number of rules, with the aim to label MOS of the study area in reasonable time, even though they are novice users. We believe that a novice user would spend less than three minutes in labeling a MOS. An expert user would label an image in thirty seconds. temperature images are more difficult to label. It is due to that the ocean temperature in the time of image capture greatly affects the MOS identification and retrieval. This is the reason why the number of knowledge rules for temperature (19) is greater than for chlorophyll (12). In chlorophyll images, three knowledge rules are enough to label all the upwellings, while in temperature images ten rules are required. upwellings are cold water structures close to coast. Thus, upwelling are harder to label when the ocean temperature is colder.

We believe that the developed tool can help oceanographers in the identification and labeling of MOS, as well as in the definition of knowledge rules for their recognition. The tool assists oceanographers in the processing of a batch of images, their visualization, and their labeling, and is very helpful in the tedious task of labeling large image databases.

Our tool has be designed to work with images from the Canary Islands and North West African coast. It means that our tool has to be modified a little in order to accept images of other places of interest for oceanographers from the point of view of global climate and primary production. This is the case, for instance, of MOS in some areas of North Atlantic and North Pacific [13–15] which have similar ocean current and wind conditions to the current study area. On the other hand, with the aim to contribute to the study of the marine environment and the coastal resource management, we are now working on adapt our tool to the detection of trends and changes in chlorophyll concentration or temperature, by analyzing the morphology, movement, and evolution of MOS.

Acknowledgments This work was funded by the EU ERDF and the Spanish Ministry of Economy and Competitiveness (MINECO) under Projects TIN2013-41576-R, TIN2013-44742-C4-4-R and CGL2013-48202-C2-2-R, and the Andalusian Regional Government (Spain) under Project P10-TIC-6114. This work also received funding from the CEiA3 and CEIMAR consortiums.

References

1. Bakun, A.: Global climate change and intensification of coastal ocean upwelling. Science **247**(4939), 198–201 (1990)
2. Change, I.P.O.C.: Climate change 2013: the physical science basis. Agenda **6**(07), 333 (2013)
3. McGregor, H., Dima, M., Fischer, H., Mulitza, S.: Rapid 20th-century increase in coastal upwelling off northwest Africa. Science **315**(5812), 637–639 (2007)
4. Gregg, W.W., Conkright, M.E., Ginoux, P., O'Reilly, J.E., Casey, N.W.: Ocean primary production and climate: global decadal changes. Geophys. Res. Lett **30**(15), 1809 (2003)
5. Angel, M.V., Fasham, M.J.R.: Eddies and biological processes. Eddies in Marine Science, pp. 492–524. Springer, Berlin (1983)
6. Rubino, A.: Fluctuating mesoscale frontal features: structures and manifestations in the real ocean. Kumulative Habilitationsschrift, Universitat Hamburg (2005)
7. Birkhoff, G., et al.: Jets, Wakes, and Cavities. Elsevier, Amsterdam (2012)
8. Chelton, D.B., Schlax, M.G., Samelson, R.M., de Szoeke, R.A.: Global observations of large oceanic eddies. Geophys. Res. Lett. **34**(15), L15606 (2007)
9. Robinson, I.S.: Ocean mesoscale features: upwelling and other phenomena. Discovering the Ocean from Space, pp. 159–193. Springer, Berlin (2010)
10. Schwartz, M.: Encyclopedia of Coastal Science. Springer Science & Business Media, New York (2006)
11. Zhang, Z., Zhang, Y., Wang, W., Huang, R.X.: Universal structure of mesoscale eddies in the ocean. Geophys. Res. Lett. **40**(14), 3677–3681 (2013)
12. Barton, E.D., Arístegui, J., Tett, P., Cantón, M., García-Braun, J., Hernández-León, S., Nykjaer, L., Almeida, C., Almunia, J., Ballesteros, S., Basterretxea, G., Escánez, J., García-Weil, L., Hernández-Guerra, A., López-Laatzen, F., Molina, R., Montero, M.F., Navarro-Pérez, E., Rodríguez, J.M., van Lenning, K., Vélez, H., Wild, K.: The transition zone of the Canary Current upwelling region. Progress Oceanogr. **41**(4), 455–504 (1998)
13. Kersalé, M., Doglioli, A., Petrenko, A.: Sensitivity study of the generation of mesoscale eddies in a numerical model of Hawaii islands. Ocean Sci. **7**(3), 277–291 (2011)
14. Lorenzo, E.D., Miller, A.J., Neilson, D.J., Cornuelle, B.D., Moisan, J.R.: Modelling observed California Current mesoscale eddies and the ecosystem response. Int. J. Remote Sens. **25**(7–8), 1307–1312 (2004)
15. Lumpkin, C.F.: Eddies and currents of the Hawaiian Islands. Ph.D. thesis, University of Hawaii (1998)
16. Oke, P.R., Griffin, D.A.: The cold-core eddy and strong upwelling off the coast of New South Wales in early 2007. Deep Sea Res. Part II: Top. Stud. Oceanogr. **58**(5), 574–591 (2011)

17. Meunier, T., Barton, E.D., Barreiro, B., Torres, R.: Upwelling filaments off Cap Blanc: interaction of the NW African upwelling current and the Cape Verde frontal zone eddy field. J. Geophys. Res.: Oceans (1978–2012) **117**, C8 (2012)
18. Tejera, A., García-Weil, L., Heywood, K., Cantón-Garbín, M.: Observations of oceanic mesoscale features and variability in the Canary Islands area from ERS-1 altimeter data, satellite infrared imagery and hydrographic measurements. Int. J. Remote Sens. **23**(22), 4897–4916 (2002)
19. Sangra, P., Pelegrí, J., Hernández-Guerra, A., Arregui, I., Martín, J., Marrero-Díaz, A., Martínez, A., Ratsimandresy, A., Rodríguez-Santana, A.: Life history of an anticyclonic eddy. J. Geophys. Res. **110**(C3), C03, 021 (2005)
20. Arístegui, J., Sangra, P., Hernández-León, S., Cantón, M., Hernández-Guerra, A., Kerling, J.: Island-induced eddies in the Canary islands. Deep Sea Res. Part I: Oceanogr. Res. Pap. **41**(10), 1509–1525 (1994)
21. Baatz, M., Hoffmann, C., Willhauck, G.: Progressing from object-based to object-oriented image analysis. Object-Based Image Analysis, pp. 29–42. Springer, Berlin (2008)
22. Musci, M., Feitosa, R.Q., Costa, G.A.: An object-based image analysis approach based on independent segmentations. In: Urban Remote Sensing Event (JURSE), 2013 Joint, pp. 275–278. IEEE (2013)
23. Drăguţ, L., Blaschke, T.: Automated classification of landform elements using object-based image analysis. Geomorphology **81**(3), 330–344 (2006)
24. Jovanovic, D., Govedarica, M., Dordevic, I., Pajic, V.: Object based image analysis in forestry change detection. In: 2010 8th International Symposium on Intelligent Systems and Informatics (SISY), pp. 231–236. IEEE (2010)
25. Rastner, P., Bolch, T., Notarnicola, C., Paul, F.: A comparison of pixel-and object-based glacier classification with optical satellite images. IEEE J. Sel. Top. Appl. Earth Obs. Remote Sens. **7**(3), 853–862 (2014)
26. Ko, B., Byun, H.: Frip: a region-based image retrieval tool using automatic image segmentation and stepwise Boolean AND matching. IEEE Trans. Multimed. **7**(1), 105–113 (2005)
27. Shrivastava, N., Tyagi, V.: A review of roi image retrieval techniques. In: Proceedings of the 3rd International Conference on Frontiers of Intelligent Computing: Theory and Applications (FICTA) 2014, pp. 509–520. Springer, Berlin (2015)
28. Vidal-Fernández, E., Piedra, J.A., Almendros-Jiménez, J.M., Cantón, M.: A location based approach to classification of mesoscale oceanic structures in SeaWiFS and MODIS-Aqua images from the North West Africa Area. Int. J. Remote Sens. **36**(24), 6135–6159 (2015)
29. Vidal-Fernández, E., Piedra, J.A., Almendros-Jiménez, J.M., Cantón, M.: OBIA system for identifying mesoscale oceanic structures in SeaWiFS and MODIS-aqua images. IEEE J. Sel. Top. Appl. Earth Obs. Remote Sens. **8**(3), 1256–1265 (2015)
30. Almendros-Jiménez, J.M., Domene, L., Piedra-Fernández, J.A.: A framework for ocean satellite image classification based on ontologies. IEEE J. Sel. Top. Appl. Earth Obs. Remote Sens. **6**(2), 1048–1063 (2013)
31. Piedra-Fernandez, J.A., Cantón-Garbín, M., Wang, J.Z.: Feature selection in AVHRR ocean satellite images by means of filter methods. IEEE Trans. Geosci. Remote Sens. **48**(12), 4193–4203 (2010)
32. Piedra-Fernández, J.A., Ortega, G., Wang, J.Z., Cantón-Garbín, M.: Fuzzy content-based image retrieval for oceanic remote sensing. IEEE Trans. Geosci. Remote Sens. **52**(9), 5422–5431 (2014)
33. NASA: Ocean Color Web. http://oceancolor.gsfc.nasa.gov (2013)
34. Liu, Z., Hou, Y.: Kuroshio front in the East China Sea from Satellite SST and remote sensing data. IEEE Geosci. Remote Sens. Lett. **9**(3), 517–520 (2012)
35. Marcello, J., Marques, F., Eugenio, F.: Automatic tool for the precise detection of upwelling and filaments in remote sensing imagery. IEEE Trans. Geosci. Remote Sens. **43**(7), 1605–1616 (2005)
36. Mityagina, M., Lavrova, O.: Dynamic phenomena in the coastal waters of the north-eastern black sea retrieved from satellite data. In: IEEE International Geoscience and Remote Sensing Symposium, 2008. IGARSS 2008, vol. 2, pp. II–347. IEEE (2008)

37. Patel, S., Balasubramanian, R., Gangopadhyay, A.: Automatic detection of oceanic eddies in SeaWiFS-derived color images using neural networks and shape analysis. Proc. IEEE IGARSS **2**, II–835–II–838 (2008)
38. Xiao, B., Hu, S., Qiang, X.: Research on the ocean primary production pattern based remote sensing. In: 2010 International Conference on Audio Language and Image Processing (ICALIP), pp. 1543–1546. IEEE (2010)
39. Sathyendranath, S., Brewin, B., Mueller, D., Doerffer, R., Krasemann, H., Mélin, F., Brockmann, C., Fomferra, N., Peters, M., Grant, M., et al.: Ocean colour climate change initiative–approach and initial results. In: IEEE International and Geoscience and Remote Sensing Symposium (IGARSS), pp. 2024–2027. IEEE (2012)
40. Saulquin, B., Gohin, F., Garrello, R.: Regional objective analysis for merging high-resolution MERIS, MODIS/Aqua, and SeaWiFS chlorophyll-a data from 1998 to 2008 on the European Atlantic shelf. IEEE Trans. Geosci. Remote Sens. **49**(1), 143–154 (2011)
41. Hu, M.K.: Visual pattern recognition by moment invariants. IRE Trans. Inf. Theory **8**(2), 179–187 (1962)
42. Maitra, S.: Moment invariants. Proc. IEEE **67**, 697–699 (1979)
43. Galvez, J.M., Cantón, M.: Normalization and shape recognition of three-dimensional objects by 3d moments. Pattern Recognit. **26**(5), 667–681 (1993)
44. Teague, M.R.: Image analysis via the general theory of moments. J. Opt. Soc. Am. **70**, 920–930 (1980)
45. Zunic, J., Sladoje, N.: Efficiency of characterizing ellipses and ellipsoids by discrete moments. IEEE Trans. Pattern Anal. Mach. Intell. **22**(4), 407–414 (2000)

Hybrid Uncertainty-Based Techniques for Segmentation of Satellite Imagery and Applications

B.K. Tripathy and P. Swarnalatha

Abstract Segmentation of an image is an essential assignment in an image examination whereby picture is divided into significant areas whose focuses have almost the same properties like; dim levels, mean qualities, or text-related characteristics. The pictures are divided into locales which best speak to the important articles in the scene. Locale parameters, for example, territory, shape, measurable parameters, and surface can be extricated and utilized for further examination of information. The examination of satellite symbolism of common scenes presents numerous one of a kind issues and it varies from an investigation and division of urban, business, or agrarian ranges. Once the division classes of a picture is obtained, it is conceivable to utilize heuristics or other area particular ways to deal with further characterize, translate, comprehend, register or extract information from the partitioned image. The applications of analysis of satellite imagery is plenty in real-life situations like weather forecasting, analysis of natural scenes, urban planning, environmental monitoring, object recognition, detection of mass wasting, etc. are well known. Several algorithms using classical approaches as well as those using uncertainty-based approaches have been proposed. The analysis shows that hybrid approaches are more efficient than the individual ones. In this chapter, we discuss on all the uncertainty-based and hybrid algorithms for segmentation of satellite imagery and their applications. Also, we propose some open problems which can be handled for future work.

Keywords Image segmentation · Satellite image · Fuzzy set · Rough set · Intuitionistic fuzzy set · Hybrid models · Data clustering

B.K. Tripathy (✉) · P. Swarnalatha
School of Computer Science and Engineering, VIT University,
Vellore 632014, India
e-mail: tripathybk@vit.ac.in

P. Swarnalatha
e-mail: pswarnalatha@vit.ac.in

S. Bhattacharyya et al. (eds.), *Hybrid Soft Computing for Image Segmentation*, DOI 10.1007/978-3-319-47223-2_7

1 Introduction

In PC vision, partitioning of an image is the procedure of apportioning a computerized picture into numerous fragments (sets of pixels, otherwise called super pixels). The objective of division is to improve and/or change the representation of a picture into something that is more significant and simpler to evaluate.

Segmenting an image into regions comes under clustering or classification process. The process can be carried out by using conventional techniques (Crisp) and uncertainty-based models (Soft). Some of the Conventional techniques used are; Canny, Sobel, Robert, and Hard C-Means where impreciseness is not taken care.

Partitioning of an image is a fundamental assignment in picture investigation whereby the picture is apportioned into significant locales whose focuses have about the same properties, e.g., dim levels, mean qualities, or textural properties. The division procedure is one of the initial phases in the remote detecting picture investigation: the picture is apportioned into districts which best speak to the pertinent articles in the scene. Locale characteristics, for example, region, shape, factual parameters and composition can be removed and utilized for further examination of the information. The division assignment can be refined in two methods:

1. Dividing up the pictures into various homogeneous locales, each having an exceptional mark,
2. Determining limits between homogeneous districts of various properties.

These division methods are known as locale-based division and edge discovery, separately. Every methodology is influenced distinctively by different elements. For a few applications edge recognition approach has not been effective. The prime cause is the nearness of little crevices in edge limits which permit converging of unique areas. Different hindrances are that these procedures are likewise frequently exceptionally delicate to nearby varieties power and the forms got are generally not shut. Along these lines, keeping in mind the end goal to yield shut limits the edges must be connected up. Then again, district-based division dependably gives shut shape areas which are a prerequisite in numerous applications. In addition, it is exceptionally straightforward and powerful in numerous applications. Mistakes in the areas limits are the principle downside of this methodology: edge pixels may be joined to any of the neighboring locales.

Image processing is among the rapidly growing technologies today, with its applications in various aspects as weather forecasting, classification of areas, etc. Picture processing shapes center exploration region inside designing and software engineering teaches as well. It incorporates three stages as, importing the picture with optical scanner or by computerized photography, examining and controlling the picture which incorporates information pressure and picture improvement and spotting designs that are not conspicuous for human eyes and thus results can be a modified picture or report that depends on picture investigation with picture preparing procedures.

The purpose of image processing is to enhance an image which undergoes segmentation for various satellite images. Processing of Analog and Digital images are the two ways to process an image for quality enhancement and segmentation based on uncertainty algorithms.The enhanced image is used for segmenting an image into required segments clusters. Image segmentation deals with segmenting the images into segments where vagueness and uncertainty has not been distinguished, where the soft computing techniques deal with vagueness. In this chapter, we primarily focus on uncertainty-based techniques in image segmentation and our focus is on satellite image segmentation.

Actually, Image segmentation is used to label each pixel in an image based on similarity of objects or its components. For segmentation methods, algorithms are categorized into dissimilar categories based on feature thresholding, template matching and region-based technique and finally clustering. These will be used according to the suitability to the applications and other metrics, such as cost, time, etc.

As general categories failed to deal with complex high-resolution satellite information, there exists necessity of novel algorithms based on computational intelligence diverse classifiers. These are effortlessly interpretable by human beings and fuzzy theory is an eye-catching methodology which is consigned as soft classifiers. As a whole in order to progress precision, all are toward, a hybrid bioinspired method which is discussed in this chapter.

The researchers are playing a prominent role in integration of various SC methods like fuzzy logic, ANN, genetic algorithms, decision trees, etc. to model a hybrid classification system which is more flexibility by exploiting tolerance and uncertainty of real-life situations.

Classification of segmentation of images is given as follows:

1.1 Edge-Based Image Segmentation

This is the essential stride of division which partitions a picture into an item and its experience. Edge recognition separates the picture by keen the adjustment in power or pixels of a photo. Dim bar outline and Gradient. Additionally manages RGB and HSV. These are superior to anything ANN.

1.2 Threshold-Based Image Segmentation [7]

Histogram thresholding is utilized to tie pre-handling and post-preparing methods required for edge division which involves Histogram Dependent Technique and Edge Maximization Technique EMT. This includes replacement threshold-based segmentation, such as optimization and Otsu thresholding.

1.3 Region-Based Image Segmentation [7]

This deals with comparison of alternative ways of noise resilient which divides an image into various regions supported by criteria, such as color, intensity, etc. The categories are region growing, region rending, and region merging.

1.4 PDE-Based Image Segmentation [7]

Halfway Differential Equations particularly in picture division which utilizes dynamic shape model or snake remodel. Wherein watershed segmentation is applicable by coupling with textural data, which leads to generate efficient results.

1.5 ANN-Based Image Segmentation [7]

Simulated Neural Network manages substantial cell that is equivalent to the segment of a photo which is mapped to the neural system. This network is trained with samples then associated between neurons. The neural networks for image segmentation are Hopfield, BPNN, SOM, etc.

1.6 Uncertainty-Based Image Segmentation [7]

It manages a dark scale picture which will be just rebuilt into a fluffy picture by utilizing a fuzzification operator that take away noise from image. The other techniques are discussed in Sect. 4.

2 Background

In this section, we shall look at the different traditional techniques briefly along with their applications. A combined method between classical and automatic approach for remote sensing image analysis is discussed in paper [11]. The algorithm deals with feature-space approach which undergo implementation of processing techniques, such as histogram thresholding and the clustering. The segmentation algorithm deals with pseudo-color images that provide excellent results if the image is composed by wide and well-defined regions resulting in an ambiguous and unsatisfactory outcome.

The paper [25] deals with the proposed method that discuss about the presence of both low radiometric contrast and moderately low-spatial determination. This will deliver a textural impact, an outskirt impact, and uncertainty in an entity/foundation qualification for satellite pictures. This basically is done by improving the morphological characteristic detection. The benefits of the new IKONOS satellite picture information for urban arranging, urban data undertakings has been discussed in paper. The classification of satellite is basically made using a pixel and a segment-based approach. A pixel-based characterization procedure has execution confinements for high-determination pictures however are perfect in usage terms. The segment-driven methodology conveys a lower extent of unclassified territories and in addition a much more homogeneous item than routine pixel-based characterization has been talked about [16].

Unsupervised factual division strategy is equipped for breaking down and ordering the satellite pictures which have nonhomogenous surface. Proposed technique demonstrated a superior in satellite picture grouping following different dispersions could be utilized for an ideal displaying and it is computationally costly [30].

The paper [27] deals with the proposed method belong to the category of causal, or feedforward, temporal segmentation techniques. The new graph-based approach performs best merge region growing, followed by the energy minimization on the image graph, where the energy consists of two terms describing the floe shape (shape term) and the gradient between the floe and the background (data term), respectively.

Graph-based image segmentation algorithm is developed in [29]. This deals with A G(V, E) an undirected graph that can be generated from an image, where nodes belongs to pixels and edges (E) connect nodes belonging to neighboring pixels. The drawback of this paper is Gaussian σ value and k calculation is not proposed.

Picture partition in paper [4] was expert utilizing the eCognition programming. A progressive grouping plan is used to wipe out regions that are not of interest and to distinguish ranges where mass developments are likely. A regulated arrangement is then directed utilizing ghostly, shape, and textural properties to recognize disappointments. There is leverage of the computerized framework which order quick mass developments that were new and more than 1 hour in a zone in a high mountain. In the meantime, the division calculation is adjusted utilizing client characterized edge of scale and heterogeneity and client characterized weightings on the different information layers.

Neighboring pixels for, e.g., comparative otherworldly values are gathered together to shape a picture fragment has been talked about in paper [13]. Portions still have the same phantom data as pixel, however now with factual qualities (mean, standard deviation, min, max, middle). The investigation of the portion's properties gives the fragment a chance to wind up an article. What's more, the likeness is characterized with fluffy rationale. At the point when characterizing mists the segregation between mists with a comparative ghastly trademark, e.g., single convective mists and mists in a frosty front, is much less demanding than on pixel premise.

The paper [9] made a proposed approach wherein the satellite picture in RGB shading space is change into YCbCr shading space and after that the changed satellite picture is part into three unique segments (stations or pictures) in light of luminance and chrominance. YCbCr Color space speaks to shading as power and endeavors the qualities of human eye furthermore our eye is more delicate to force than tone.

The paper [15] discuss about the implemented services that correspond to the various levels of remote sensing data processing that needs enhancement phase, extraction of cloud at base level and estimation of fractal dimension at top level. The cloud extraction is made using Markovian segmentation of infrared data which leads to high computation complexity and eradicated by segmentation parallel algorithm [32].

The paper [8] deals with segmentation and classification of remote sensing images. The classified image is given to K-Means Algorithm and Back propagation Algorithm of ANN to calculate the density count. In this method, there is no need of training (supervised learning). But accuracy is not good, and multiple land covers, data have not been used. In future different neural network algorithms can be used to classify the satellite images and the classification results of those images will be compared with results of existing classification methods [31].

The paper [9] discuss about enhancement of the satellite imagery using color separation of satellite imagery using color transformation. It process the regions grouped into a set of FCM clustering algorithm using ERDAS IMAGING software. Here also, training (supervised learning) is not essential for segmentation process. Some more applications of clustering techniques in image segmentation can be found in [22, 23].

Partitionutilization and overcomes way to deal with outline a progressive arrangement of characterization. Bolster vector machine (SVM)-based order procedure has been embraced for the errand of recognizing an information picture as having a place with one of the Desertic, Coastal, or Fluvial landform super-group classes. This paper proposed a strategy that gives better arrangement results in instances of rises, bull bow, and Plains when contrasted with Unsupervised technique [28] and takes lesser time than the unsupervised strategy. In any case, their technique takes additional time than the strategy proposed in [8].

The paper [12] talk about around a technique that treats every point in the information set, which is the guide of all conceivable shading blends in the given picture, as a potential group focus and gauges its potential concerning the other information components. The point with the greatest estimation of potential is thought to be a bunch focus. On the off chance that we need to have more number of groups, then consideration of one or all the more new bunch does not represent any issue in the proposed approach, though in fuzzy c-means grouping, the grouping must be done everywhere. In the meantime for little pictures, this methodology improves results.

In [18] a technique is suggested that uses the residuals of morphological opening and shutting changes in light of a geodesic metric. The proposed strategy performs well within the sight of both low radiometric contrast and generally low-spatial determination. Yet, for pictures with substantial and homogeneous areas, this methodology is computationally costly.

The paper [33] portrays a division calculation which consolidates old picture handling calculations with methods which are satisfactory from learning revelation in databases (KDD) and information mining to breakdown. These are the fragmented unstructured satellite pictures of common scenes which decide the quantity of classes in the picture naturally.

The paper [24] talks about around a programmed cloud-sort arrangement frameworks that has restricted by ambiguities in multispectral cloud-sort marks. The calculation introduces a cloud-sort arrangement that settles disarrays in infrared cloud-sort marks by a correlation of textural measures on known and obscure cloud-sort fragments. This arrangement with two stages, for example, the division technique and the grouping methodology. However, this has an issue of grouping of 107 windows of infrared satellite information brought about an order exactness of roughly 95 %. The paper [14] deals with a new method of performing multiscale, hierarchical segmentation of images using texture properties that has been shown to perform consistently better than other untrained, unsupervised texture segmentation algorithms. The images are first quantized using contiguity-enhanced K-Means clustering. This method outperforms other segmentation algorithms on radar and satellite weather images and result some merits of the multiscale technique in spite of some problems that are yet to be resolved on weather satellite imagery [19].

Urban arranging requires opportune securing and investigation of spatial and fleeting data for settling on educated choices is made in paper [6]. The paper portrays a methodology utilizing both per-pixel and item-based order techniques.The cloud free orthorectified Ikonos [20] very high-resolution satellite is classified into per-pixel and object-based categories. Per-pixel is a supervised classification and object based is an object-oriented classification that leads to GIS data integration and modeling. As a result, the precision of these two classes utilizing an item arranged characterization strategy further enhanced from 89 to 97 %. The joined methodology utilizing per-pixel and article situated arrangement strategies may demonstrate helpful in the investigation of VHR satellite information like Ikonos likewise as a future work [3].

3 Uncertainty-Based Techniques

Modern day data sets have inherent imprecision in them and so in order to handle them models dealing with uncertainty have become essential. Some of the existing models of uncertainty like fuzzy sets [38], rough sets [24] and intuitionistic fuzzy sets [7] have already been used by researchers to develop algorithms dealing with data clustering based upon them.

A fuzzy set A can be defined to be associated with a membership function μ_A such that every element x in U is associated with its membership value in [0, 1] through this function.

This notion was further extended to define the notion of intuitionistic fuzzy set in [7], where the nonmembership function is defined as not being ones complement of the membership function but another such function V_A such that every element x in U is associated with its membership value in [0, 1] through this function.

This notion was further extended to define the notion of intuitionistic fuzzy set in [7], where the nonmembership function is defined as not being ones complement of the membership function but another such function V_A such that the sum of the membership and nonmembership values of every element always lies in [0, 1]. The hesitation function π_A is defined as ones complement of the sum of V_A and V_A. So, for a fuzzy set there is no hesitation value for any element in U. Almost at the same time as that of intuitionistic fuzzy sets another uncertainty-based model called the rough set was introduced in [24]. This notion is based upon the concept of equivalence relations defined over the universes. Any subset X of U is associate with two crisp sets $\underline{R}X$ and $\overline{R}X$ with respect to an equivalence relation R, called the lower and upper approximations of X with respect to R [24]. If $\underline{R}X \neq \overline{R}X$ then X is said to be rough with respect to R. Else, it is said to be R-definable. In case X is R-rough we take the difference between $\underline{R}X$ and $\overline{R}X$ as the boundary region of X with respect to R and denote it by $BN_R(X)$.

Data clustering is an important topic under data mining and it has wide applications. Under this we accumulate similar elements under a group called a cluster and these clusters are substantially dissimilar from each other. A class of algorithms called the C-means algorithms deal with data clustering.

In case of crisp clustering the clusters generated using a c-means algorithm are disjoint. But the disjoint clusters generated are not of much use in real-life applications. So, uncertainty-based C-means algorithms have been developed where the clusters generated may not be disjoint. This provides a generality and improved applicability to the clustering algorithms.

The fields of application for the clustering techniques vary over a wide area of spectrum and the algorithms developed are more suitable to one or more of the fields like information retrieval, satellite image analysis, machine learning, bioinformatics and pattern recognition.

3.1 Fuzzy C-Means

This algorithm was proposed in [37] and for the first time the objective function approach came into existence. As it is well known now the solution space for a fuzzy c-means algorithm is of infinite dimension and no exact algorithm can find out a solution in real time. As a solution to this the objective function approach was introduced and optimization techniques came into force in data clustering.

To be more precise, the clusters generated by a fuzzy c-means are fuzzy sets and hence have overlapping by nature. Every element has degrees of belongingness to the clusters. This belongingness to all the clusters sums to 1.

3.2 Rough C-Means

The rough C-means was put forth in [15]. Here, as expected each of the clusters generated is a rough set. Hence, each one of them has a lower approximation comprising of the certain elements and an upper approximation comprising of uncertain values, which belong to the boundary of more than one cluster.

Algorithm 1 Fuzzy C-Means

1: Each of the clusters be initialized with a center value
2: The centres be updated with the formula

$$v_i = \frac{\sum_{k=1}^{N} (\mu_{ik})^p Y_k}{\sum_{k=1}^{M} (\mu_{ik})^p} \quad (1)$$

3: The distance d_{ik} between the centre v_i and element x_k be obtained by using the Euclidean distance formula.
4: The partition matrix V be generated with its elements being given by
$\mu_{ik} = \frac{1}{\sum_{l=1}^{D} \left(\frac{d_{ik}}{d_{lk}}\right)^{\frac{2}{(p-1)}}}$
if $d_{ij} > 0$, otherwise = 1.
5: Calculate the new partition matrix V' using steps -3 and 4.
6: If $||V^{(r)} - V^{(r+1)}|| < \varepsilon$, stop. Else, go to step-2

While computing the new centers, two weight functions w_{low} and w_{up} are used with their suffixes providing the indication as to which factor they are to be multiplied. In fact, w_{low} is the weight factor for the lower approximation and w_{up} is the weight factor for the boundary region, where it is assumed that $w_{low} > w_{up}$ such that $w_{low} + w_{up} = 1$.

Algorithm 2 Rough C-Means

1: Each of the clusters be initialized with a center value
2: Compute the distances of each element x_k from the cluster centres and find the two lowest distances say d_{pk} and d_{qk} from the pth and q^{th} cluster centres.
3: If $d_{pk} - d_{qk}$ is less than a preassigned value (ε) then assign x_k to the upper approximations of both the pth and q^{th} clusters. Otherwise, assign it to the lower approximation of the cluster from which it has the minimum distance.
4: Calculate the fresh cluster centres by using the formula

$$v_i = \begin{cases} w_{low}\frac{\sum_{x_k \in \underline{B}V_i} x_k}{|\underline{B}V_i|} + w_{up}\frac{\sum_{x_k \in \overline{B}U_i - \underline{B}U_i} x_k}{|\overline{B}U_i - \underline{B}U_i|} & \text{if } |\underline{B}V_i| \neq 0 \text{ and } |\overline{B}V_i - \underline{B}V_i| \neq 0 \\ \frac{\sum_{x_k \in \overline{B}V_i - \underline{B}V_i} x_k}{|\overline{B}V_i - \underline{B}V_i|} & \text{if } |\underline{B}V_i| = 0 \text{ and } |\overline{B}V_i - \underline{B}V_i| \neq 0 \\ \frac{\sum_{x_k \in \underline{B}V_i} x_k}{|\underline{B}V_i|} & \text{Else} \end{cases} \quad (2)$$

5: Repeat from step 2 until there are no more assignment.

3.3 Intuitionistic Fuzzy C-Means

Algorithm 3 Intuitionistic Fuzzy C-Means

1: Step-1, Step-2 and Step-3 are same as in fuzzy C-means (Algorithm 2).
2: Compute the hesitation value by using the formula:

$$\pi_A x = 1 - \mu_A(x) - \frac{1 - \mu_A(x)}{1 + \lambda \mu_A(x)}, \quad x \in X \tag{3}$$

3: Compute the fresh matrix V by taking .
$\mu'_{ik} = \mu_{ik} + \pi_{ik}$
4: Step-6, Step-7 and Step-8 are as in FCM (Algorithm 2).

4 Hybrid Techniques

The combination of two models can provide better models than their components. In fact, at the time of inception of rough sets there was an apprehension that this new model will compete with fuzzy sets. But, in [1] it was established that the models complement each other and provide better models as also has been found later. To this extent the rough fuzzy c-means (RFCM) [16] is a positive example to the claim above.

Once again, the clusters generated by RFCM are rough fuzzy sets and the advantage of these clusters is that the lower approximations contain the certain elements of a cluster whereas the boundary elements which belong to more than one cluster have degrees of membership also. The convergence is faster in the case of crisp clustering [35] and this is not true for uncertainty-based c-means algorithms [36]. However, the real-life situations are mostly uncertainty based and hence the uncertainty-based means are more useful. Using fuzzy sets the fuzzy c-means [38] was developed. This has been analyzed for edge techniques in [15].

Pure rough set-based clustering algorithms and their hybrid models with fuzzy sets are more prevalent now a day. Some rough fuzzy c-means algorithmic techniques are proposed in [16, 17, 20, 21]. Using the intuitionistic fuzzy sets [10] an extension of FCM was developed in [5] and because of the presence of the hesitation values this algorithm has been found to be more effective which is establishes by taking MRI scanned images.

It is difficult to obtain smooth edges for satellite images due to abrupt change in their brightness levels. This is a reason why satellite image segmentation is preferred to be done by using uncertainty-based clustering algorithms. As the vagueness and uncertainty are different in the rough set nomenclature and also direct two different types of imprecision, the hybrid models like RFCM and the rough intuitionistic fuzzy c-means algorithms are used.

There are many hard edge detection techniques like the zero-based detection, Sobel detector, Canny edge detector and Robert cross detector. But non-smoothness of satellite images make these techniques less efficient to the uncertainty-based techniques.

The emphasis is given on satellite imagery in this chapter. Because of the above, the chapter has discussed with hybrid soft computing techniques which is supposed and has been established as the most suitable among all the uncertainty-based algorithms for segmentation of images.

4.1 *The Hybrid Algorithms*

In this section, we shall discuss on some algorithms which has been obtained as a composition of RCM and FCM or RCM and IFCM.

Algorithm 4 RFCM

1: Each of the clusters be initialized with a cluster value..

2: The Euclidean distances d_{ik} of x_k from the ith cluster center are computed. If $d_{ik} = 0$ then compute the membership value μ_{ik} of the kth element in the ith cluster using
$\mu_{ik} = \frac{1}{\sum_{j=1}^{C} \left(\frac{d_{ik}}{d_{jk}}\right)^{\frac{2}{m-1}}}$

3: Let μ_{ik} and μ_{jk} be the lowest and next lowest membership values of the kth element in clusters V_i and V_j, respectively.
If $\mu_{ik} - \mu_{jk} < \varepsilon$ then $x_k \in \overline{B}U_i$ and $x_k \in \overline{B}U_i$
Else $x_k \in \underline{B}V_i$

4: The new cluster centers are computed by using the formula,

$$v_i = \begin{cases} w_{low} \frac{\sum_{x_k \in \underline{B}U_i} x_k}{|\underline{B}U_i|} + w_{up} \frac{\sum_{x_k \in \overline{B}U_i - \underline{B}U_i} (\mu'_{ik})^m x_k}{\sum_{x_k \in \overline{B}U_i - \underline{B}U_i} (\mu'_{ik})^m} & \text{if } |\underline{B}U_i| \neq 0 \text{ and } |\overline{B}U_i - \underline{B}U_i| \neq 0 \\ \frac{\sum_{x_k \in \overline{B}U_i - \underline{B}U_i} (\mu'_{ik})^m x_k}{\sum_{x_k \in \overline{B}U_i - \underline{B}U_i} (\mu'_{ik})^m} & \text{if } |\underline{B}U_i| = 0 \text{ and } |\overline{B}U_i - \underline{B}U_i| \neq 0 \\ \frac{\sum_{x_k \in \underline{B}U_i} x_k}{|\underline{B}U_i|} & \text{Else} \end{cases}$$

5: Steps starting with the second one are repeated until there are no more assignment of elements remains or the terminating condition is satisfied.

4.1.1 RFCM

The algorithm was proposed in the year 2006 in [18] and was improved by Maji and Pal in 2007.

4.1.2 RIFCM

The RFCM has been extended to propose and study the RIFCM in 2013 in the paper [2]. As in previous cases, here the clusters are rough intuitionistic fuzzy sets. Hence, each cluster has a lower approximation comprising of certain elements and has a boundary, which comprises of uncertain elements with both membership and non-membership values to the cluster boundary. In fact, the elements may belong to the boundary of more than one cluster boundaries with different membership and non-membership values. The parameters used here are having similar meanings as in RFCM unless otherwise stated.

Algorithm 5 The RIFCM algorithm

1: Each of the clusters be initialized with a cluster value.
2: The Euclidean distances d_{ik} of x_k from the ith cluster center are computed. If $d_{ik} = 0$ then compute the membership value μ_{ik} of the kth element in the ith cluster using
3: The hesitation values π_{ik} are computed by using the formula

$$\pi_A x = 1 - \mu_A(x) - \frac{1 - \mu_A(x)}{1 + \lambda\mu_A(x)} \mid x \in X$$

4: The modified membership values μ'_{ik} using the relation $\mu'_{ik} = \mu_{ik} + \pi_{ik}$
5: Stepes 5 and 6 are same as steps 3 and 4 of RFCM algorithm (Algorithm 4) with μ_{ik} being replaced by μ'_{ik}
6: If neither the termination condition is met nor there are no elements to be assigned the repeat steps from step 2.

4.1.3 The Measuring Indices

Here, we introduce two of the indices which measure the efficiency of a clustering algorithm from their values. These are the Davies–Bouldin (DB) index and the Dunn (D) index.

Davies–Bouldin (DB) Index

The DB index is obtained as the ratio of sum of within cluster distance to between-cluster distance. It is formulated as given

$$DB = \frac{1}{c}\sum_{i=1}^{c} max_{k \neq i}\left\{\frac{S(v_i) + S(v_k)}{d(v_i, v_k)}\right\} \text{for } i > 1, k < c$$

We shall express below the within cluster distances for the RCM, RFCM and RIFCM later. Lower value of DB index for a clustering algorithm in comparison to those for other algorithms shows that the algorithm is better than the compared algorithms.

Dunn (D) Index

This value is used to identify the clusters which are compact and separated [1]. The expression for its computation is

$$Dunn = min_i\left\{min_{k \neq i}\left\{\frac{d(v_i, v_k)}{max_1 S(v_1)}\right\}\right\} \text{for } k > 1, i, l < c$$

Unlike the DB index value, the higher the D index value for an algorithm it is supposed to be better.

Within Cluster Distance for RCM

$$S(v_i) = \begin{cases} w_{low}\dfrac{\sum_{x_k \in \underline{B}U_i} \|x_k - v_i\|^2}{|\underline{B}U_i|} + w_{up}\dfrac{\sum_{x_k \in \overline{B}U_i - \underline{B}U_i} \|x_k - v_i\|^2}{|\overline{B}U_i - \underline{B}U_i|} & \text{if } |\underline{B}U_i| \neq 0 \text{ and } |\overline{B}U_i - \underline{B}U_i| \neq 0 \\ \dfrac{\sum_{x_k \in \overline{B}U_i - \underline{B}U_i} \|x_k - v_i\|^2}{|\overline{B}U_i - \underline{B}U_i|} & \text{if } |\underline{B}U_i| = 0 \text{ and } |\overline{B}U_i - \underline{B}U_i| \neq 0 \\ \dfrac{\sum_{x_k \in \underline{B}U_i} \|x_k - v_i\|^2}{|\underline{B}U_i|} & \text{if } |\underline{B}U_i| \neq 0 \text{ and } |\overline{B}U_i - \underline{B}U_i| = 0 \end{cases}$$

Within Cluster Distance for RFCM

$$S(v_i) = \begin{cases} w_{low} \dfrac{\sum_{x_k \in \underline{B}U_i} \|x_k - v_i\|^2}{|\underline{B}U_i|} + w_{up} \dfrac{\sum_{x_k \in \overline{B}U_i - \underline{B}U_i} (\mu_{ik})^m \|x_k - v_i\|^2}{\sum_{x_k \in \overline{B}U_i - \underline{B}U_i} (\mu_{ik})^m} & \text{if } |\underline{B}U_i| \neq 0 \text{ and } |\overline{B}U_i - \underline{B}U_i| \neq 0 \\ \dfrac{\sum_{x_k \in \overline{B}U_i - \underline{B}U_i} (\mu'_{ik})^m \|x_k - v_i\|^2}{\sum_{x_k \in \overline{B}U_i - \underline{B}U_i} (\mu_{ik})^m} & \text{if } |\underline{B}U_i| = 0 \text{ and } |\overline{B}U_i - \underline{B}U_i| \neq 0 \\ \dfrac{\sum_{x_k \in \underline{B}U_i} \|x_k - v_i\|^2}{|\underline{B}U_i|} & \text{if } |\underline{B}U_i| \neq 0 \text{ and } |\overline{B}U_i - \underline{B}U_i| = 0 \end{cases}$$

Within Cluster Distance for RIFCM

The formulae for this is obtained from those of RFCM by replacing μ_{ik} with μ'_{ik}

4.1.4 Testing

A comparative analysis of the effectiveness of the above uncertainty-based methods and hybrid methods were conducted by taking four types of different satellite images. The images were filtered by using a filtering technique called the refined bit plane [34] filtering method. After that the filtered images were segmented by using the different images. The results show that the RIFCM algorithm provides the best results among all the segmentation methods under consideration.

4.1.5 The Index Values

In two ways of measuring indices such as the DB index and the D index provide a measure of the efficiencies of different clustering algorithms. The table below provides the Db and the D values for the above four satellite images. Lower value for the DB measure and higher values for the D measure indicates that the algorithm is of better efficiency. As per this estimation the RIFCM algorithm comes out to be the most efficient among all the above algorithms considered (Table 1).

Table 1 The values of the DB and D measures for all the methods for the different satellite images

	FCM	RCM	IFCM	RFCM	RIFCM
HILLS					
DB - INDEX	0	0.184944	15.70054	0.193631	0.193631
DUNN INDEX	0	8.525729	0.211887	0.021204	0.016745
FRESHWATER					
DB - INDEX	14.70054	0	0.40423	0.399804	19.70054
DUNN INDEX	0.125887	0	2.211887	4.910032	0.02116
FRESHWATERVALLEY					
DB - INDEX	0.201419	0.281419	0.4	0.399804	0.281655
DUNN INDEX	4.03718	4.43718	0.021228	4.910032	0.02001
DROUGHT					
DB-INDEX-WRBP	0.90423	0.428179	0.429886	0.429886	1.00423
DUNN INDEX-WRBP	0.90116	4.208979	0.021067	0.017018	0.08116

5 Case Study

This case study is a step further in image segmentation in the sense that we have used segmentation in depth reconstruction. A filter called geometric correction is used and compared the performance when it is used with the case when it is not used. A technique called anaglyph approach for this purpose and the images are again those obtained from the satellites [26] (Fig. 1).

It has been observed that the images obtained through remote sensing do not have high precision and so are not suitable for high precision algorithms. The reason is being the occurrence of distortions like geometric errors. Also the parameters like atmospheric propagation, sensor response, and illuminations have their effect on the images. In order to deal with these problems in [26], an approach was proposed where in the first phase a preprocessing is done so that the images obtained from the satellites are free from the above errors. This step is followed by clustering using the algorithms discussed in this chapter. The process is done through some geometric correction techniques. The image is reconstructed using depth dimension/depth map generation. For all these anaglyph images are used for better interpretation of satellite imagery. The experiments performed show that the imprecise algorithms described in this chapter are extremely useful for segmentation of satellite images after passing through the process mentioned.

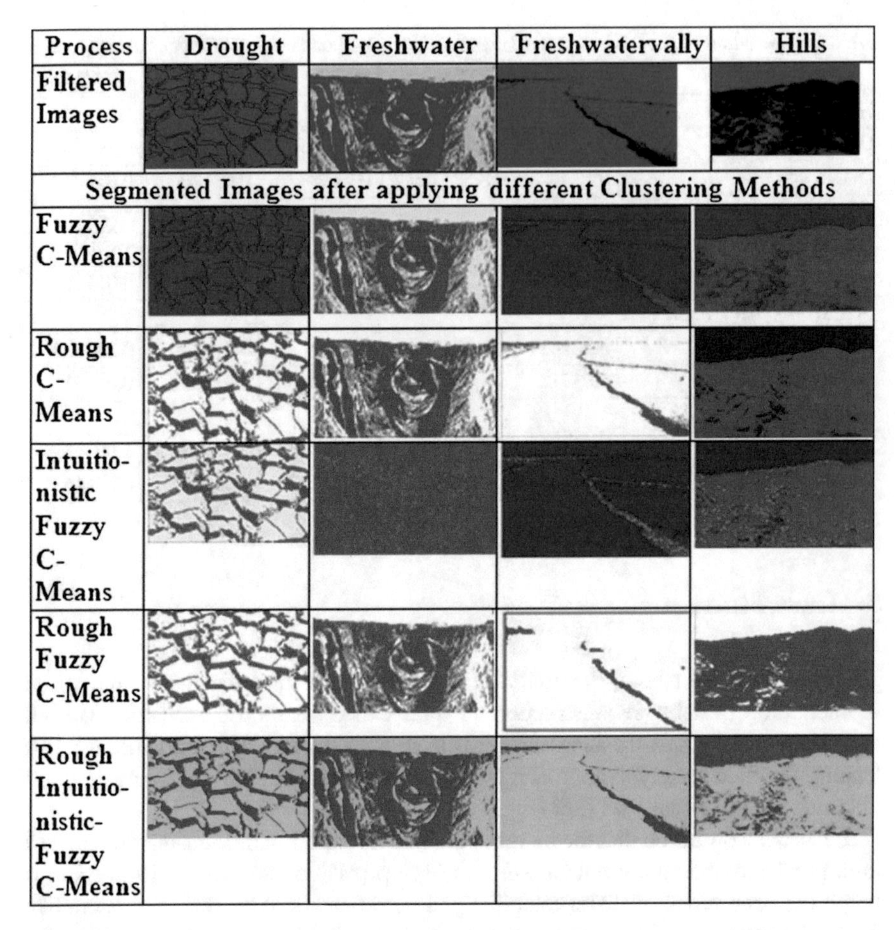

Fig. 1 The segmented images of the four satellite images obtained asoutput from the different uncertainty-based algorithms

Depth Map Generation Algorithm (Depth Reconstruction) and Anaglyph Technique

The third dimension map generation (Z-dimension) is a grayscale image and is to possess the same resolution as that of the original input image. It is used for generation of Depth Map.The following is the algorithm used for this purpose.

Figure 2 provides a comparison of original image with its depth maps, first without geometric correction and next with geometric correction. It is clearly evident that the depth map generation out of the filtered image provides much better result than the one without using the filter.

In Figs. 3 and 4, images have been generated from the third dimension map.The Figs. 3c and 4c. depict the resulting anaglyph 3D image, from a set of generated frames. It is worth noting that the Anaglyph 3D images can be viewed through the

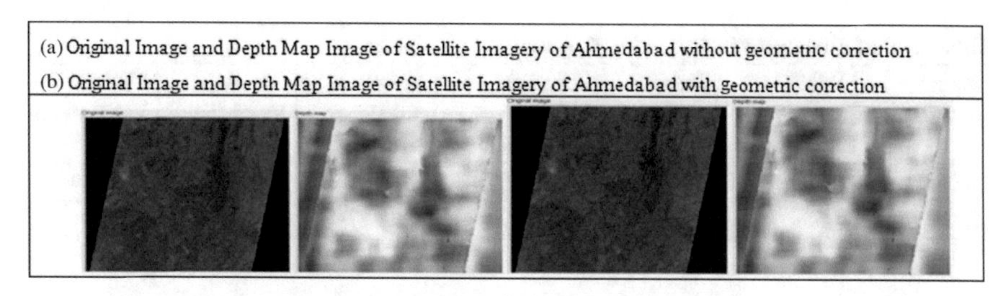

Fig. 2 Original image followed by its depth map without geometric correction and the procedure repeated for depth map with geometric correction

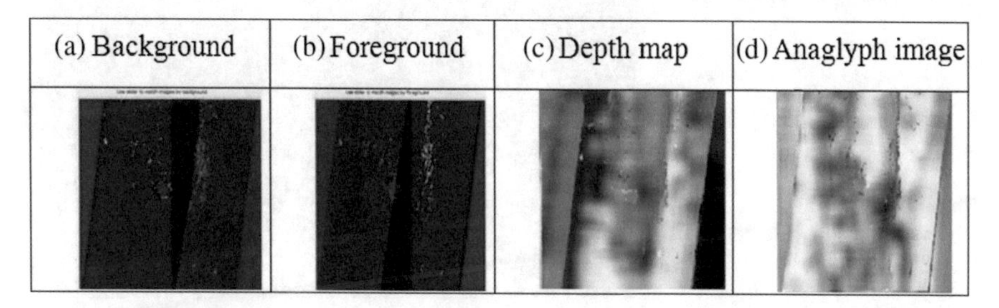

Fig. 3 Depth images of **a** background, **b** foreground, **c** depth map and **d** anaglyph images of without geometric correction (WOGC-C-D)

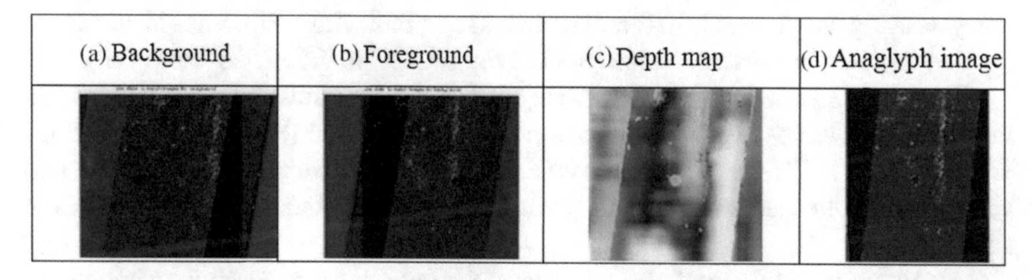

Fig. 4 Depth images of **a** background, **b** foreground, **c** depth map and **d** anaglyph images of with geometric correction(WGC-C-D)

color-coded anaglyph glasses.Both these images reveal an integrated stereographic image and reach one eye. As is evident from Figs. 3d and 4d, the visual cortex of the brain fuses this into perception of a three dimensional scene or composition.

5.1 Discussion on the Results

Each of the methods WOGC-C D and WGC-C-D involves three steps dealing with depth map without geometric correction and depth map with geometric correction respectively [26]. In the first phase either the geometric correction is used or not. The

Table 2 Performance of PSNR and RMSE values

Input images	PSNR Value-WOGC	PSNR Value-WGC	PSNR Value-clustered	RMSE Value-clustered	PSNR Value-depth map	RMSE Value-depth map
WOGC	8.3992	13.7734	11.0707	13.7734	6.9112	10.3854
WGC	7.5597	12.9627	10.8370	12.9627	6.5697	8.9570

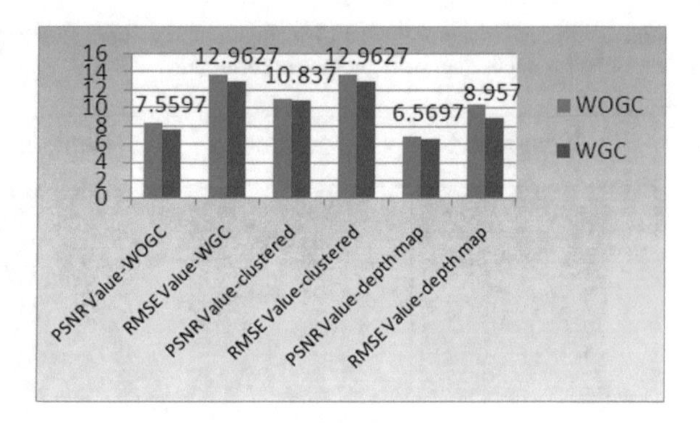

Fig. 5 The bar chart showing PSNR and RMSE values without GC and with GC

corresponding results are represented in the tabular form in Table 2. The performance analysis using the metrics PSNR and RMSE as per Table 2 and Fig. 5 has been carried out and it yields improvised value (8.3992 for WOGC to 7.5597dB for WGC).

The analysis shows that the clustering shows better results when WGC is used than when it is not used as is evident from Table 2 and Fig. 5 (8.3992-11.0707dB for WOGC) and (7.5597-10.8370dB for WGC). The clustered images result in minimum values for the measures PSNR and RMSE when GC is used than when GC is not used [26].

The output in the form of clusters is used for depth map generation using reconstruction techniques and stereo pair images shows improved visualization of satellite imagery as illustrated in Fig. 6 and has an efficient unsupervised true color composite. Again the measures PSNR and RMSE are computed to measure the performance of the approach and it provides much better results.

6 Scope for Future Work

There are several other image segmentation methods like the family of kernel-based algorithms KFCM, KIFCM, KRCM, KRFCM and KRIFCM; the possibilistic image segmentation methods like the PFCM, PIFCM, PRFCM and PRIFCM; their combinations like PKFCM, PKIFCM, PKIFCM, and PKRIFCM. These methods have been applied on many MRI images, cancer cell images and metal coin images. However, it has been established recently that the different kernels are suitable for different

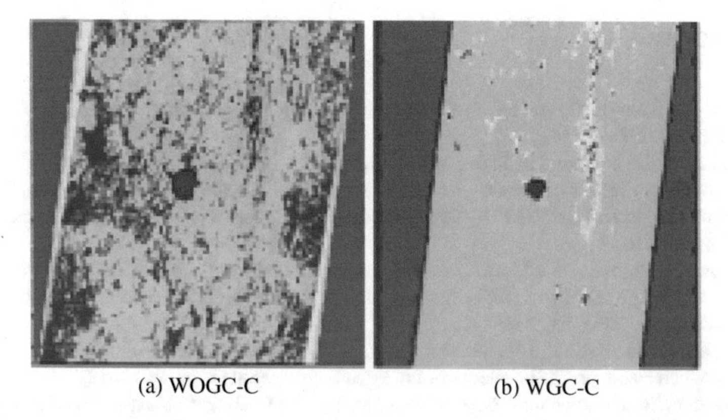

(a) WOGC-C (b) WGC-C

Fig. 6 Measures of accuracy for Original satellite imagery when clustered with true color composite

images. That is no single kernel which provides better results for all types of images. It would be an interesting study to apply all these methods to satellite images and find their efficiencies. Also, the proposed model will be applied for reconstruction of satellite imagery for better interpretation of results which resolves varied societal issues.

7 Conclusions

The important area of segmentation of satellite images is studied in this chapter. A presentation was made on different crisp segmentation techniques and also the individual uncertainty-based algorithms like the FCM, IFCM, and RCM. However, it has been experimentally concluded that the hybrid techniques like the RFCM and the RIFCM are superior to the individual methods even for satellite images. Four different types of satellite images were taken to illustrate that RIFCM is the best among all the segmentation methods dealt with the measurement of two indices; namely the DB and the D index were taken to establish this observation. We provide two sets of case studies. The first one being different types of satellite images and the second one is for the depth reconstruction using geometric correction using anaglyph approach for satellite images. The overall conclusion is that the hybrid clustering algorithms provide better segmentation and when used for depth reconstruction the images with geometric correction generates better results than the cases when this filtering technique is not used. To evaluate the approaches the accuracy measures PSNR and RMSE are used. Finally, at the end of the chapter we have suggested some research directions for further work on this topic.

References

1. Adak, C.: Rough clustering based unsupervised image change detection. arXiv preprint arXiv:1404.6071 (2014)
2. Atanassov, K.T.: Intuitionistic fuzzy sets. Fuzzy Sets Syst. **20**(1), 87–96 (1986)
3. Baboo, S.S., Thirunavukkarasu, S.: Image segmentation using high resolution multispectral satellite imagery implemented by fcm clustering techniques. Int. J. Comput. Sci. Issues (IJCSI) **11**(3), 154 (2014)
4. Barlow, J., Franklin, S., Martin, Y.: High spatial resolution satellite imagery, dem derivatives, and image segmentation for the detection of mass wasting processes. Photogramm. Eng. Remote Sens. **72**(6), 687–692 (2006)
5. Bermúdez, J.D., Segura, J.V., Vercher, E.: A multi-objective genetic algorithm for cardinality constrained fuzzy portfolio selection. Fuzzy Sets Syst. **188**(1), 16–26 (2012)
6. Bhaskaran, S., Paramananda, S., Ramnarayan, M.: Per-pixel and object-oriented classification methods for mapping urban features using ikonos satellite data. Appl. Geogr. **30**(4), 650–665 (2010)
7. Farooque M., YasirRaeen, S.: Self organizing map based improved color image segmentation. International Journal of Advanced Research in Computer Science and Software Engineering **6**(3), 456–462 (2016)
8. Gagrani, A., Gupta, L., Ravindran, B., Das, S., Roychowdhury, P., Panchal, V.: A hierarchical approach to landform classification of satellite images using a fusion strategy. In: Computer Vision, Graphics and Image Processing, pp. 140–151. Springer, New York (2006)
9. Ganesan, P., Rajini, V., Sathish, B., Kalist, V., Basha, S.K.: Satellite image segmentation based on ycbcr color space. Indian J. Sci. Technol. **8**(1), 35–41 (2015)
10. Garai, A., Mali, K.: Density based fuzzy c means (DBFCM) image segmentation
11. Guarnieri, A., Vettore, A.: Automated techniques for satellite image segmentation. Int. Arch. photogramm. Remote Sens. Spat Info. Sci. **34**(4), 406–410 (2002)
12. Hanmandlu, M., Jha, D., Sharma, R.: Segmentation of Satellite Images by Modified Mountain Clustering (2000). https://www.cse.iitb.ac.in/~sharat/icvgip.org/icvgip00/I-47.pdf
13. Huckle, R., Olesen, F.: Cloud analysis from meteosat data using image segmentation for climate model verification. In: Proccessings of the EUMETSAT Meteorological Satellite Conference. Citeseer (2008)
14. Lakshmanan, V., Rabin, R., DeBrunner, V.: Hierarchical texture segmentation of weather radar and satellite images. Use of Meteosat WV data for monitoring moisture changes in the environment of a tornado-producing storm, p. 37 (2001)
15. Lingras, P., West, C.: Interval set clustering of web users with rough k-means. J. Intell. Info. Syst. **23**(1), 5–16 (2004)
16. MACQUEEN, J.: Some methods for classification and analysis of multivariate observations. In: Proceedings of the fifth Berkeley symposium on mathematical statistics and probability, 1967, vol. 1, pp. 281–297. University of California Press (1967)
17. Maji, P., Pal, S.K.: Rfcm: A hybrid clustering algorithm using rough and fuzzy sets. Fundam. Informs. **80**(4), 475–496 (2007)
18. Maji, P., Pal, S.K.: Rough set based generalized fuzzy-means algorithm and quantitative indices. IEEE Trans. Syst. Man Cybern. Part B Cybern. **37**(6), 1529–1540 (2007)
19. Mangai, U.G., Samanta, S., Das, S., Chowdhury, P.R., Varghese, K., Kalra, M.: A hierarchical multi-classifier framework for landform segmentation using multi-spectral satellite images-a case study over the indian subcontinent. In: Fourth Pacific-Rim Symposium on Image and Video Technology (PSIVT), pp. 306–313. IEEE (2010)
20. Meinel, G., Lippold, R., Netzband, M.: The potential use of new high resolution satellite data for urban and regional planning. IMAGE (1997)
21. Mitra, S.: An evolutionary rough partitive clustering. Pattern Recognit. Lett. **25**(12), 1439–1449 (2004)

22. Mukhopadhyay, A., Bandyopadhyay, S., Maulik, U.: Clustering using multi-objective genetic algorithm and its application to image segmentation. In: IEEE International Conference on Systems, Man and Cybernetics, 2006. SMC'06. vol. 3, pp. 2678–2683. IEEE (2006)
23. Mukhopadhyay, A., Maulik, U., Bandyopadhyay, S.: Multiobjective genetic clustering with ensemble among pareto front solutions: Application to mri brain image segmentation. In: Seventh International Conference on Advances in Pattern Recognition, 2009. ICAPR'09. pp. 236–239. IEEE (2009)
24. Parikh, J.A., Rosenfeld, A.: Automatic segmentation and classification of infrared meteorological satellite data. IEEE Trans. Syst. Man Cybern. Part B Cybern. **8**(10), 736–743 (1978)
25. Pesaresi, M., Benediktsson, J.A.: A new approach for the morphological segmentation of high-resolution satellite imagery. IEEE Trans. Geosci. Remote Sens. **39**(2), 309–320 (2001)
26. Prabu, S., Tripathy, B.K., Swarnalatha, P., Ramakrishna, R., Moorthi S.M.: Depth Reconstruction using Geometric Correction with Anaglyph Approach for Satellite Imagery, pp. 5–14 (2014)
27. Price, C., Tarabalka, Y., Brucker, L.: Graph-based method for multitemporal segmentation of sea ice floes from satellite data. In: Latin American Remote Sensing Week (2013)
28. Rao, S.G., Puri, M., Das, S.: Unsupervised segmentation of texture images using a combination of gabor and wavelet features. In: ICVGIP, pp. 370–375 (2004)
29. Ravali, K., Kumar, M.R., Rao, K.V.: Graph-based high resolution satellite image segmentation for object recognition. Int. Arch. Photogramm. Remote Sens. Spat. Info. Sci. **40**(8), 913 (2014)
30. Rekik, A., Zribi, M., Hamida, A.B., Benjelloun, M.: An optimal unsupervised satellite image segmentation approach based on pearson system and k-means clustering algorithm initialization. methods **8**, 9
31. Sathya, P., Malathi, L.: Classification and segmentation in satellite imagery using back propagation algorithm of ann and k-means algorithm. Int. J. Mach. Learn. Comput. **1**(4), 422 (2011)
32. Shelestov, A., Kravchenko, O., Korbakov, M.: Services for Satellite Data Processing (2005)
33. Soh, L.K., Tsatsoulis, C.: Segmentation of satellite imagery of natural scenes using data mining. IEEE Trans. Geosci. Remote Sens. **37**(2), 1086–1099 (1999)
34. Swarnalatha, P., Tripathy, B.: A novel fuzzy c-means approach with bit plane algorithm for classification of medical images. In: International Conference on Emerging Trends in Computing, Communication and Nanotechnology (ICE-CCN), pp. 360–365. IEEE (2013)
35. Tou, J.T., Gonzalez, R.C.: Pattern Recognition Principles (1974)
36. Vieira, S.M., Sousa, J.M., Kaymak, U.: Fuzzy criteria for feature selection. Fuzzy Sets Syst. **189**(1), 1–18 (2012)
37. Xiong, H., Wu, J., Chen, J.: K-means clustering versus validation measures: a data-distribution perspective. IEEE Trans. Syst. Man Cybern. Part B Cybern. **39**(2), 318–331 (2009)
38. Zadeh, L.A.: Fuzzy sets. Info. Control **8**(3), 338–353 (1965)

Improved Human Skin Segmentation Using Fuzzy Fusion Based on Optimized Thresholds by Genetic Algorithms

Anderson Santos, Jônatas Paiva, Claudio Toledo and Helio Pedrini

Abstract Human skin segmentation has several applications in computer vision beyond its main purpose of distinguishing between skin and nonskin regions. Despite the large number of methods available in the literature, accurate skin segmentation is still a challenging task. Many methods rely only on color information, which does not completely discriminate the image regions due to variations in lighting conditions and ambiguity between skin and background color. This chapter extends upon a self-contained method for skin segmentation that outlines regions from which the overall skin color can be estimated and such that the color model is adjusted to a particular image. This process is based on thresholds that were empirically defined in a first approach. The proposed method has three main contributions over the previous one. First, genetic algorithm (GA) is applied to search for better thresholds that will be used to extract appropriate seeds from the general probability and texture maps. Next, the GA is also applied to define thresholds for edge detectors aiming to improve edge connections. Finally, a fuzzy method for fusion is included where its parameters are optimized by GA during a learning phase. The improvements added to the skin segmentation method are evaluated on a set of hand gesture images. A statistical analysis is conducted over the computational results achieved by each evaluated method, indicating a superior performance of our novel skin segmentation method.

Keywords Human skin segmentation · Fuzzy fusion · Self-adaptation skin segmentation · Genetic algorithms

A. Santos (✉) · H. Pedrini
Institute of Computing, University of Campinas, Campinas,
SP 13083-852, Brazil
e-mail: anderson.santos@ic.unicamp.br

H. Pedrini
e-mail: helio@ic.unicamp.br

J. Paiva · C. Toledo
Institute of Mathematics and Computer Science, University of São
Paulo, São Carlos, SP 13566-590, Brazil
e-mail: jlpaiva@gmail.com

C. Toledo
e-mail: claudio@icmc.usp.br

S. Bhattacharyya et al. (eds.), *Hybrid Soft Computing for Image Segmentation*, DOI 10.1007/978-3-319-47223-2_8

1 Introduction

The process of human skin segmentation can be defined as a two-class labeling problem, where the main objective is to determine if a pixel belongs to a skin or nonskin category, according to a set of extracted features. The input is an image and the output corresponds to a map indicating which class each pixel belongs to.

Skin segmentation plays an important role in a diversity of applications, such as face detection [1, 2], gesture analysis [3, 4], person tracking [5, 6], nudity detection [7, 8], content-based image retrieval [9, 10], human–computer interaction [11, 12], among others.

The skin detection configures a preprocessing stage for such applications, demanding certain requirements in terms of speed and simplicity. The term simplicity here means that complex parameter configuration, fixed threshold selection, and manual adaptation should be avoided. In other words, the skin segmentation process should enable real-time applications and be amenable to different purposes.

There are many challenges associated with the automatic segmentation of human skin in uncontrolled environment. Images acquired through varying illumination conditions typically present different skin characteristics. Shadows, light intensity variation, reflections, noise, and person's pose may cause discrepancy in skin regions of the same image. Image resolution and dimensions, as well as compression techniques, are also significant aspects.

In addition to digital imaging aspects, there are inherent difficulties caused by natural characteristics of the human skin: its tones vary across different ethnicities, its texture changes with age (babies have a soft skin while elders have a more coarse skin), its elasticity depends on facial expressions and pose.

Although not robust to all these challenges, color presents a meaningful evidence for skin detection. In fact, the vast majority of researches found in the literature focus on color information to determine whether a pixel belongs to a skin region or not. Nevertheless, there is an intrinsic problem associated with the use of color. It does not provide a clear separability between the pixels that belong to skin and one portion of nonskin, referred to as skin-like pixels. The process to identify a transition between skin and skin-like pixels is complex and can frequently leads to a number of segmentation errors.

The main purpose of this chapter is to describe an approach that sets parameters for a skin segmentation process through genetic algorithms(GA). The new technique extends upon prior work developed by Santos and Pedrini [13], where the skin segmentation method is self-adaptive since it generates a skin color model specific for each image. This model reduces color ambiguity by decreasing the number of skin-like pixels. Although there are methods that perform adaptive segmentation, they usually rely on face detection or a prior knowledge, which can be both unavailable or difficult to determine. Additionally, a fuzzy fusion (FF) technique is also introduced in this chapter to combine different fuzzy classifications and assign to each pixel a membership grade related to the skin set.

The novelty of the method relies on its self-contentedness by combining a previous self-adaptation skin segmentation (SASS) method with GA and FF technique. It uses spatial analysis to obtain true skin regions from which a specific color model is derived. Instead of empirically selecting thresholds [13], multiple GAs are applied to find more appropriate values for these parameters. Fuzzy parameters are also optimized by a genetic algorithm during FF learning phase.

This text is organized as follows: Sect. 2 presents some relevant works related to human skin segmentation. Section 3 describes the proposed methodology. Section 4 presents and evaluates the experiments results obtained with our method. Section 5 discusses the general contributions of the work.

2 Background

Many approaches found in the literature have been proposed to address the problem of human skin segmentation [14–17]. The simplest and earlier strategies for classifying a pixel are based on static decision rules that restrict skin to some specific intervals on a chosen color space. Sobottka et al. [18] developed a skin detection method based on the HSV color space. A transformation of RGB color space into a single-channel was proposed by Cheddad et al. [19] for skin detection purpose. Hsu et al. [20] adopted various thresholds that partition the HSI color space into three zones that define the skin pixels. A rule based on two quadratic functions for the normalized RG space was proposed by Soriano et al. [21].

A more sophisticated scheme, proposed by Jones and Rehg [5], is based on modeling the statistical distribution of color. There are two main approaches: parametric and non-parametric techniques. Both approaches calculate the probability of a given color (c) to be skin ($P(skin|c)$), which generates a probability map such that the segmentation can be performed through a threshold. However, parametric approaches assume that the skin distribution fits some explicit model.

In order to have a more accurate model, it is possible to suppose that there is an overlap between skin and non-skin colors [15], such that many researchers have adapted the mentioned methods according to the context. For instance, Kovac et al. [22] defined different rules depending on lighting conditions, whereas Phung et al. [23] created an iterative method for determining an optimal threshold for the probability map of a particular image.

Nevertheless, the most significant results are obtained by content-based adaptation, more specifically for face detection. The first of such approaches [24] uses the region acquired by a face detector to update a unimodal Gaussian previously determined. Taylor and Morris [25] used only the facial skin in normalized RG to construct a Gaussian model, discarding any previous training. A more robust technique [26] uses the face region to build a local skin histogram and a $P_{face}(skin|c)$ is derived and combined with the general probability for the final map.

Another strategy, known as spatial analysis, considers the structural alignment in the neighborhood of pixels classified as skin, generally with a probability map, such that it refines the segmentation process by removing false positives. Most of these techniques perform an expansion of seeds found by a high threshold. This expansion can be performed through different criteria, such as energy accumulation [26], cost propagation [27], and threshold hysteresis [28]. Although cost propagation is complex, it usually provides superior results skin segmentation, where the Dijkstra's algorithm [29] is used to calculate the shortest routes in a combined domain composed of hue, luminance, and skin probability.

Wang et al. [30] used fixed rules for RGB and YCbCr color spaces, then combined the result of both and applied the gray-level co-occurrence matrix (GLCM) [31, 32] to extract texture features and classify the found skin regions. Although the false positive rate decreased, the true positive rate also decreased.

Ng and Chi-Man [33] combined both color and texture features for skin segmentation. The texture features were extracted through 2D Daubechies wavelets, whereas a Gaussian mixture model was used to classify the skin regions. Nonskin regions were discarded by using K-means. The method is dependent on the number of clusters and the improvement was not significant since the decrease in true skin detection is approximately the same as false skin detection.

Jiang et al. [34] employed a histogram-based skin probability map to find initial skin candidates. A lower threshold was used as a second stage to discard skin-like pixels. Gabor wavelets were used to extract texture features and combined to produce an untrained texture map. Therefore, a threshold on this map was required to eliminate nonskin texture. Similarly to other methods, this approach also compromises the true skin detection. Then, the authors used color and texture information to select markers of watershed segmentation [31] to grow skin regions.

A common problem found in most of the works mentioned is their lack of adaptability to particular conditions present in the images. The parameters set on those methods are the same when applied over different images. They usually are fixed by considering an average score for a previous set of images. Even content-based adaptation approaches, which attempt to define a color model for each image, apply previously defined parameters. Furthermore, some methods rely on accurate detection of faces in the images, which is not a simple task. To overcome such problems, a self-contained adaptive segmentation was proposed by Santos and Pedrini [13]; however, the definition of appropriate threshold values is still a challenge for an effective adaptation.

Evolutionary approaches can be applied to estimate thresholds in image problems. Gupta et al. [35] proposed a differential evolution strategy to perform wavelet shrinkage for noisy images. Thavavel et al. [36] employed a multi-objective genetic algorithm to determine wavelet denoising threshold values for a tomographic reconstruction. Mukhopadhyaya and Mandal [37] used genetic algorithm to find threshold values for medical image denoising.

A hybrid approach is proposed by Xie and Bovik [38] for dermoscopy image segmentation. A GA is applied to select seeds (leaf neurons) and combined with a self-generating neural network (SGNN) to optimize and stabilize clustering results. Another hybrid artificial neural network (ANN) method is introduced by Razmjooy et al. [39] to address the skin classification problem. This approach applies an evo-

Table 1 Comparison of seed-based skin detection methods

Method	Seed detection	Propagation	Color model	Fusion models
Ref. [34]	Fixed threshold	Watershed	Global	No
Ref. [26]	Face detection	Energy transference	Local and global	Weighted mean of empirically defined weights
Ref. [27]	Fixed threshold	Minimum cost	Global	No
Ref. [28]	Fixed threshold	Fixed threshold on neighbor differences	Global	No
Ours	Adaptive threshold by GA optimization	Control based on adaptive edge detection with GA	Local and global	Fuzzy fusion of texture and color models with GA parameter optimization

lutionary algorithm, named Imperialist Competitive Algorithm (ICA), to optimize initial weights of the feedforward neural network.

Fuzzy approaches have also been applied to the skin detection problem. Charir and Elmoataz [40] proposed a method based on fuzzy C-means clustering algorithm for skin color segmentation. Hamid and Jemma [41] developed a fuzzy inference system for face detection. Kim et al. [42] proposed a linear matrix inequality fuzzy clustering to estimate the rule structure and parameters for the membership function.

The method proposed in this chapter is related to those developed in [35–37] in the sense that they also apply a GA to optimize threshold values. In our method, we extend the self-contained adaptive segmentation (SASS) method described in [13] by replacing thresholds empirically defined with those optimized using GA. In contrast with approaches described in [40–42], this work employs the Fuzzy Fusion (FF) technique to generate a fuzzy map for final image segmentation. The overall combination of methods leads to the development of a hybrid approach that combines SASS, GAs, and FF in a different way from that proposed in [38, 39].

Table 1 presents a summary of relevant characteristics present in our method and in other seed-based skin detection approaches available in the literature.

3 Improved Human Skin Segmentation

In our methodology, a genetic algorithm is applied to optimize thresholds for the skin segmentation approach developed by Santos and Pedrini [13]. The method combines spatial analysis and adaptive models for better skin probability estimation. The proposed method can be divided into three major steps:

(i) seeds are extracted from the general probability map and a texture map through a combination of two thresholds. These thresholds are optimized by a genetic algorithm designed to maximize the skin probability, the number of seeds, and region homogeneity.
(ii) a seed fill [43] is applied to grow the seeds into a region, determined by the image edges obtained with Canny detectors [31]. The thresholds used by the edge detectors are provided by a genetic algorithm in order to make the edges more connected. The resulting regions are used to estimate the local skin color probability.
(iii) a fusion of texture, global, and local color probabilities is performed with fuzzy integral. In this step, a genetic algorithmis applied to optimize parameters during the learning phase of the fuzzy method.

Figure 1 illustrates the main components of our skin segmentation process. Algorithm 1 summarizes the overall method, whose details are explained in the next subsections.

Algorithm 1: Proposed skin segmentation method.

input : color image I
histograms of skin (H_{skin}) and nonskin ($H_{nonskin}$) colors
texture filter f
fuzzy measures F_{uzzm}
output: final probability map M_{final}

1 Build general probability map (M_{global}) using H_{skin} and $H_{nonskin}$
2 $T_{map} \leftarrow rgb2gray(I) * f$
3 $t_1, t_2 \leftarrow$ *Genetic Algorithm 1*
4 **for** $x \in I$ **do**
5   **if** $M_{global}(x) \geq t_1 \wedge T_{map}(x) \leq t_2$ **then**
6         $Seeds \leftarrow x$
7 $H, S, V \leftarrow rgb2hsv(I)$
8 $[t_l^1, t_h^1, t_l^2, t_h^2, t_l^3, t_h^3] \leftarrow$ *Genetic Algorithm 2*
9 $H_{edge} \leftarrow Canny(H, t_l^1, t_h^1)$
10 $S_{edge} \leftarrow Canny(S, t_l^2, t_h^2)$
11 $V_{edge} \leftarrow Canny(V, t_l^3, t_h^3)$
12 $Edges \leftarrow H_{edge} \vee S_{edge} \vee V_{edge}$
13 $C \leftarrow$ *FloodFill(Seeds, Edges)*
14 **for** $x \in I$ **do**
15   **if** $x \in C$ **then**
16         $H_{skin\_local}(color(x))$++
17 Build local probability map (M_{local}) using H_{skin_local} and $H_{nonskin}$
18 $M_{final} \leftarrow FuzzyFusionT_{map}, M_{global}, M_{local}, F_{uzzm})$
19 **return** M_{final}

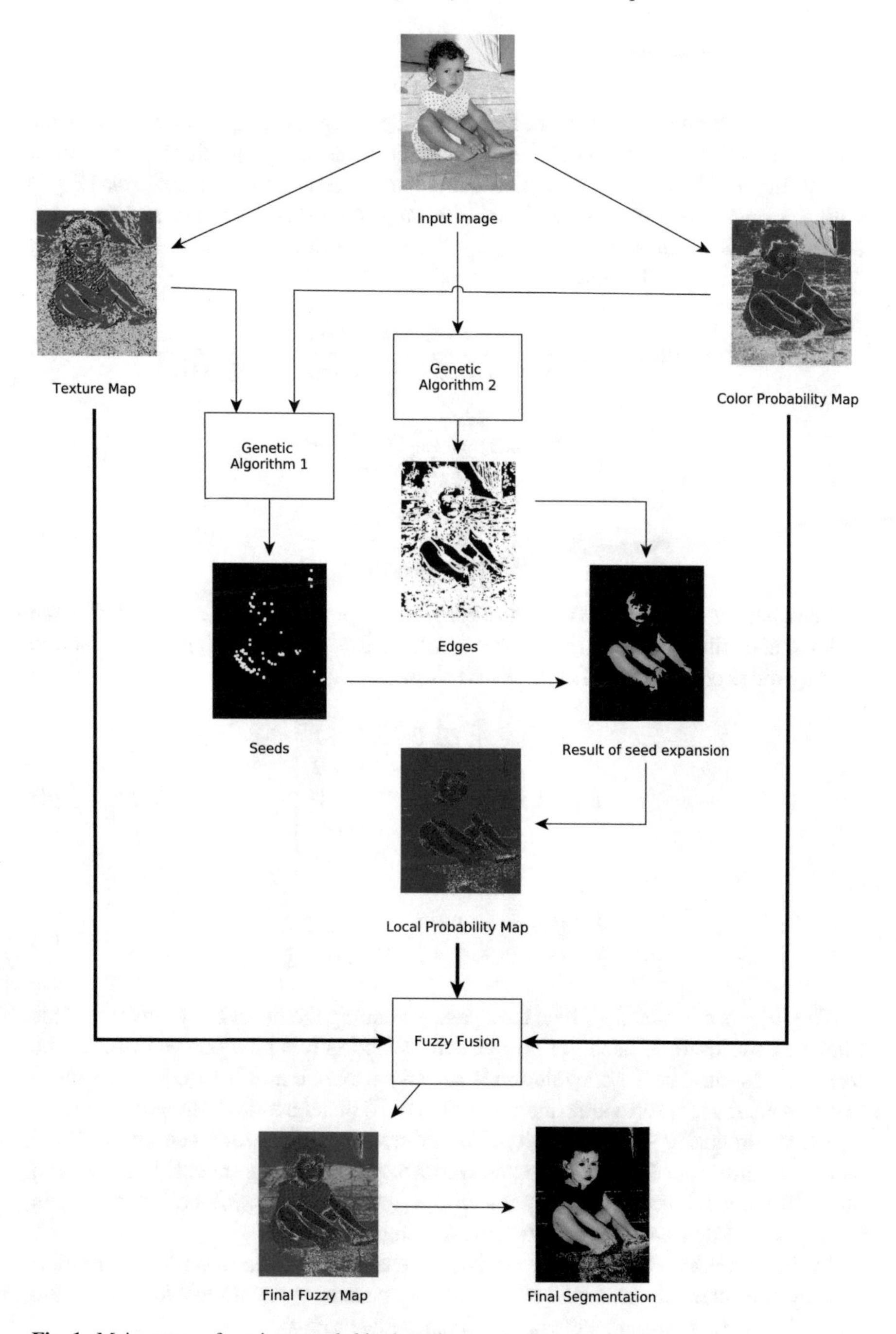

Fig. 1 Main stages of our improved skin detection methodology

3.1 Seed Detection

The most important step in a seed growing algorithm is the proper choice of the seeds. Before the current seed detection, it is necessary to produce a skin color probability map and a texture map. The first map is generated by Bayes' rule (Eq. 1) with the posterior probabilities defined through histograms of skin and non-skin colors collected from a training set (Eqs. 2 and 3). Therefore, it is possible to map the color of each pixel in the image to a skin probability.

$$P(skin|c) = \frac{P(c|skin)P(skin)}{P(c|skin)P(skin) + P(c|\neg skin)P(\neg skin)} \tag{1}$$

$$P(c|skin) = \frac{H_{skin}(c)}{\sum(H_{skin}(i))} \tag{2}$$

$$P(c|\neg skin) = \frac{H_{\neg skin}(c)}{\sum(H_{\neg skin}(i))} \tag{3}$$

The texture map is created through a convolution process between grayscale image and a spatial filter mask. This mask constitutes a 5×5 Laws filter [44], which is built by the product of two 1D masks, as shown in Eqs. 4 and 5.

$$E5^t \cdot S5 = \begin{bmatrix} 1 & 0 & -2 & 0 & 1 \\ 2 & 0 & -4 & 0 & 2 \\ 0 & 0 & 0 & 0 & 0 \\ -2 & 0 & 4 & 0 & -2 \\ -1 & 0 & 2 & 0 & -1 \end{bmatrix} \tag{4}$$

$$\begin{aligned} E5 \text{ (Edge)} &= [\,-1\ -2\ \ 0\ 2\ 1\,] \\ S5 \text{ (Spot)} &= [\,-1\ \ \ 0\ 2\ 0\ 1\,] \end{aligned} \tag{5}$$

The filter is combined with its transpose producing the mean between them. This filter was successfully used to improve skin detection in a previous work [45]. The resulting response of the convolution is scaled between 0 and 1, where values close to 1 mean a more texture area and those close to 0 mean a more homogeneous area.

After computing the probability and texture maps, the seeds can be extracted using one threshold for each map and retrieving the resulting intersection between them. The desired seeds are the ones with a high skin color probability, placed in homogeneous regions, and that are spread along the image.

In the proposed approach, we applied a genetic algorithm to optimize simultaneously the threshold values. This is done by searching for thresholds within the interval [0,1] that minimize the fitness function from Eq. 6.

$$f_{seed}(Seeds^{[t_1,t_2]}) = \begin{cases} \frac{1}{\min(M_{global} \wedge Seeds^{[t_1,t_2]})} + \frac{1}{|CC(Seeds^{[t_1,t_2]})|} + \\ \quad + \max(T_{map} \wedge Seeds^{[t_1,t_2]}) & \text{if } Seeds \neq \emptyset \\ +\infty & \text{otherwise} \end{cases} \tag{6}$$

where M_{global} represents the skin color probability map, $CC(n)$ corresponds to the connected components of n, T_{map} refers to the texture map, and $Seeds^{[t_1,t_2]}$ denotes the set of pixel seeds found with thresholds t_1 and t_2 following the steps in lines 4–6 of Algorithm 1. The first term imposes that every pixel in *Seeds* will have a high probability, whereas the second maximizes the number of components, which consequently causes a spatial dispersion of the pixels. The last term is associated with the homogeneity of the seed pixels, minimizing the texture value of every seed pixel. Additionally, there is a penalty if no seed pixel is found, forcing it to retrieve at least one pixel.

3.2 *Edge Detection*

The seeds must be expanded into skin regions; however, the main drawback of seed expansion methods is the occurrence of "leakages," that is, when a seed grows to a region of nonskin. To overcome such problem, we also apply a genetic algorithm in this step to optimize the threshold values required for the Canny edge detector [31] by considering both the connectivity and continuity as goals.

In order to benefit from color information, the edge detection is performed in each channel of the image transformed to the HSV color space. The resulting three edge maps are then combined through a union operation. Therefore, we need to evolve six parameters, once the Canny edge detector employs a low and a high threshold for each channel in the threshold hysteresis. As typically occurs, the low threshold (t_l) is considered as a fraction of the higher one (t_h). We then optimize t_h and a fraction α that produces $t_l = t_h * \alpha$. We limited α ranging from 0.01 to 0.5, and t_h between 0.1 and 0.9. Extreme values are undesirable since they generate either too many or only few edges.

There are many methods for evaluating the result of edge detection. A widely used strategy is the Pratt's figure of merit [46], however, it requires an edge map as reference. Unsupervised methods have been developed by analyzing edges locally [47] and comparing them against patterns [48]. These methods can be suitable for comparing two different edge detectors, however, they do not consider the amount of information needed. For instance, a circle will have the highest connectivity regardless of its size or quantity. Therefore, this type of metric is not appropriate for the genetic algorithm since it gives different outcomes for similar results.

Due to these factors, we adopt here a modification of a connected component (CC) based edge quality evaluation [49]. It relates connected components in an

8- and 4-neighborhood, as expressed in Eq. 7.

$$f_{edge}(Edge^{[A,T_h]}) = \frac{|CC_8(Edge^{[A,T_h]})|}{|CC_4(Edge^{[A,T_h]})|\, H} \quad (7)$$

where H is the image entropy and $Edge^{[A,T_h]}$ is the resulting edge mask found. We have $A = [\alpha^1, \alpha^2, \alpha^3]$ and $T_h = [t_h^1, t_h^2, t_h^3]$ to define the set of thresholds $t_l^i = t_h^i * \alpha^i$ and t_h^i, for $i = 1, 2, 3$. They are applied to Canny detector as presented in lines 9–12 of Algorithm 1.

Since every component belonging to a 4-neighborhood is also a 8-neighborhood, the best way to minimize its ratio would be to produce as many CC_4 as possible that are connected by diagonals forming only one CC_8. This imposes a line connectivity, penalizing edge gaps. Furthermore, we also include the maximization of the entropy of the binary edge image to cope with insufficient information that may be caused by the detection of too simple connected geometric forms.

3.3 Seed Expansion and Local Probability Map

Given the seed locations and the edge mask, we can use such information to expand the seeds into a significant skin region. In order to do that, we apply the flood fill algorithm [43] to the binary image representing the edge mask. The algorithm recursively aggregates the seed neighboring pixels until they reach a boundary (edge or image border).

Once we have generated these regions, we use them to build a local statistical model that adapts to the particular conditions of the image. From the histogram of these resulting skin regions, we obtain a $P_{local}(c|skin)$. As for nonskin, we assume that the local distribution follows the global one, expressed in Eq. 8.

$$P_{local}(skin|c) = \frac{P_{local}(c|skin)}{P_{local}(c|skin) + P(c|\neg skin)} \quad (8)$$

Therefore, we generate a skin color probability map in the same way as the global one, but using adaptive colors.

3.4 Fuzzy Fusion

Given the texture map and the local and global skin color probability, they can be seen as different sources of information. The texture map will only account for the homogeneity of the regions, the local map is in favor of colors similar to the region found, and the global gives a more general information that can be useful, if the colors in the found region do not represent the skin in the entire image.

The goal of the information fusion is to transform the data of multiple sources into one representation form [50]. Our sources can be perceived as fuzzy membership functions. Moreover, we can apply the fuzzy integral as a fusion operator to obtain a final map representing the membership of the pixels to skin.

Fuzzy integral generalizes other common fusion operators, such as average, maximum, and order weighted average. It is characterized by the use of fuzzy membership functions ($h(x_i)$) as integrands, whereas fuzzy measures as weights and the type of applied fuzzy connectives.

Fuzzy measures serve as the a priori importance for the integrands. They define coefficients for each source, which are denoted as fuzzy densities, and also for the union of different sources characterizing the level of agreement between them. Thus, the coefficients $\mu(A_j)$ are defined for all subsets of the set of integrands (χ) in the interval [0, 1] and they need to satisfy the monotonicity condition expressed in Eq. 9.

$$A_j \subset A_k \implies \mu(A_j) \leq \mu(A_k) \quad \forall A_j, A_k \in \chi \tag{9}$$

There are two main fuzzy integrals that differentiate on the fuzzy connectives used: Sugeno Fuzzy Integral (Eq. 10) [51], which uses minimum ($\wedge$) and maximum ($\vee$), as well as Choquet Fuzzy Integral (Eq. 11) [52] that employs product and addition.

$$S_\mu[h_1(x_i), \ldots, h_n(x_n)] = \bigvee_{i=1}^{n} [h_{(i)}(x_i) \wedge \mu(A_{(i)})] \tag{10}$$

$$C_\mu[h_1(x_i), \ldots, h_n(x_n)] = \sum_{i=1}^{n} h_{(i)}(x_i)[\mu(A_{(i)}) - A_{(i-1)}] \tag{11}$$

where the enclosed subindex $_{(i)}$ refers to a previous sorting on the integrands, $h_{(1)}(x_1)$ is the source with the highest value, and $A_{(k)}$ is the subset with the k highest values, such that $A_{(n)} = \chi$. To perform our fusion process, we apply the Choquet Fuzzy Integral, as it has shown to be superior for classification purpose [50, 53].

In order to establish the fuzzy measures, we used a genetic algorithmduring a learning phase of the fuzzy method. To avoid dealing with the monotonicity condition and also narrow down the search space, we used the particular fuzzy-λ measures [54] that define the coefficients of the union of subsets based on the individual subsets, as shown in Eq. 12.

$$\mu(A_i \cup A_j) = \mu(A_i) + \mu(A_j) + \lambda\mu(A_i)\mu(A_j) \qquad \forall A_i, A_j \in \chi \tag{12}$$

where λ is found by considering that the fuzzy measure for the interaction of all sources is equal to 1 ($\mu(\chi) = 1$). Therefore, the genetic algorithm evolves only three parameters corresponding to the fuzzy densities of the texture map, the local and the global skin color probability maps, all of them in the interval of [0, 1].

To evaluate the fitness, we selected a validation dataset and applied the method by retrieving the elected information sources, conducting the fusion and evaluating the F_{score} of the resulting fuzzy map with respect to a threshold of 0.5. Equation 13 summarizes the fitness function as the mean of the result for each image I_i in the validation set, given by

$$f_{\mu}(\mu^f = [\mu_1, \mu_2, \mu_3]) = \frac{\sum_{i=1}^{N} F_{score}(GT(I_i), FzFs(T^i_{\text{map}}, M^i_{global}, M^i_{local}, \mu^f) > 0.5)}{N} \tag{13}$$

where $GT(I_i)$ is the hand labeled mask (ground truth) of image I_i and $FzFs$ is the fuzzy fusion function.

3.5 Genetic Algorithms

The genetic algorithms(GAs) applied by our work are developed with the Global Optimization Toolbox in Matlab [55]. For all cases where the GAs are employed, as shown in Fig. 1 and Algorithm 1 in Sect. 3, the same steps are performed. These basic stages are described in Algorithm 2.

Algorithm 2: Genetic Algorithm.

input : number of variables; lower and upper bounds for each variable.
output: best solution found.

1 Create a random initial population.
2 *generationsBestChanged* $\leftarrow 0$
3 **while** (*Number of Generations* $\leq$ *Generation Limit*) *and* (*generationsBestChanged* $\leq$ *Max Stall Generations*) **do**
4   Calculate the fitness of each individual.
5   Select parents in the population based on their fitness.
6   Choose the individuals in the current population with best fitness to pass to the next population.
7   Create child individuals from the chosen individuals.
8   Replace the population to one formed by the best individuals and the newly created children.
9 **return** *best individual.*

The three versions of GAs proposed have as input a set of lower and upper bounds. These bounds are used to set values for genes in each individual, since the individuals are represented as a list of real-coded values. The optimized real values achieved by the best individual is the GAs output.

Initially, a population is created randomly with values within the lower and upper bounds. Afterwards, the evolutionary steps take place with the best individuals from the previous population being passed to the next on each generation. The next population is also compounded by the new individuals created during this generation. In our approach, the next population has 20 % of the best elements of the current population, and the remaining 80 % from the best new individuals created.

New individuals are created by following two steps. In the first one, members of the current population (parents) are chosen based on their fitness. This selection for reproduction is performed by a stochastic universal sampling (SUS) technique [56]. In this technique, individuals are mapped to contiguous segments of a line where the size of the segments are proportional to the fitness of each element. Moreover, the better the fitness, the larger the segment. The individuals are sorted sequentially, based on the ranking provided by their fitness. The selection for reproduction is executed through this line by choosing the elements of the section where the selection steps on. A fixed step size is calculated from the number of parents being chosen. For example, if five parents are needed, then $step\ size = 1/5$. The initial step is a random number less than the step size.

In the second step, the new individuals are effectively created by performing crossover and mutation operations. For crossover, the technique applied is the scattered crossover, where a binary random vector with the same size as the parents is created. The new individuals will be formed by real values from the first parent, when we have 0's in the random binary vector, and with values inherited from the second parent, when we have 1's in the same vector. Figure 2 illustrates the scattered crossover. The applied mutation operator is the Gaussian, which adds a random number (within the bound limits) taken from a Gaussian distribution to each gene.

The evolutionary part of the algorithm ends either when a maximum number of generations is reached (*Generation Limit*) or the best individual of the population does not change for a number of generations (*Max Stall Generations*). At the end, the values encoded in the best individual of the population is returned. In both cases

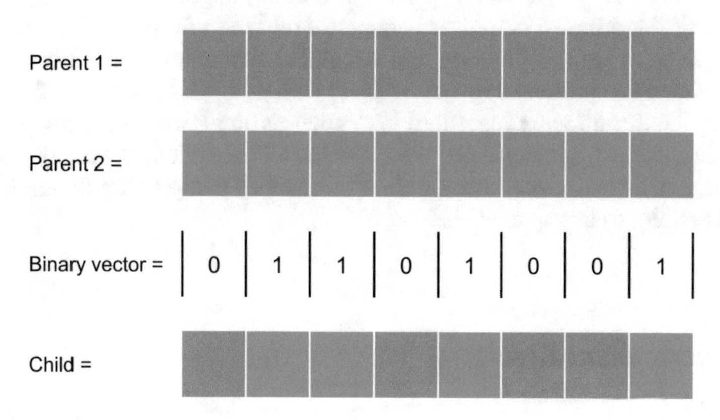

Fig. 2 Example of scattered crossover

where the GA was used in this work, the GAs parameters were empirically set as *Population Size* = 50, *Generation Limit* = 100, *Max Stall Generations* = 10, *Crossover Rate* = 0.8 and *Mutation Rate* = 0.2, based on previous tests over some images.

From the basic steps presented in Algorithm 2, the main differences among the GAs employed in Algorithm 1 are explained as follows:

(a) Genetic Algorithm 1 optimizes threshold values for texture and skin color probability maps. An individual is represented by (t_1, t_2) with $t_1, t_2 \in [0, 1]$. The fitness function employed to evolve a population with such individuals is expressed by Eq. 6 and it is calculated following the steps described in Algorithm 3. The best optimized thresholds (t_1, t_2) are returned to Algorithm 1, where t_1 is applied to set M_{global}, the skin color probability map, whereas t_2 sets T_{map}, the texture map.

Algorithm 3: Fitness calculation from Eq. 6

input : color image I
output: f_{Value}

1 **for** $x \in I$ **do**
2 **if** $M_{global}(x) \geq t_1 \wedge T_{map}(x) \leq t_2$ **then**
3 $Seeds^{[t1,t2]} \leftarrow x$

4 $f_{Value} \leftarrow f_{seed}(Seeds^{[t1,t2]})$ **return** f_{Value}

(b) Genetic Algorithm 2 optimizes threshold values for the Canny edge detectors. The individual is represented by $[\alpha^1, t_h^1, \alpha^2, t_h^2, \alpha^3, t_h^3]$ with $\alpha \in [0.01, 0.5]$, $t_h \in [0.1, 0.9]$ and $t_l^i = \alpha^i * t_h^i$ for $i = 1, 2, 3$. The fitness function employed evolves individuals from Eq. 7 following the steps in Algorithm 4. The best optimized thresholds (t_l^i, t_h^i) for $i = 1, 2, 3$ are sent to Canny edge detectors in Algorithm 1.

(c) Genetic Algorithm 3 optimizes parameters related to fuzzy densities of the texture map, as well as the local and the global skin color probability maps. An individual is represented by (μ_1, μ_2, μ_3) with $\mu_i \in [0, 1]$ for $i = 1, 2, 3$. In this case, the evolutionary algorithm is applied during a learning phase of Fuzzy Fusion method, previously to the Algorithm 1 execution. During the learning phase, μ_i is optimized over a training dataset of images. The fitness function follows Eq. 13 as explained in Sect. 3.4.

4 Experimental Results

This section presents and evaluates the computational results obtained with the proposed method. Initially, the image sets are described, then the evaluation metrics and

Algorithm 4: Fitness calculation from Eq. 7

input : color image I
$A = [\alpha^1, \alpha^2, \alpha^3]$
$T_h = [t_h^1, t_h^2, t_h^3])$
output: f_{Value}

1 $H, S, V \leftarrow rgb2hsv(I)$
2 $H_{edge} \leftarrow Canny(H, t_l^1, t_h^1)$
3 $S_{edge} \leftarrow Canny(S, t_l^2, t_h^2)$
4 $V_{edge} \leftarrow Canny(V, t_l^3, t_h^3)$
5 $Edge^{[A,T_h]} \leftarrow H_{edge} \vee S_{edge} \vee V_{edge}$
6 $f_{Value} \leftarrow f_{edge}(Edge^{[A,T_h]})$
7 **return** f_{Value}

methodology are defined. Finally, the comparative results are presented, including a statistical analysis and a qualitative assessment.

4.1 Data Sets

Three available skin detection databases were used throughout this research. All of them provide a manually annotated ground truth that makes possible to identify the pixel class (skin or nonskin) for the training and quantitatively evaluate the detection output.

The Compaq database [5] was used to train the Bayesian model, which allows us to acquire the histograms of skin and nonskin. These histograms are used as inputs to Algorithm 1. The database contains images acquired from the Internet in a variety of settings. There are 8963 nonskin images and 4666 skin images, which sum approximately 1 billion pixels. It is one of the largest databases for skin segmentation and it is extensively used in the literature. The smallest image containing skin has 38×39 pixels, the largest one has 1068×848 pixels, whereas the median has 254×266 pixels.

The ECU database [16] was used during the learning phase by Fuzzy Fusion method. The database is composed of images collected manually from the Internet and another small portion captured by researchers. The collection provides a diversity in terms of background scenes, lighting conditions, and skin types.

The third set of images comes from the HGR1 database [57] and it is composed of images registered for hand gesture recognition. A total of 12 different individuals placed in uncontrolled background and lighting conditions is featured. The images vary from 174×131 to 640×480 and present mostly hands and arms representing the skin regions. The proposed method is validated over these 899 images from HGR1. Figure 3 illustrates some images from these databases.

(a) Compaq (b) ECU (c) HGR1

Fig. 3 Image samples from databases used in our experiments

4.2 Evaluation Settings

The performance of the skin segmentation process was evaluated primarily by the F_{score}, expressed in Eq. 14, which is the harmonic mean of Precision and Recall

$$F_{score} = 2\,\frac{\text{Precision} \times \text{Recall}}{\text{Precision} + \text{Recall}} \tag{14}$$

where Precision and Recall for a classified image I and the expected classification G are defined in Eqs. 15 and 16, respectively.

$$\text{Precision} = \frac{\sum\limits_{i \in I} I(i) \in Skin \wedge G(i) \in Skin}{\sum_{i \in I} I(i) \in Skin} \tag{15}$$

$$\text{Recall} = \frac{\sum\limits_{i \in I} I(i) \in Skin \wedge G(i) \in Skin}{\sum_{j \in G} G(j) \in Skin} \tag{16}$$

Precision corresponds to the percentage of pixels correctly classified as skin out of all the pixels classified as skin, whereas Recall is the percentage of the pixels correctly classified as skin out of all the pixels expected to be skin.

In order to analyze our method and validate its improvements, we compare it against (i) the standard Bayesian approach to skin segmentation [5] (Bayesian), which considers only the skin color global probability, (ii) the previous unoptimized version [13] (SASS - Self Adaptive Skin Segmentation), (iii) the proposed improvement approach using genetic algorithms (SASS+GA), (iv) the use of fuzzy fusion (SASS+Fuzzy), and (v) the combined method (SASS+GA+Fuzzy). They are all evaluated by using the same threshold set to 0.5 and, when the fuzzy fusion is not presented, the simple average is used for combination, except for the Bayesian approach that outputs only one result.

4.3 Statistical Analysis

We evaluate the F_{score} of each image from the test set of 899 images for all methods. Table 2 shows the average result for each evaluated method. The score values increase according to the proposed improvements. The best F_{score} is achieved when genetic algorithms and fuzzy method are added to SASS.

A statistical analysis is performed to assess the significance of the obtained results and to perform a more systematic comparison among the methods. If there is such a difference, the analysis also reports which methods present the best results. The type of statistical test used in the comparisons is defined according to the fulfillment of certain criteria. If all criteria are met, a parametric test is used, otherwise a non-parametric is applied.

The first criterion is to determine whether the variances of the samples (results) for the different methods are uniform. The assessment used in this case is the Levene's test [58], whose results (Table 3) show that, since the p-value is less than α, we can reject the null hypothesis that the variances are identical.

Since the criterion for homogeneity of variances between the samples was not met, a parametric test should not be used. Hence, the Friedman test [59], which considers paired comparisons of the same instances (results for a same image), was used to determine if there were significant differences among the results obtained with the evaluated methods. To determine which methods presented the best results, a post-hoc analysis was conducted with the Nemenyi's test [59] for paired comparisons.

As it can be observed from Table 4, there are significant differences among the results obtained with the different evaluated methods (p-value $< \alpha$). The pairwise comparisons among the methods is conducted with the Nemenyi's test (Table 5), which groups the methods with respect to their performance. Methods in the Group

Table 2 Segmentation results for all evaluated methods

Method	F_{score} (%)
Bayesian	77.29
SASS	59.45
SASS + Fuzzy	78.07
SASS + GA	83.76
SASS + GA + Fuzzy	**84.90**

Table 3 Levene's test for the results obtained with different methods

F (observed)	F (critical)	DF	p-value	α
17.3856	2.3779	4	<0.0001	0.05

Table 4 Friedman's test for the results of the different methods

Q (observed)	Q (critical)	DF	p-value	α
1924.3145	9.4877	4	<0.0001	0.05

Table 5 Nemenyi's test for paired comparisons

Method	Groups				
SASS + GA + Fuzzy	A				
SASS + GA		B			
SASS + Fuzzy			C		
Bayesian				D	
SASS					E

Table 6 Number of times each method presented the best F_{score} value for each image

Method	Wins
SASS + GA + Fuzzy	613
SASS + GA	88
SASS + Fuzzy	111
Bayesian	86

A achieved better results than the ones in Group B, and so on. Methods within the same group have no significant difference to each other.

Based on the results for the Nemenyi's test, we can state that SASS combined with GA and Fuzzy presented the best results in comparison to all the other methods. Furthermore, SASS combined only with GA achieved results that outperformed the other three methods under evaluation, showing the improvement given by the evolutionary approach when optimizing parameters for the skin segmentation method.

A second type of comparison takes into account the absolute number of times where each method outperformed the others. There is a tie if two methods reach the same results with $|F_{score}^{method\ 1} - F_{score}^{method\ 2}| < 0.0001$. In this case, a win is computed for both methods. For the 899 images, SASS combined with GA and Fuzzy presented the best results 613 times (68 %), SASS combined only with GA was the best method on 88 images (10 %), whereas SASS combined with Fuzzy method was better than the others 111 times (12 %). Moreover, SASS was the best method for 3 images (less than 1 %) and the Bayesian approach had 86 best results (10 %). These results show that the use of both GA and Fuzzy clearly provides a quality increase for the SASS method. Table 6 summarizes all the previously mentioned results.

4.4 Qualitative Analysis

In order to perform a qualitative analysis of the results, Fig. 4 illustrates the output from the evaluated methods on four different image samples from the test set. By comparing the segmentation obtained through the method with and without fuzzy, it is possible to infer that the use of fuzzy fusion operates mainly on the removal of false positive error. This is more evident on the images in Rows 2 and 3, where the undesired background is removed.

The use of genetic algorithms clearly produced better results on the featured images. Even for the image shown in Row 1, where the segmentation is poor, the result is superior compared to the other methods. This segmentation improvement can be explained by the optimization of the intermediary step. Figure 5 shows the resultant detection of edges and seeds based on the threshold obtained with original SASS [13] and with the method proposed in this work.

The seeds found with the SASS method are considered good as they are placed in true skin regions; however, the approach identified too many edges which prevent the flood fill to detect consistent regions. On the other hand, the edges and seeds found with SASS combined with genetic algorithms were sufficiently accurate to provide a good segmentation.

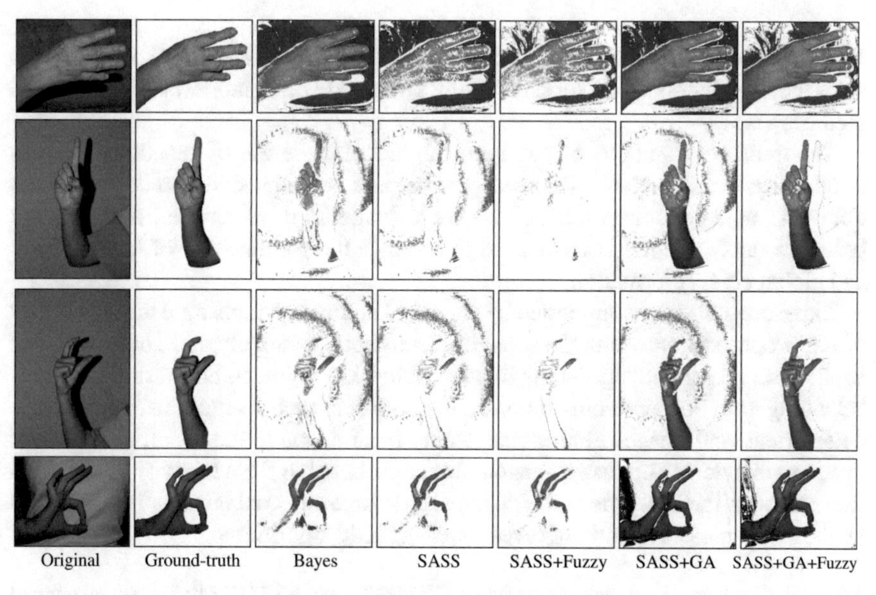

Fig. 4 Examples of skin segmentation obtained with the different methods evaluated in our experiments

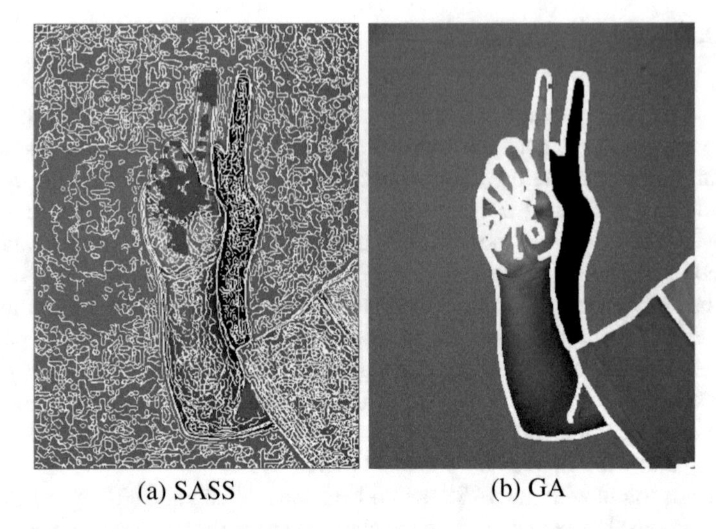

Fig. 5 Edges and seeds obtained with original SASS method [13] and with the use of genetic algorithm. Edges are shown in white and seeds in *red*

5 Conclusions

This work extends upon a self-adaptive human skin segmentation method that eliminates the need for object detection by using a local skin color model obtained through seed growing.

The main contributions of our approach includes the use of genetic algorithms to optimize simultaneously the seed growing and edge detection processes, where different parameters are evaluated for each image. Furthermore, a fusion method based on fuzzy integral is introduced to enhance the combination of texture, local and global color information.

Experimental results supported by statistical analysis demonstrate the superiority of the proposed improvements when executed both separately and combined. The application of genetic algorithms improved the skin segmentation results, counterbalancing common problems found in the data sets such as illumination changes, noise effect, and images at different scaling factors, due to the adaptive search for image parameters. The fuzzy integral also contributed to the adaptivity of the algorithm through the best type of fusion between source and goal information, achieved by the fuzzy measures and improved by the genetic algorithms.

Acknowledgments The authors are thankful to FAPESP (grant #2011/22749-8) and CNPq (grant #307113/2012-4) for their financial support.

References

1. Hu, X., Peng, S., Yan, J., Zhang, N.: Fast face detection based on skin color segmentation using single Chrominance Cr. In: 7th International Congress on Image and Signal Processing, pp. 687–692. IEEE (2014)
2. Ji, S., Lu, X., Xu, Q.: A fast face detection method combining skin color feature and adaboost. In: International Conference on Multisensor Fusion and Information Integration for Intelligent Systems, pp. 1–5. IEEE (2014)
3. Palacios, J.M., Sagüés, C., Montijano, E., Llorente, S.: Human-computer interaction based on hand gestures using RGB-D sensors. Sensors **13**(9), 11842–11860 (2013)
4. Wachs, J.P., Kölsch, M., Stern, H., Edan, Y.: Vision-based hand-gesture applications. Commun. ACM **54**(2), 60–71 (2011). Feb
5. Jones, M.J., Rehg, J.M.: Statistical color models with application to skin detection. Int. J. Comput. Vis. **46**(1), 81–96 (2002)
6. Vo, D.M., Jiang, L., Zell, A.: Real time person detection and tracking by mobile robots using RGB-D images. In: IEEE International Conference on Robotics and Biomimetics, pp. 689–694. IEEE (2014)
7. Jeong, C.-Y., Kim, J.-S., Hong, K.-S.: Appearance-based nude image detection. In: 17th International Conference on Pattern Recognition, vol. 4, pp. 467–470. IEEE (2004)
8. Platzer, C., Stuetz, M., Lindorfer, M.: Skin sheriff: a machine learning solution for detecting explicit images. In: 2nd International Workshop on Security and Forensics in Communication Systems, pp. 45–56. ACM, New York (2014)
9. Acton, S.T., Rossi, A.: Matching and retrieval of tattoo images: active contour CBIR and glocal image features. In: IEEE Southwest Symposium on Image Analysis and Interpretation, pp. 21–24. IEEE (2008)
10. Choraś, R.S.: CBIR System for detecting and blocking adult images. In: 9th WSEAS International Conference on Signal Processing, pp. 52–57. World Scientific and Engineering Academy and Society, Stevens Point (2010)
11. Manresa-Yee, C., Varona, J., Mas, R., Perales, F.J.: Hand tracking and gesture recognition for human-computer interaction. Progress In Computer Vision And Image, Analysis, pp. 401–412 (2010)
12. Ren, Z., Meng, J., Yuan, J.: Depth camera based hand gesture recognition and its applications in human-computer-interaction. In: 8th International Conference on Information, Communications and Signal Processing, pp. 1–5. IEEE (2011)
13. Santos, A., Pedrini, H.: A Self-adaptation method for human skin segmentation based on seed growing. In: 10th International Conference on Computer Vision Theory and Applications, pp. 455–462. Berlin (2015)
14. Kakumanu, P., Makrogiannis, S., Bourbakis, N.: A survey of skin-color modeling and detection methods. Pattern Recognit. **40**(3), 1106–1122 (2007)
15. Kawulok, M., Nalepa, J., Kawulok, J.: Skin detection and segmentation in color images. Advances in Low-Level Color Image Processing, pp. 329–366. Springer, Berlin (2014)
16. Phung, S.L., Bouzerdoum, A., Chai, D.: Skin segmentation using color pixel classification: analysis and comparison. IEEE Trans. Pattern Anal. Mach. Intell. **27**(1), 148–154 (2005)
17. Zarit, B.D., Super, B.J., Quek, F.K.: Comparison of five color models in skin pixel classification. In: International Workshop on Recognition, Analysis, and Tracking of Faces and Gestures in Real-Time Systems, pp. 58–63 (1999)
18. Sobottka, K., Pitas, I.: Face localization and facial feature extraction based on shape and color information. In: International Conference on Image Processing, vol. 3, pp. 483–486. IEEE (1996)
19. Cheddad, A., Condell, J., Curran, K., Mc Kevitt, P.: A skin tone detection algorithm for an adaptive approach to steganography. Signal Process. **89**(12), 2465–2478 (2009)
20. Hsu, R.-L., Abdel-Mottaleb, M., Jain, A.K.: Face detection in color images. IEEE Trans. Pattern Anal. Mach. Intell. **24**(5), 696–706 (2002)

21. Soriano, M., Martinkauppi, B., Huovinen, S., Laaksonen, M.: Skin detection in video under changing illumination conditions. In *15th International Conference on Pattern Recognition*, vol. 1, pp. 839–842. IEEE (2000)
22. Kovac, J., Peer, P., Solina, F.: Human skin color clustering for face detection. In: International Conference on Computer as a Tool. vol. 2, pp. 144–148. IEEE (2003)
23. Phung, S.L., Chai, D., Bouzerdoum, A.: Adaptive skin segmentation in color images. In: International Conference on Multimedia and Expo, vol. 3, pp. 111–173 (2003)
24. Fritsch, J., Lang, S., Kleinehagenbrock, M., Fink, G.A., Sagerer, G.: Improving adaptive skin color segmentation by incorporating results from face detection. In: 11th IEEE International Workshop on Robot and Human Interactive Communication, pp. 337–343 (2002)
25. Taylor, M.J., Morris, T.: Adaptive skin segmentation via feature-based face detection. In: SPIE Photonics Europe, p. 91390P. International Society for Optics and Photonics (2014)
26. Kawulok, M.: Energy-based blob analysis for improving precision of skin segmentation. Multimed. Tools Appl. **49**(3), 463–481 (2010)
27. Kawulok, M.: Fast propagation-based skin regions segmentation in color images. In: 10th IEEE International Conference and Workshops on Automatic Face and Gesture Recognition, pp. 1–7 (2013)
28. Ruiz-del Solar, J., Verschae, R.: Skin Detection using Neighborhood Information. In: Sixth IEEE International Conference on Automatic Face and Gesture Recognition, pp. 463–468. IEEE (2004)
29. Dijkstra, E.W.: A note on two problems in connection with graphs. Numer. Math. **1**, 269–271 (1959)
30. Wang, X., Zhang, X., Yao, J.: Skin color detection under complex background. In: International Conference on Mechatronic Science, Electric Engineering and Computer, pp. 1985–1988 (2011)
31. Gonzalez, R., Woods, R., Eddins, S.: Digital Image Processing using MATLAB. Gatesmark Publishing, Knoxville (2009)
32. Schwartz, W.R., Pedrini, H.: Color textured image segmentation based on spatial dependence using 3D co-occurrence matrices and Markov random fields. 15th International Conference in Central Europe on Computer Graphics. Visualization and Computer Vision, pp. 81–87. Czech Republic (2007)
33. Ng, P., Chi-Man, P.: Skin color segmentation by texture feature extraction and K-means clustering. In: Third International Conference on Computational Intelligence, Communication Systems and Networks, pp. 213–218. IEEE (2011)
34. Jiang, Z., Yao, M., Jiang, W.: Skin detection using color, texture and space information. Fourth Int. Conf. Fuzzy Syst. Knowl. Discov. **3**, 366–370 (2007)
35. Gupta, V., Chan, C.C., Sian, P.T.: A differential evolution approach to PET image de-noising. In: 29th Annual International Conference of the IEEE Engineering in Medicine and Biology Society, pp. 4173–4176 (2007)
36. Thavavel, V., Basha, J.J., Krishna, M., Murugesan, R.: Heuristic wavelet approach for low-dose EPR tomographic reconstruction: an applicability analysis with phantom and in vivo imaging. Expert Syst. Appl. **39**(5), 5717–5726 (2012)
37. Mukhopadhyay, S., Mandal, J.: Wavelet based denoising of medical images using sub-band adaptive thresholding through genetic algorithm. Procedia Technol. **10**, 680–689 (2013) (First International Conference on Computational Intelligence: Modeling Techniques and Applications)
38. Xie, F., Bovik, A.C.: Automatic segmentation of dermoscopy images using self-generating neural networks seeded by genetic algorithm. Pattern Recognit. **46**(3), 1012–1019 (2013)
39. Razmjooy, N., Mousavi, B.S., Soleymani, F.: A hybrid neural network imperialist competitive algorithm for skin color segmentation. Math. Comput. Model. **57**(3–4), 848–856 (2013)
40. Chahir, Y., Elmoataz, A.: Skin-color detection using fuzzy clustering. Int. Symp. Commun. Control Signal Process. **3**(1), 1–4 (2006)
41. Hmida, M.B., Jemaa, Y.B.: Fuzzy classification, image segmentation and shape analysis for human face detection. In: 13th IEEE International Conference on Electronics, Circuits and Systems, pp. 640–643. IEEE (2006)

42. Kim, M.H., Park, J.B., Joo, Y.H.: New fuzzy skin model for face detection. Advances in Artificial Intelligence, pp. 557–566. Springer, Berlin (2005)
43. Soille, P.: Morphological Image Analysis: Principles and Applications. Springer Science & Business Media, Berlin (2013)
44. Laws, K.I.: Rapid texture identification. In: 24th Annual Technical Symposium, International Society for Optics and Photonics, pp. 376–381 (1980)
45. Santos, A., Pedrini, H.: Human skin segmentation improved by texture energy under superpixels. In: Pardo, A., Kittler, J. (eds.) Progress in Pattern Recognition, Image Analysis, Computer Vision, and Applications. Volume 9423 of Lecture Notes in Computer Science, pp. 35–42. Springer International Publishing, Berlin (2015)
46. Pratt, W.K.: Digital Image Processing. Wiley-Interscience, New York (2001)
47. Kitchen, L., Rosenfeld, A.: Edge evaluation using local edge coherence. IEEE Trans. Syst. Man Cybern. **11**(9), 597–605 (1981)
48. Zhu, Q.: Efficient evaluations of edge connectivity and width uniformity. Image Vis. Comput. **14**(1), 21–34 (1996) (Image and Vision Computing Journal on Vision-Based Aids for the Disabled)
49. Tao, C., Xiankun, S., Hua, H., Xiaoming, Y.: Image edge detection based on ACO-PSO algorithm. Int. J. Adv. Comput. Sci. Appl. **6**(7), 47–54 (2015)
50. Soria-Frisch, A.: Soft Data Fusion for Computer Vision. Fraunhofer-IRB-Verlag (2004)
51. Murofushi, T., Sugeno, M.: Fuzzy measures and fuzzy integrals. In: Grabisch, M., Murofushi, T., Sugeno, M. (eds.) Fuzzy Measures and Integrals - Theory and Applications, pp. 3–41. Physica Verlag, Heidelberg (2000)
52. Murofushi, T., Sugeno, M.: An interpretation of fuzzy measures and the Choquet integral as an integral with respect to a fuzzy measure. Fuzzy Sets Syst. **29**(2), 201–227 (1989)
53. Soria-Frisch, A., Verschae, R., Olano, A.: Fuzzy fusion for skin detection. Fuzzy Sets Syst. **158**(3), 325–336 (2007)
54. Tahani, H., Keller, J.M.: Information fusion in computer vision using the fuzzy integral. IEEE Trans. Syst. Man Cybern. **20**(3), 733–741 (1990)
55. Global Optimization Toolbox. http://www.mathworks.com/products/global-optimization/ (2016). Accessed 24 Feb 2016
56. Baker, J.E.: Reducing bias and inefficiency in the selection algorithm. In: Second International Conference on Genetic Algorithms on Genetic Algorithms and Their Application, pp. 14–21. L. Erlbaum Associates Inc., Hillsdale (1987)
57. Kawulok, M., Kawulok, J., Nalepa, J.: Spatial-based skin detection using discriminative skin-presence features. Pattern Recognit. Lett. **41**, 3–13 (2014)
58. Sheskin, D.J.: Handbook of Parametric and Nonparametric Statistical Procedures. CRC Press, Boca Raton (2003)
59. Demšar, J.: Statistical comparisons of classifiers over multiple data sets. J. Mach. Learn. Res. **7**, 1–30 (2006)

Uncertainty-Based Spatial Data Clustering Algorithms for Image Segmentation

Deepthi P. Hudedagaddi and B.K. Tripathy

Abstract Data clustering has been an integral and important part of data mining. It has wide applications in database anonymization, decision making, image processing and pattern recognition, medical diagnosis, and geographical information systems, only to name a few. Data in real-life scenario are having imprecision inherent in them. So, early crisp clustering techniques are very less efficient. Several imprecision-based models have been proposed over the years. Of late, it has been established that the hybrid models obtained as combination of these imprecise models are far more efficient than the individual ones. Several clustering algorithms have been put forth using these hybrid models. It is also found that conventional fuzzy clustering algorithms fail in incorporating the spatial information. This chapter focuses on discussing some of the spatial data clustering algorithms developed so far and their applications mainly in the area of image segmentation.

Keywords Clustering · Uncertainty · Image segmentation · Spatial data

1 Introduction

Enormous data is created everyday and is being available across several resources. It is manually impossible to explore this ocean of data and choose data relevant to a specific task. Therefore, a mechanism to group this data was required. It was then that clustering was found to be the best technique. Man has been doing clustering intuitively. It is been several years that clustering has evolved to be one of the most easy and reliable technique. Several clustering algorithms as and when required for a particular application are developed in literature.

Clustering has always had a major role in applications of data mining. It is basically a grouping technique. The elements in same cluster are highly identical and possess

D.P. Hudedagaddi (✉) · B.K. Tripathy
SCOPE, VIT University, Vellore 632014, Tamil Nadu, India
e-mail: deepthiph@gmail.com

B.K. Tripathy
e-mail: tripathybk@vit.ac.in

S. Bhattacharyya et al. (eds.), *Hybrid Soft Computing for Image Segmentation*, DOI 10.1007/978-3-319-47223-2_9

Fig. 1 Clustering stages

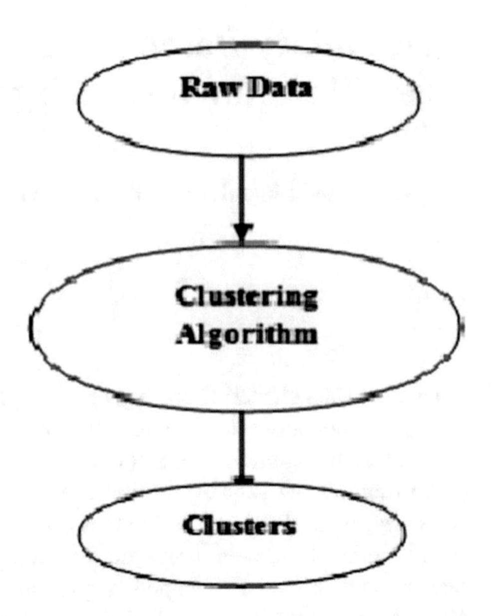

the same features. But the elements are highly unidentical with those in the other cluster [2]. A clustering technique is considered to be the best if it is able to find out most or all of the patterns hidden. It should produce clusters that are extremely similar within the cluster and dissimilar with other clusters. Usually, distance function is being used to measure cluster similarity. Data mining possesses challenges on data such as dealing with noise, interpretability, scalability, dimensionality, and dealing with different data attributes and many more. Clustering has been applied on data that has interval scaled, binary, nominal, ordinal, ratio, and mixed type variables [1]. Clustering can be mainly of five types. They are center-based, contiguous, density-based, shared property-based and well-separated clusters. Data is well described by its attributes. Attributes can be huge for large dimensional data [13]. Clustering has built a reputation of its own in the fields of information retrieval, image segmentation, and web data mining (Fig. 1).

2 Based Models

Sets have always been thought to be a notion of mathematics. However, the lesser known fact is that it has its own role to play in natural language. Sets are usually considered to be collection of related things, like paintings, books, etc. However, in real world, the data is more vague and imprecise. To handle such kind of data, the concept of sets was modified. This led to the development of uncertainty-based models as there was a need to handle this uncertain data.

2.1 *Fuzzy Sets*

Zadeh [24] came up with theory of fuzzy sets. This was the best way to handle vagueness. He introduced the membership function in which an element could belong to a certain set to a certain degree. The range of it's membership is 0 to 1.This was in complete contrast to conventional set theory where an element either belonged or did not belong to the set.

2.2 *Rough Sets*

Pawlak's theory of rough sets [17, 18] came as another alternative approach to vagueness. Rough set theory is a precise implementation of vagueness, in which imprecision is given by boundary region of a set. Membership is not a primary function in this theory. This theory provides a mathematical view to vagueness and imprecision. In rough sets, a set is defined by the perception and knowledge about it in the universe. Hence, two dissimilar elements can be indiscernible with respect to the knowledge, information and perception.

2.3 *Hybrid Models*

The fusion of both rough and fuzzy sets along with intuitionistic fuzzy sets has been successful in giving better results than individual theories. Several hybrid clustering algorithms are developed and have found their applications in several fields.

3 Uncertainty-Based Data Clustering Algorithms

Many clustering algorithms are developed keeping in consideration the uncertainty involved in databases. One of the first developed algorithm was Hard C-Means (HCM) [21]. Attanasov developed intuitionistic fuzzy sets. Dubois and Prade introduced rough fuzzy sets [12], whereas Saleha et al. [19] introduced rough intuitionistic fuzzy sets. Development of HCM and making modifications to it to incorporate uncertainty led to introduction of uncertainty-based clustering algorithms. Rough C-Means (RCM) [14], Rough Fuzzy C-Means (RFCM) [7, 13, 18, 21], Fuzzy C-Means (FCM) [14], and Intuitionistic fuzzy c-Means (IFCM) [8] are few of them. The hybrid versions of fuzzy and intuitionistic fuzzy along with rough sets incorporates the concepts of graded and non-graded membership and hence uncertainty is taken care by the boundary regions for all the elements in the cluster. The approximations of rough fuzzy sets handle data uncertainty, while the detection accuracy is improved by intuitionistic fuzzy sets.

3.1 Fuzzy C-Means Algorithm (FCM)

Bezdek developed FCM. This algorithm allows data items to belong to multiple clusters. Every data item is given by a membership value.

1. Initialise c cluster centers.
2. Compute the Euclidean distance d_{ik} between data items x_k and centroids v_i using

$$d(x, y) = \sqrt{(x_1 - y_1)^2 + (x_2 - y_2)^2 + \cdots + (x_n - y_n)^2} \tag{1}$$

3. Create partition matrix U: If $d_{ij} > 0$ then

$$\mu_{ik} = \frac{1}{\sum_{j=1}^{C} \left(\frac{d_{ik}}{d_{jk}}\right)^{\frac{2}{m-1}}} \tag{2}$$

Else

$$\mu_{ik} = 1$$

4. The cluster centroids are computed using

$$\mathrm{V_i} = \frac{\sum_{\mathrm{j=1}}^{\mathrm{N}} (\mu_{\mathrm{ij}})^{\mathrm{m}} \mathrm{x_j}}{\sum_{\mathrm{j=1}}^{\mathrm{N}} (\mu_{\mathrm{ij}})^{\mathrm{m}}} \tag{3}$$

5. Recompute new U using step 2 and 3.
6. If $\left\| U^{(r)} - U^{(r+1)} \right\| < \varepsilon$ then stop. If not, start process from step 4.

3.2 RFCM

RFCM algorithm is developed using a combination of rough set and fuzzy set by Mitra and Maji [9, 13] and it handles uncertainty and vagueness efficiently. The membership of elements in the overlapping regions is the highlight of this algorithm [4].

1. Initialise c cluster centers.
2. Calcluate μ_{ik} using Eq. (2).
3. If μ_{ik} and μ_{jk} are the two lowest memberships of an element y_k in that order and the difference between these two is less than ε then y_k is put in both $\overline{\mathrm{B}}\mathrm{U_i}$ and $\overline{\mathrm{B}}\mathrm{U_j}$ else y_k is put in $\underline{\mathrm{B}}\mathrm{U_i}$.

4. The fresh centers are computed.
5. Steps from 2 onwards are repeated till termination condition is satisfied.

3.3 IFCM

This was introduced by T. Chaira and is based upon the hesitation component of IFS.

1. Initialise cluster centers.
2. Calculate Euclidean distance d_{ik} between data items y_i and cluster centers w_k.
3. Compute membership matrix $V = \mu_{ik}$ such that $\mu_{ik} = 1$ if $d_{ik} = 0$; else μ_{ik} is computed using Eq. (2).
4. Compute the hesitation matrix π.
5. Calculate the modified membership matrix V'.

$$\mu'_{ik} = \mu_{ik} + \pi_{ik} \tag{4}$$

6. Calculate cluster centroids using

$$V_i = \frac{\sum_{j=1}^{N} (\mu'_{ij})^m y_j}{\sum_{j=1}^{N} (\mu'_{ij})} \tag{5}$$

7. Repeat steps from 2 to 5 for computing new partition matrix.
8. Stop if $\left\| V'^{(r)} - V'^{(r+1)} \right\| < \varepsilon$ else repeat steps 4–8.

3.4 RIFCM [5]

1. Initialize cluster centers.
2. Calculate Euclidean distance d_{ik} between data items y_i and cluster centers w_k.
3. Compute membership matrix $V = \mu_{ik}$ as in step 3 of IFCM algorithm.
4. Compute π_{ik}.
5. Compute μ'_{ik} and normalize

$$\mu'_{ik} = \mu_{ik} + \pi_{ik}.$$

6. Same as step 3 in RFCM algorithm.
7. New centers are computed.
8. Same as step 5 in RFCM algorithm.

Process repeats from step 2 either till condition for termination is met or till all the objects are assigned [16].

4 Image Segmentation

Segmentation of images has been the need of the hour specially in major fields like medical diagnosis and space related images. Getting quality and realistic segments of the image is extremely challenging and is also very important. Different perspectives of images can be got by different segments of images. Magnetic resonance imaging (MRI) is the technique which analyzes the alterations in the volume, shape and distribution of tissues occurred due to several neurological conditions. Image segmentation is a way to obtain reliable measurements of these variations. Researchers have found ways for automatic analysis of these segments. But many of these techniques fail to use the spatial information of MRI signals.

5 Spatial Data Clustering Algorithms in Image Segmentation

Clustering is mainly required for studying of data. Hierarchical and partitioning clustering are two major classifications of clustering. Partitioning clustering does not allow overlapping in clusters. K-means and self-organizing maps are some of the partitioning clustering techniques. Hierarchical clustering, has a tree like structure with several nested partitions in sequence. Ward's method, single, average and complete linkage clustering are some of the hierarchical clustering methods.

Three types of clustering are studied to incorporate spatial information. They are regionalization, spatial, and point pattern analysis. In spatial clusters, spatial properties like distances and locations are used to define the similarity in clusters. Spatial clustering methods could either be hierarchical or partitioning, or grid or density based. Regionalization technique involves optimizing the objective function while grouping spatial objects into regions. Several geographic applications like analysis of landscape, zoning of climate, and remote sensing image segmentation require clusters to be in serial. Hot-spot or point pattern analysis lays its focus on detecting unusual concentrations of events in space. Geographic analysis machine (GAM) is a scan statistic tool to determine spatial clusters.

GAM uses the number of points in a specified area as a test statistic. A Monte Carlo technique can be followed for determining if the number of points in the specified area is significant. GAM is unsuitable for multiple-testing problem. The computational workload also adds as a disadvantage. However, to find and examine local clusters, all scan statistics need a sufficient amount of computational power.

Spatial clustering and analysis of these clusters play a vital role in quantifying patterns of geographic variations. Few of the major applications include surveillance of diseases, population genetics, analysis of crime, spatial epidemiology, landscape ecology, and many other fields. Spatial data mining algorithms largely rely on processing of neighborhood relations since the neighbors of many objects have to be investigated in a single run of a distinctive algorithm [15].

Chuang et al. [7] explained that a traditional FCM algorithm does not utilize the image's spatial information. They modified present FCM algorithm and developed spatial FCM (sFCM). This technique yields regions that are more similar. It is successful in removing noisy spots. Hence, it is a strong technique for noisy image segmentation. On similar lines, spatial IFCM was also developed by Tripathy et al. by introducing the intuitionsitic feature in the objective function [20]. They have proved the algorithm to be better by giving results using DB and D indices.

5.1 Spatial Information

In images, the neighboring pixels have high degree of correlation among the neighboring pixels [6]. These exhibit similar characteristic values, and hence the probability that they belong to the same cluster is high. This relationship has not been considered in earlier methods. A spatial function exploits the spatial information:

$h_{ij} = \sum_{k \varepsilon NB(x_j)} u_{ik}$

where $NB(x_j)$ is the pixels in neighborhood of x_j. Spatial function h_{ij} provides likeliness degree that x_j is in i^{th} cluster. Spatial function value of a pixel will be higher if the neighborhood pixels of the cluster in considerations also belong to the same cluster. The membership function incorporates it as

$$u_{ij}^{'} = \frac{u_{ij}^{p} h_{ij}^{q}}{\sum_{k=1}^{c} u_{kj}^{p} h_{kj}^{q}} \tag{6}$$

p denotes weightage of the initial membership and q gives spatial function. If image is noisy, the spatial function brings down the number of misclassified pixels by taking neighboring pixels into account.

5.1.1 Fuzzy Clustering for Image Thresholding with Spatial Constraints [23]

Yang et al. proposed a spatially weighted fuzzy C-means (SWFCM) which incorporated spatial neighboring details into conventional FCM. They had realized that ambiguity and indistinguishable histogram are an issue in segmenting real world images, and hence provided a solution fixing these issues. The principle of their technique is to involve the neighborhood information. Considering influence of neighboring pixels on central pixel, they extended fuzzy membership function to:

$u*_{ik} = u_{ik} p_{ik}$ where k = 1, 2, ..., n (n gives number of image data); p_{ik} is spatial constraint or weight. Thus, objective function of SWFCM is given by

$$J_q\left(U^*, V^*\right) = \sum_{k=1}^{n} \sum_{i=1}^{c} (u_{ik}^*)^q d^2\left(x_k, v_i^*\right) \tag{7}$$

Like FCM, membership degree u* and cluster centers v_i were updated.

$$(u_{ik}^*)^b = \frac{p_{ik}}{\sum_{j=1}^{c} (\frac{d_{ik}}{d_{jk}})^{2/(q-1)}} \quad (8)$$

and

$$v_i^{*(b+1)} = \frac{\sum_{k=1}^{n} [u_{ik}^{*\,(b)}]^q \; x_k}{\sum_{k=1}^{n} [u_{ik}^{*\,(b)}]^q} \quad (9)$$

The motto here was to come up with p_{ik}: the weight variable, which is a priori information to monitor process of clustering.

$$p_{ik} = \frac{\sum_{x_n \in N_k^i} 1/d^2(x_n, k)}{\sum_{x_n \in N_k} 1/d^2(x_n, k)} \quad (10)$$

Here N_k^i denotes the whole data set in class I and N_k represents a subset of having N_k^i the neighbors of the kth pixel. Once the computation of weights is over the variable $p_i k$ is used to initiate a fresh iteration. A fast FCM and the representatives and memberships obtained from it are used to avoid any premature stoppage of SWFCM, which is executed after that of FCM gets over. After that the maximum membership defuzzification technique is used in order to get the crisp partition matrix. By this, an object gets classified with the largest membership to its class. This method was called soft thresholding scheme (Fig. 2).

5.1.2 FCM for Segmentation of Image Having Spatial Information [7]

A segmentation technique with reduced noise and providing homogeneous clusters was proposed by Chuang et al. This method utilizes the spatial information and cluster weighting by altering these from the distribution of cluster information in neighborhood.

The characteristic that pixels in a single neighborhood are similar and they have better chance of being put in the same cluster is high is used by Chuang et al.

It is a two-pass process where the first pass computes the memberships as in FCM. The next pass these memberships are transformed to spatial domain leading to the computation of spatial functions. Again FCM is used for the new outputs. Taking the threshold value as 0.02, it is checked if the maximum difference between two cluster centers is less than it. If so, it is stopped and defuzzification is performed.

The evaluation of the results is done through validity functions like the entropy V_{pe} and coefficient V_{pc} of the partitioning. These validity functions indicate that in the new approach the performance is better. The optimal clustering is obtained for highest V_{pe} and lowest V_{pc}.

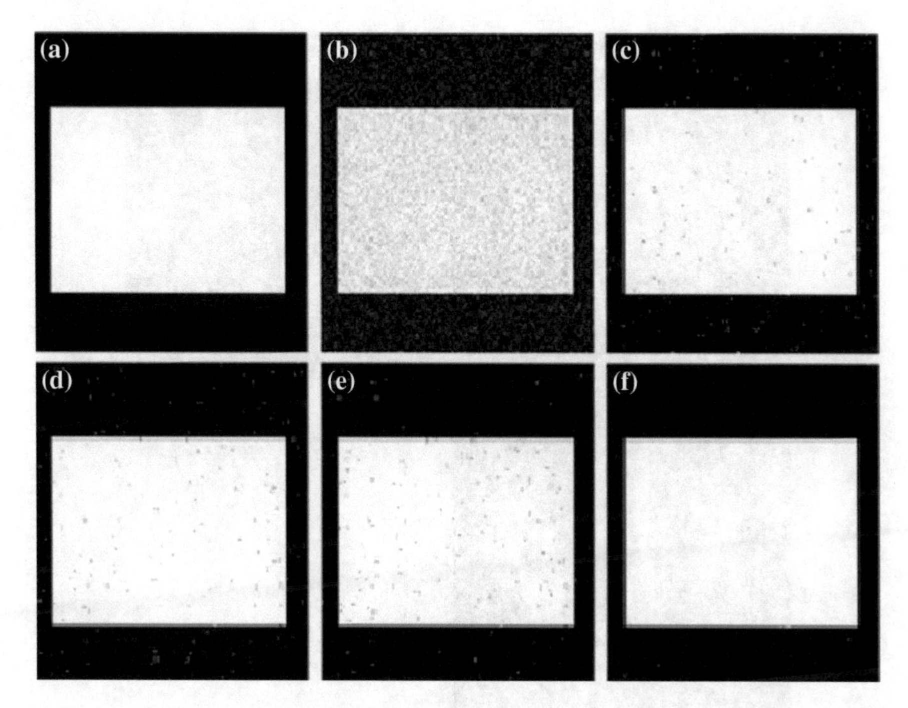

Fig. 2 Thresholding results: **a** original image **b** noisy image with SNR = 5 **c** Jawahar's method **d** Otsu's method **e** Kapur's method **f** Yang's method

$$V_{pc} = \frac{\sum_j^N \sum_i^c u_{ij}^2}{N} \tag{11}$$

and

$$V_{pe} = -\frac{\sum_j^N \sum_i^c [u_{ij} \log u_{ij}]}{N} \tag{12}$$

A good clustering generates entities that are compact within one cluster and entities that are separated between different clusters. Minimum V_{xb} indicates good clustering (Fig. 3).

$$V_{xb} = \frac{-\sum_j^N \sum_i^c u_{ij} \vee x_j - v_i \vee^2}{N * \left(min_{i \neq k} \left\{ ||v_k - v_i||^2 \right\}\right)} \tag{13}$$

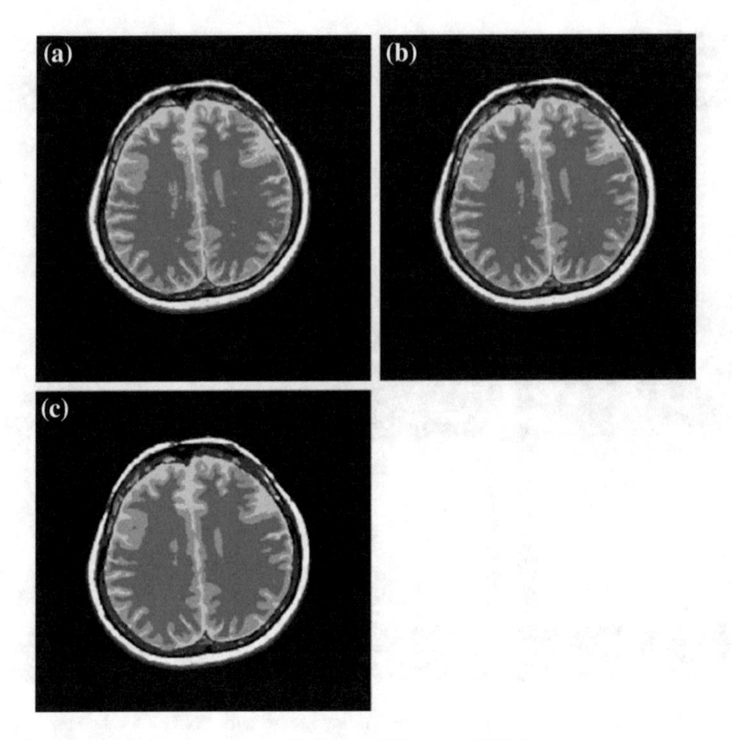

Fig. 3 MRI segmented images using **a** FCM; **b** $\text{sFCM}_{1,1}$; **c** $\text{sFCM}_{0,2}$

5.1.3 Fast and Robust Image Segmentation with Spatial Information Using [22]

Xiang et al. developed a modified FCM using high inter-pixel correlation and local context information. Initially, a similarity model in spatial domain is established. The initial centers and memberships of the clusters are evaluated based on this model. Then, membership function is changed as per correlation among pixels. Finally, segmentation of image is done using this modified FCM. Figure 4 results showed that developed technique is successful in achieving better segmentation results and also, is faster.

The major steps for image segmentation are as follows.

1. Initialise randomly c clusters and set $\varepsilon > 0$. (Usually c = 2, m = 2, $\varepsilon = 0.0001$.)
2. Calculate the local image feature F_{ij} for all neighbor windows of the image.
3. Calculate the visual weights w_i for pixel i over image.

$$w_i = \sum_{j \in \Omega_i} (F_{ij} . g(x_j, y_j) / \sum_{j \in \Omega_i} F_{ij}$$

where Ω_i is a square window on pixel i in spatial domain. $g(x_j, y_{j)}$ is gray value of the pixel j, (x_j, y_j) is a spatial coordinate of the pixel $_j$. F_{ij} gives local image feature.

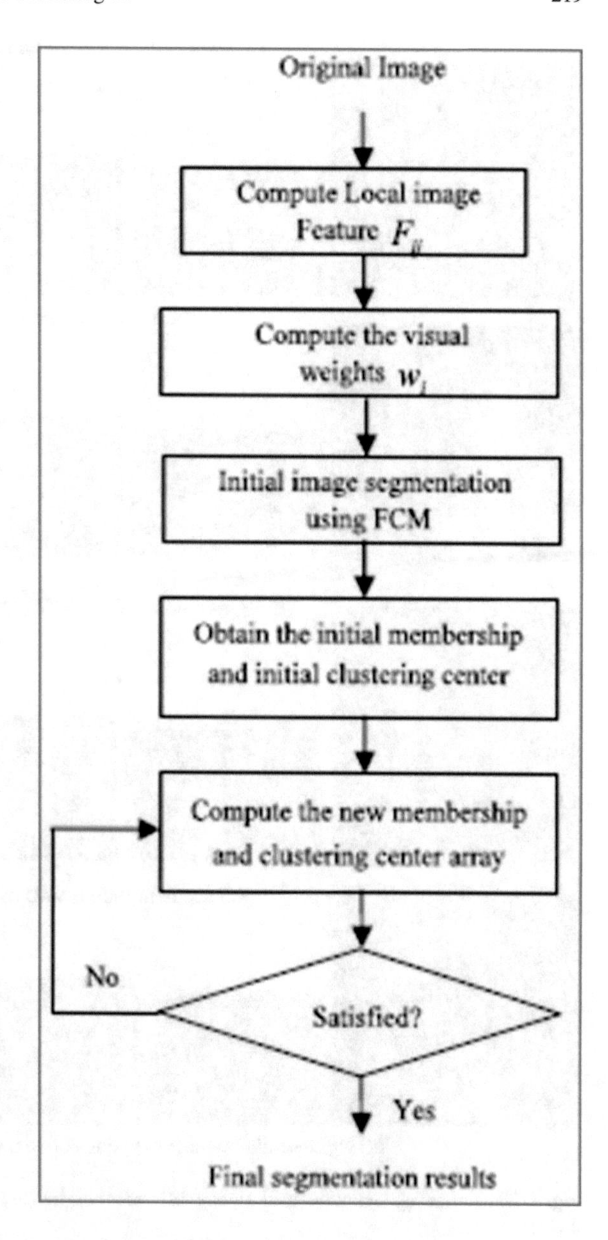

Fig. 4 Fast and robust FCM using spatial information for image segmentation

4. Segment image using equations in step 3, the membership and clustering center array are obtained after convergence as initial parameters.

$$\mu_k(x_i, y_i) = \frac{(w_i - v_k)^{-2/m-1}}{\sum_{j=0}^{c-1}(w_i - v_j)^{-2/m-1}}$$

Fig. 5 The image segmentation results using fast and robust image segmentation method (the noisy image)

$$v_k = \frac{\sum_{i=0}^{q-1} r_i \mu_k(x_i, y_i)^m w^i}{\sum_{i=0}^{q-1} r_i \mu_k(x_i, y_i)^m}$$

Here, r_i denotes gray value pixels equal to i, and $\sum_{i=0}^{q-1} r_i = M \times N$.

5. Calculate new membership and clustering center array using equations in step 4 based on clustering center results and initial membership.
6. Repeat Step 5 till the termination criterion is met $|V_{new} - V_{old}| < \varepsilon$.

Fig. 6 SIFCM algorithm

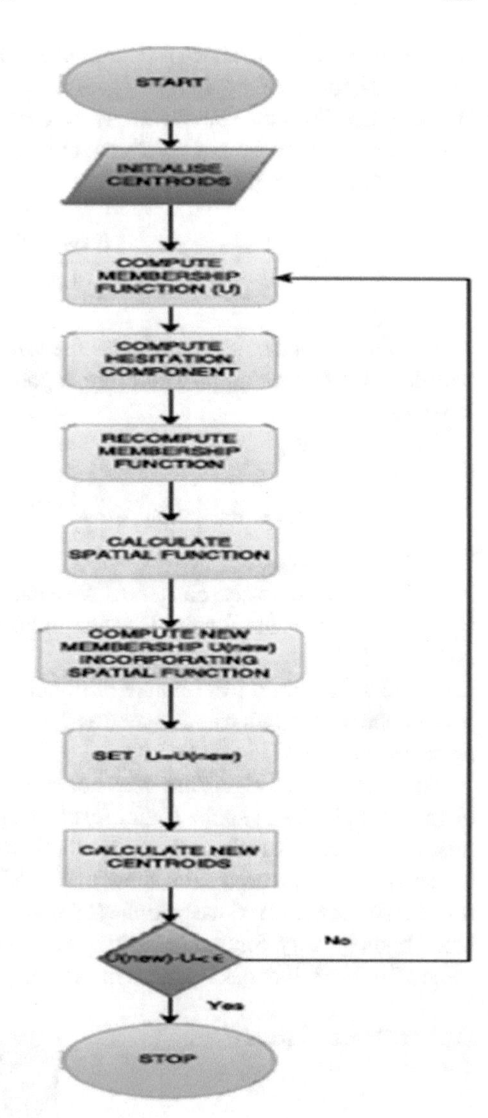

The final segmentation results are obtained.

The Fig. 5 provide the segmentation results on a noisy image by applying FCM and also the fast and robust image segmentation technique.

5.1.4 Image Segmentation Using Spatial IFCM [20]

Tripathy et al. developed the intuitionistic fuzzy C-means with spatial information (sIFCM). This was an extension to Chang's work.

The sIFCM algorithm is (Fig. 6).

Speckle noise of variance 0.04 and mean 0 was induced on the image. FCM and sFCM are applied to the image. V_{pc} and V_{pe} is calculated. They used DB (Davies–Bouldin) and D (Dunn) indices are used to measure the cluster quality in addition to the evaluation metrics used by Chung et al. [7].

The DB index is given by

$$\mathrm{DB} = \frac{1}{c}\sum_{i=1}^{c} \max_{k!=i}\left\{\frac{S(v_i) + S(v_k)}{d(v_i, v_k)}\right\} \quad \text{for } 1 < i, k < c \tag{14}$$

The aim of this index is to minimize the within-cluster distance and maximize the between-cluster separation. Therefore, a good clustering should have minimum DB index [3].

The D index is given by

$$\mathrm{Dunn} = \min_i\left\{\min_{k!=i}\left\{\frac{d(v_i, v_k)}{\max_l S(v_l)}\right\}\right\} \quad \text{for } 1 < k, i, l < c \tag{15}$$

It maximizes the between-cluster distance and minimizes the within-cluster distance. Hence, a higher value for the D index indicates better efficiency of the clustering algorithm [24].

A 225 × 225 brain MRI image of dimension was used for proving their results. The number of clusters was assigned as c = 3. The results of MRI image with speckle noise are shown in (Figs. 7 and 8).

Traditional FCM algorithm does not cluster the image efficiently in the presence of spurious blobs and spots. Better results can be obtained by increasing the spatial function degree. This allows the membership function to be incorporated in the spatial information. However, results show that sIFCM approach produces more desirable results. This method reduces number of spurious spots. The image is segmented with better homogeneity. Smoother segmentation can be obtained by considering a higher value of q; but it induces blurness in the image (Table 1).

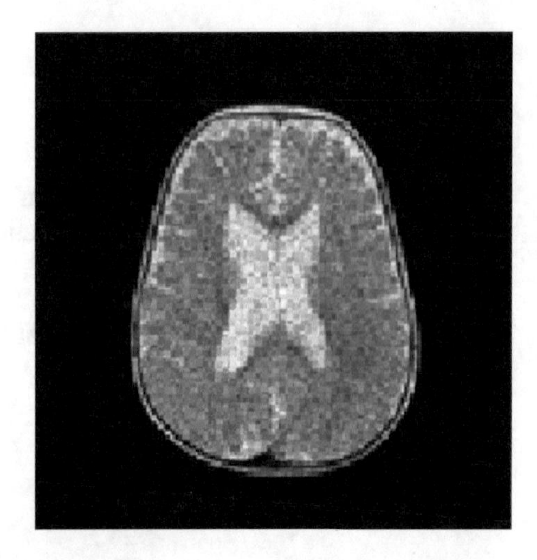

Fig. 7 MRI image-speckle noise

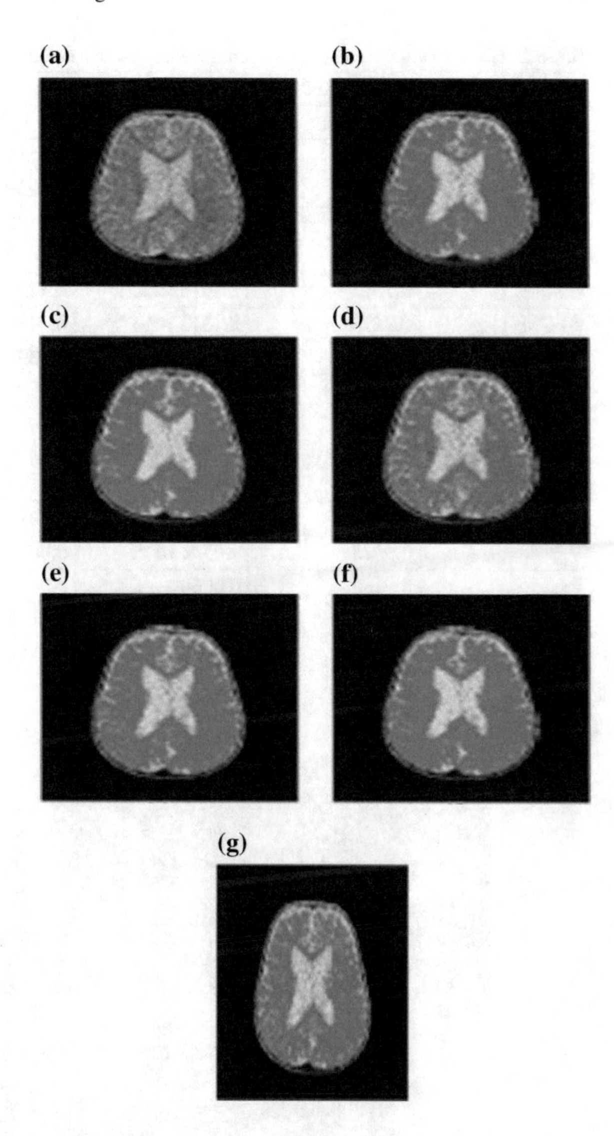

Fig. 8 Noisy image segmentation. **a** FCM. **b** sFCM1,1. **c** $sFCM_{1,2}$. **d** $sFCM_{2,1}$. **e** $sIFCM_{1,1}$. **f** $sIFCM_{1,2}$. **g** $sIFCM_{2,1}$

5.1.5 Applications of Spatial FCM and IFCM on Leukemia Images

Deepthi et al. have applied the spatial clustering algorithms on leukemia images and have found the following results.

The results in Table 2 show that the sFCM succeeds in providing better results than conventional FCM.

The segmented images with the application sFCM provide better clarity and understanding of presence of leukemia cells than that of conventional FCM (Fig. 9).

Table 1 Cluster validity measures on image with speckle noise

Method	Results on the noisy image			
	V_{pc}	V_{pe}	DB Index	D Index
FCM	0.6975	2.8195×10^{-4}	0.4517	3.4183
$sFCM_{1,1}$	0.7101	5.9541×10^{-9}	0.4239	3.6734
$sFCM_{2,1}$	0.6922	7.7585×10^{-12}	0.4326	3.4607
$sFCM_{1,2}$	0.6874	4.2711×10^{-12}	0.4412	3.6144
$sIFCM_{1,1}$	0.7077	1.1515×10^{-08}	0.4254	3.6446
$sIFCM_{2,1}$	0.7135	8.1312×10^{-13}	0.4276	3.7472
$sIFCM_{1,2}$	0.713	4.6770×10^{-13}	0.4393	3.4968

Table 2 Performance indices of sFCM on leukemia image

Index	FCM	$sFCM_{2,1}$	$sFCM_{1,1}$	$sFCM_{1,2}$
V_{pc}	0.2118	0.2300	0.2219	0.4963
V_{pe}	0.0178	2.05×10^{-6}	0.0003	7.69×10^{-8}
V_{xb}	0.0168	0.0060	0.0074	0.0522
DB	0.4670	0.4185	0.4197	0.4173
D	2.2254	2.9951	2.9318	3.4340

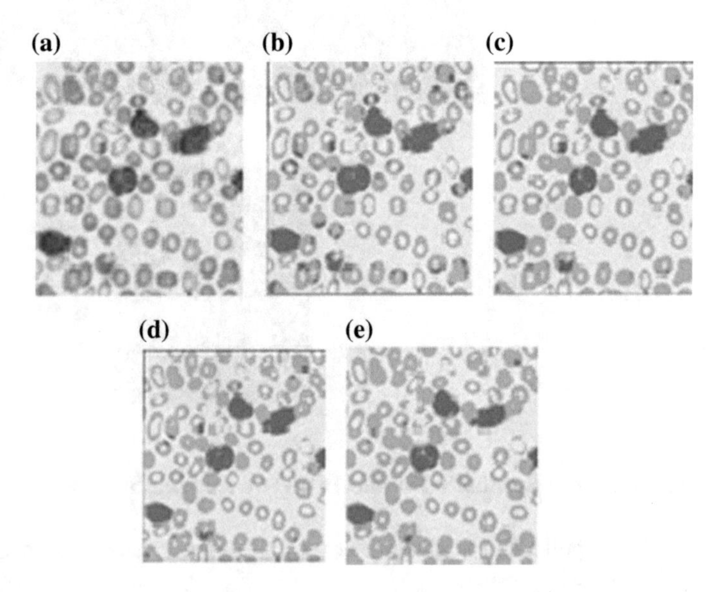

Fig. 9 **a** Original image, segmented images of leukemia using **b** FCM; **c** $sFCM_{2,1}$; **d** $sFCM_{1,1}$; **e** $sFCM_{1,2}$

Table 3 Performance indices of sIFCM on leukemia image

Index	FCM	IFCM	$sIFCM_{1,1}$	$sIFCM_{1,2}$	$sIFCM_{2,1}$
V_{pc}	0.2118	0.2074	0.4762	0.2241	0.2281
V_{pe}	0.0178	0.02585	3.56E-005	1.68E-005	5.90E-006
V_{xb}	0.0168	0.02096	0.05929	0.0069	0.0063
DB	0.467	0.4906	0.4126	0.4208	0.4188
D	2.2254	2.0529	3.5866	3.0203	2.9655

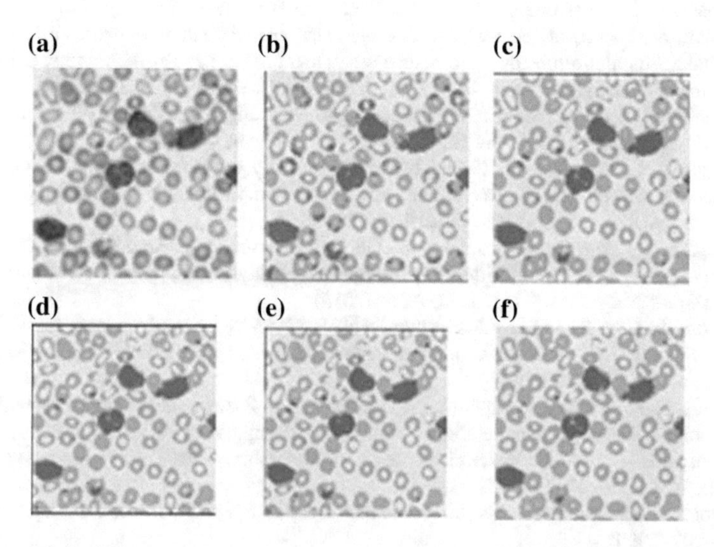

Fig. 10 **a** Original image, segmented images of leukemia using **b** FCM; **c** IFCM **d** $sFCM_{2,1}$; **e** $sFCM_{1,1}$; **f** $sFCM_{1,2}$

The results in Table 3 show that the sIFCM succeeds in providing better results than conventional FCM.

The segmented images with the application sIFCM provide better clarity and understanding of presence of leukemia cells than that of conventional FCM and IFCM [10, 11] (Fig. 10).

6 Conclusion

The conventional hybrid algorithms are not capable of handling the spatial data. The spatial data has to be considered for image segmentation to obtain more accurate results. Hence, researchers have now been working on modifying conventional uncertain data clustering algorithms to incorporate spatial data. The chapter has put forth several modifications of fuzzy C-means with respect to spatial information. Extension of these algorithms with other uncertainty-based hybrid models and studying their behavior is the area that is not explored to the complete extent by the researchers.

References

1. Atanassov, K.T.: Intuitionistic fuzzy sets. Fuzzy sets Syst. **20**(1), 87–96 (1986)
2. Azadeh, A., Saberi, M., Anvari, M.: An integrated artificial neural network fuzzy c-means-normalization algorithm for performance assessment of decision-making units: the cases of auto industry and power plant. Comput. Ind. Eng. **60**(2), 328–340 (2011)
3. Bezdek, J.C., Pal, N.R.: Some new indexes of cluster validity. IEEE Trans. Syst., Man, Cybern., Part B: Cybern. **28**(3), 301–315 (1998)
4. Bhargava, R., Tripathy, B.: Kernel based rough-fuzzy c-means. In: Maji, P., Ghosh, A., Narasimha Murty, M., Ghosh, K., Pal, S.K. (eds.) Pattern Recognition and Machine Intelligence, pp. 148–155. Springer, Heidelberg (2013)
5. Bhargava, R., Tripathy, B., Tripathy, A., Dhull, R., Verma, E., Swarnalatha, P.: Rough intuitionistic fuzzy c-means algorithm and a comparative analysis. In: Proceedings of the 6th ACM India Computing Convention, p. 23. ACM (2013)
6. Buades, A., Coll, B., Morel, J.M.: A non-local algorithm for image denoising. In: IEEE Computer Society Conference on Computer Vision and Pattern Recognition, 2005. CVPR 2005. vol. 2, pp. 60–65. IEEE (2005)
7. Chuang, K.S., Tzeng, H.L., Chen, S., Wu, J., Chen, T.J.: Fuzzy c-means clustering with spatial information for image segmentation. Comput. Med. Imaging Gr. **30**(1), 9–15 (2006)
8. Davies, D.L., Bouldin, D.W.: A cluster separation measure. IEEE Trans. Pattern Anal. Mach. Intell. **2**(2), 224–227 (1979)
9. Dubois, D., Prade, H.: Rough fuzzy sets and fuzzy rough sets*. Int. J. Gen. Syst. **17**(2–3), 191–209 (1990)
10. Hudedagaddi, D., Tripathy, B.: Application of spatial FCM on cancer cell images. In: Proceedings of the National Conference on Recent Trends in Mathematics and Information Technology, pp. 96–100 (2016)
11. Hudedagaddi, D., Tripathy, B.: Application of spatial IFCM on leukaemia images. Soft Comput. Med. Data Satell. Image Anal. (SCMSA) **7**(5), 33–40 (2016)
12. Lingras, P., West, C.: Interval set clustering of web users with rough k-means. J. Intell. Inf. Syst. **23**(1), 5–16 (2004)
13. Maji, P., Pal, S.K.: RFCM: A hybrid clustering algorithm using rough and fuzzy sets. Fundam. Inf. **80**(4), 475–496 (2007)
14. Maji, P., Pal, S.K.: Rough set based generalized fuzzy-means algorithm and quantitative indices. IEEE Trans. Syst. Man Cybern. Part B: Cybern. **37**(6), 1529–1540 (2007)
15. Mennis, J., Guo, D.: Spatial data mining and geographic knowledge discovery-an introduction. Comput., Environ. Urban Syst. **33**(6), 403–408 (2009)
16. Mittal, D., Tripathy, B.: Efficiency analysis of kernel functions in uncertainty based c-means algorithms. In: International Conference on Advances in Computing, Communications and Informatics (ICACCI), 2015, pp. 807–813. IEEE (2015)
17. Pawlak, Z.: Rough sets. Int. J. Comput. Inform. Sci. **11**(5), 341–356 (1982)
18. Pawlak, Z., Skowron, A.: Rudiments of rough sets. Inf. Sci. **177**(1), 3–27 (2007)
19. Rizvi, S., Naqvi, H.J., Nadeem, D.: Rough intuitionistic fuzzy sets. In: Proceedings of the 6th Joint Conference on Information Sciences (JCIS), pp. 101–104. Durham, NC (2002)
20. Tripathy, B., Basu, A., Govel, S.: Image segmentation using spatial intuitionistic fuzzy c means clustering. In: IEEE International Conference on Computational Intelligence and Computing Research (ICCIC), 2014, pp. 1–5. IEEE (2014)

21. Tripathy, B., Tripathy, A., Rajulu, K.G.: Possibilistic rough fuzzy c-means algorithm in data clustering and image segmentation. In: IEEE International Conference on Computational Intelligence and Computing Research (ICCIC), 2014, pp. 1–6. IEEE (2014)
22. Wang, X.Y., Bu, J.: A fast and robust image segmentation using FCM with spatial information. Digit. Signal Process. **20**(4), 1173–1182 (2010)
23. Yang, Y., Zheng, C., Lin, P.: Fuzzy clustering with spatial constraints for image thresholding. Opt. Appl. **35**(4), 943 (2005)
24. Zadeh, L.A.: Fuzzy sets. Inf. Control **8**(3), 338–353 (1965)

Coronary Artery Segmentation and Width Estimation Using Gabor Filters and Evolutionary Computation Techniques

Fernando Cervantes-Sanchez, Ivan Cruz-Aceves and Arturo Hernandez-Aguirre

Abstract This paper presents a novel method based on single-scale Gabor filters (SSG) consisting of three steps for vessel segmentation and vessel width estimation of X-ray coronary angiograms. In the first stage, a comparative analysis of genetic algorithms, and two estimation of distribution algorithms in order to improve the vessel detection rate of the SSG, while reducing the computational time of the training step is performed. The detection results of the SSG are compared with those obtained by four state-of-the-art detection methods via the area (A_z) under the receiver operating characteristic (ROC) curve. In the second stage, a comparative analysis of five automatic thresholding methods is performed in order to discriminate vessel and non-vessel pixels from the Gabor filter response. In the last step, a procedure to estimate the vessel width of the segmented coronary tree structure is presented. The experimental results using the SSG obtained the highest vessel detection performance with $A_z = 0.9584$ with a training set of 40 angiograms. In addition, the segmentation results using the interclass variance thresholding method provided a segmentation accuracy of 0.941 with a test set of 40 angiograms. The performance of the proposed method consisting of the steps of vessel detection, segmentation, and vessel width estimation shows promising results according to the evaluation measures, which is suitable for clinical decision support in cardiology.

Keywords Automatic segmentation · Coronary arteries · Estimation of distribution algorithms · Gabor filters · Genetic algorithms · Vessel width estimation

---

F. Cervantes-Sanchez · A. Hernandez-Aguirre
Centro de Investigación en Matemáticas (CIMAT), A.C., Jalisco S/N, Col. Valenciana, 36000 Guanajuato, GTO, Mexico
e-mail: fernando.cervantes@cimat.mx

A. Hernandez-Aguirre
e-mail: artha@cimat.mx

I. Cruz-Aceves (✉)
CONACYT - Centro de Investigación en Matemáticas (CIMAT), A.C., Jalisco S/N, Col. Valenciana, 36000 Guanajuato, GTO, Mexico
e-mail: ivan.cruz@cimat.mx

S. Bhattacharyya et al. (eds.), *Hybrid Soft Computing for Image Segmentation*, DOI 10.1007/978-3-319-47223-2_10

1 Introduction

Automatic segmentation and width estimation of coronary arteries represent an important and challenging problem for systems that perform computer-aided diagnosis in cardiology. In clinical practice, the process carried out by specialists consists on a visual examination followed by a manual delineation or inspection of affected regions along coronary arteries, which can be subjective, time consuming, and labor intensive. Due to this, the development of computational methods to obtain an efficient segmentation, and subsequently an accurate vessel width estimation plays an essential role.

In literature, different strategies for the detection and segmentation of blood vessels have been proposed. Most of them, are based on mathematical morphology [1–4], Gaussian filters [5–10], Hessian matrix [11–13], and Gabor filters [14–16]. In general, these four types of methods require a training stage to select the most appropriate parameter values. The morphology-based method of Qian et al. [3] uses the size S of the structuring elements in order to perform the top-hat operator to detect vessel-like structures of different calibers. The Gaussian filters [7] are defined by four parameters; length and width of the Gaussian template, the spread of the intensity profile, and number of oriented filters. The method of Wang et al. [13] uses a range of values to define the spread of the Gaussian profile (σ), and a discrete step size (δ) to perform multiscale analysis. The most widely used strategy to determine the best parameter values for these methods is based on the definition of a training set of images and a global search method to find the best set of parameters using the area under the receiver operating characteristic (ROC) curve as performance metric [17].

Moreover, the Gabor filters introduced by Rangayyan et al. [14, 15] use three parameters, which have to be tuned for each particular application. These parameters represent the elongation of the Gabor kernel, the average thickness of the vessels to be detected, and the number of directional filters in the range $[-\pi/2, \pi/2]$. The exhaustive global search was used in both works to determine the optimal parameter values over a predefined search space. In general, the Gabor filters obtain superior performance than the spatial methods discussed above; however, since these filters are performed in the frequency domain, the training stage by an exhaustive global search is computationally expensive.

In order to solve the parameter optimization problem by reducing the computational time of the exhaustive global search in a training stage, some strategies using population-based methods such as genetic algorithms [18] and estimation of distribution algorithms [19, 20] have been introduced. In this chapter, a novel unsupervised method for vessel segmentation and vessel width estimation in X-ray coronary angiograms is presented. In the first step, vessel detection is performed using single-scale Gabor filters, which are tuned by a procedure involving a comparative analysis of three evolutionary computation techniques. In the second step, five thresholding methods are compared to classify vessel and nonvessel pixels from the Gabor filter response in terms of accuracy measure. Finally, in the third step, the segmented coronary artery is processed to obtain the vessel centerline by using mathematical morphology techniques in order to estimate the vessel width of the entire coronary tree structure.

The remainder of this paper is organized as follows. In Sect. 2, the fundamentals of the single-scale Gabor filters and three evolutionary computation techniques are described in detail. In Sect. 3, the proposed method for parameter optimization and vessel width estimation is introduced. The experimental results are analyzed in Sect. 4, and conclusions are given in Sect. 5.

2 Background

The present section introduces the fundamentals of the single-scale Gabor filters used for the enhancement of vessel-like structures in medical images, and three evolutionary computation techniques of the state-of-the-art employed for solving discrete and continuous optimization problems.

2.1 *Single-Scale Gabor Filters (SSG)*

The Gabor filter is defined as a Gaussian function modulated by a sinusoid [21], which can be rotated at different orientations by using a geometric transformation [14]. Consequently, the Gabor filter can be applied to enhance vessel-like structures at different orientations by using a directional filter bank [22]. The main kernel of a Gabor filter can be defined as follows:

$$g(x, y) = \frac{1}{2\pi\sigma_x\sigma_y} \exp\left[-\frac{1}{2}\left(\frac{x^2}{\sigma_x^2} + \frac{y^2}{\sigma_y^2}\right)\right] \cos(2\pi f_o x), \quad (1)$$

where σ_x and σ_y are the standard deviation of the Gaussian function, and f_o represents the frequency of the modulating sinusoid. The single-scale Gabor filter designed by Rangayyan et al. [14, 15] is governed by three different parameters. The first parameter τ is used to approximate the average thickness (in pixels) of the vessel-like structures to be detected, which is defined as follows:

$$\tau = \sigma_x 2\sqrt{2 \ln 2}. \quad (2)$$

The second parameter (l) is introduced to control the elongation of the Gabor kernel as $\sigma_y = l\sigma_x$, and the last parameter κ determines the number of oriented filters in the range $[-\frac{\pi}{2}, \frac{\pi}{2}]$ as $\kappa = 180/\theta$, where (θ) is the angular resolution. These oriented kernels are convolved with the input image, and for each pixel the maximum response over all orientations is preserved in order to acquire the filter response.

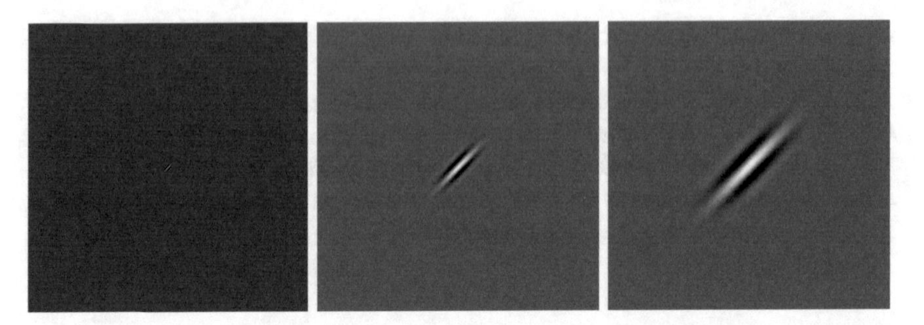

Fig. 1 Gabor filter kernels of size 300 × 300 pixeles with $l = 2.5$, $\theta = 45°$ and using $\tau = 5$, $\tau = 10$, and $\tau = 20$ pixels, respectively

To obtain the highest performance of the single-scale Gabor filters, the continuous parameter of elongation (l), and the discrete parameters of average thickness (τ) and number of oriented filters (κ) have to be determined. In Fig. 1, a subset of Gabor kernels with different parameter values is presented.

2.1.1 Training Stage Using ROC Curve Analysis

Since the single-scale Gabor filters are governed by three main parameters (κ, τ, l), an optimization process to select the most suitable value for each parameter is required. This optimization process has been commonly performed through a training stage calculating the area (A_z) under the ROC curve, and by applying an exhaustive global search using different range of values for each parameter.

The ROC curve represents a measure to assess the performance of a classification system. To obtain this curve from the Gabor filter response, a sliding threshold is applied to the gray-scale filter response in order to plot the true-positive fraction (TPF) against false-positive fraction (FPF). The area A_z under this curve can be approximated by using the Riemann-sum method.

In the exhaustive global search of the training stage, for each combination of parameters (κ, τ, l), the filter response is evaluated in terms of the A_z value. The set of parameters with the highest A_z over a predefined training set of images is used to be directly applied in the test set of images. The main disadvantage of the exhaustive global search is the fact that it is computationally expensive; therefore, the application of stochastic optimization methods can be introduced to reduce the number of evaluations and computational time of the training stage, which is highly desirable for medical applications.

To illustrate the importance of the optimal parameter selection for the single-scale Gabor filters, in Fig. 2, a subset of coronary angiograms is introduced along with the ground-truth images, where by visual inspection, and in terms of detecting vessels at different diameters, superior performance was acquired with low values of the Gabor parameter τ.

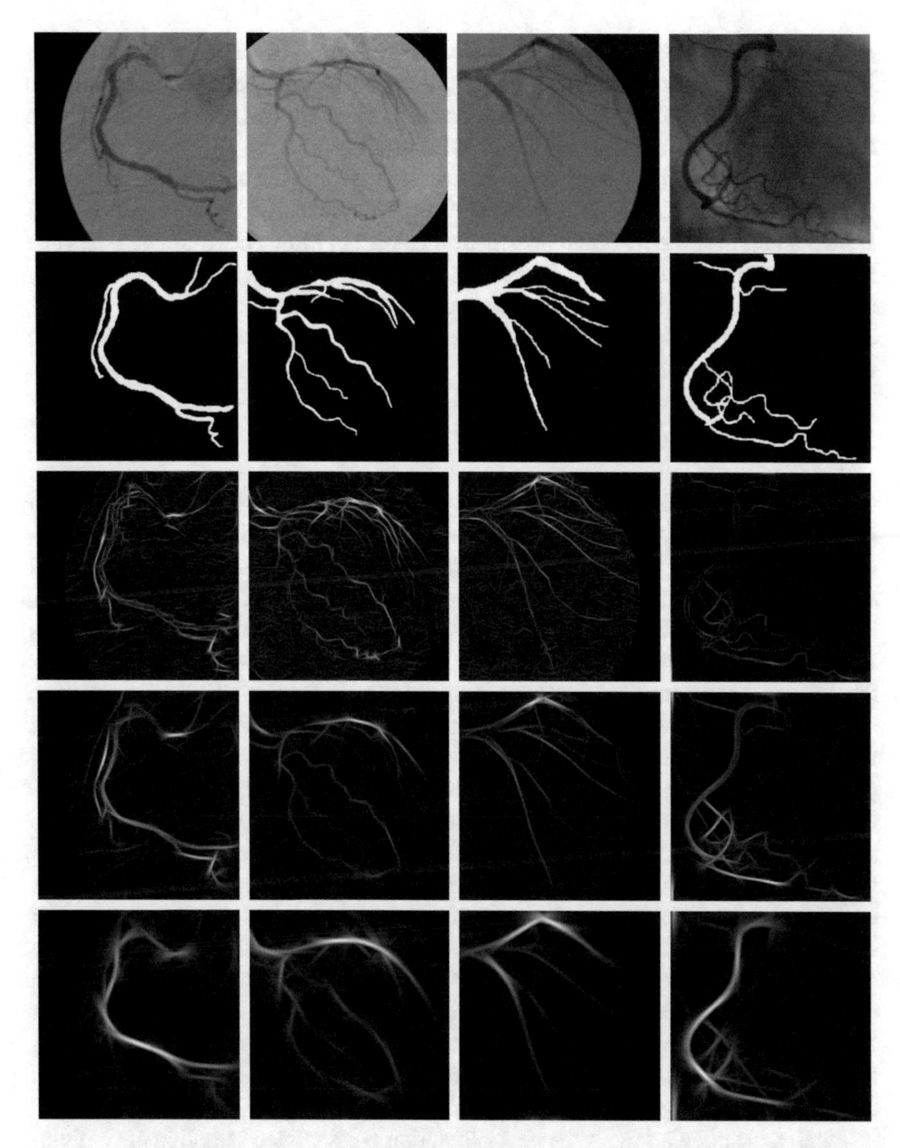

Fig. 2 *First row* subset of coronary angiograms. *Second row* ground-truth images. The remaining *three rows* present the Gabor filter response for the angiograms in the *first row* with $l = 2.5$, $\kappa = 45$, and using $\tau = 5$, $\tau = 10$, and $\tau = 20$ pixels, respectively

2.2 Evolutionary Computation Techniques

Evolutionary computation (EC) techniques are used to solve optimization problems based on different strategies, most of them Nature-inspired. The optimal solution to the problem is searched into a set of *individuals* with same features but different values. For each iteration also known as *generation*, a new set of individuals called *population*, is obtained from modifying features of the *individuals* following nature-inspired processes such as selection, crossover, and mutation. The best solutions during generations are kept and labeled as *elite individuals*, where the final solution for the optimization problem is chosen between them.

In the present work, we focus on genetic algorithms(GA) and two estimation of distribution algorithms [23–25] called univariate marginal distribution algorithm and Boltzmann univariate marginal distribution algorithm [26]. The difference between those algorithms is how *individuals* features are modified. GA modify them directly through mutation and crossover processes. On the other hand, UMDA and BUMDA estimate the probability distribution of the features, then each value is sampled from its respective marginal distribution.

2.2.1 Genetic Algorithm

Genetic algorithm(GA) is an EC technique that simulates the genetic evolution [27, 28]. GA codes the *individuals* features in a chain of *genes* that can be evaluated on the objective function. A selection process assures strong individuals will survive and inherit the genes in a crossover process. The main idea behind the GA, is that strong individuals produce stronger offspring. Then a mutation process is included to enhance the search space. Multiple techniques have been developed for the different nature-inspired processes, some of them will be briefly discussed below.

GA commonly start with a population randomly generated, which is evaluated through generations using an objective function. The best individual found with this process is saved as potential solution to the optimization problem. For the following generations, the process will be repeated from the selection step. In this work, roulette-wheel method is used for the selection step. This selection strategy forms the selected set of individuals randomly from the current population. The sum of the evaluation values of the population is computed, and each individual probability of being chosen is the fraction that its evaluation value contributes to the global sum.

The crossover operator combines the features of the selected set trying to form stronger offspring. Two children are formed taking randomly two individuals from the selected set as parents. In the uniform crossover, the genes that will be taken from each parent are chosen with uniform probability, then they are mixed to form the new individuals, as it is shown in Fig. 3.

Finally, the population for the next generation is assembled sampling the individuals from the offspring and the previous selected set. The new population is assessed using the objective function. If an individual is better on its evaluation than the current solution, it is now considered the potential solution.

According to the above description, the GA can be implemented as follows:

1. Initialize number of generations G, number of individuals N_p, crossover rate CR, and mutation rate MR.

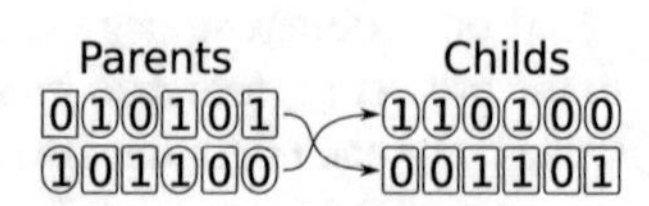

Fig. 3 Crossover using the uniform genes selection from each parent

2. Initialize the individuals into the search space.
3. Evaluate individuals in the objective function.
4. For each generation $g = \{1, \ldots, G\}$:

   a. Apply selection operator;
   b. Perform crossover and mutation steps to generate new individuals;
   c. Evaluate new individuals in the objective function;
   d. Replace individuals with the worst fitness.

5. Stop if the convergence criterion is satisfied (e.g., number of generations).

Example: Optimization of the 2-D Rastrigin Function

The main application of the evolutionary computation techniques is the optimization of mathematical functions. In order to illustrate the implementation of a GA in an optimization process, the *Rastrigin* function in two dimensions is introduced, which

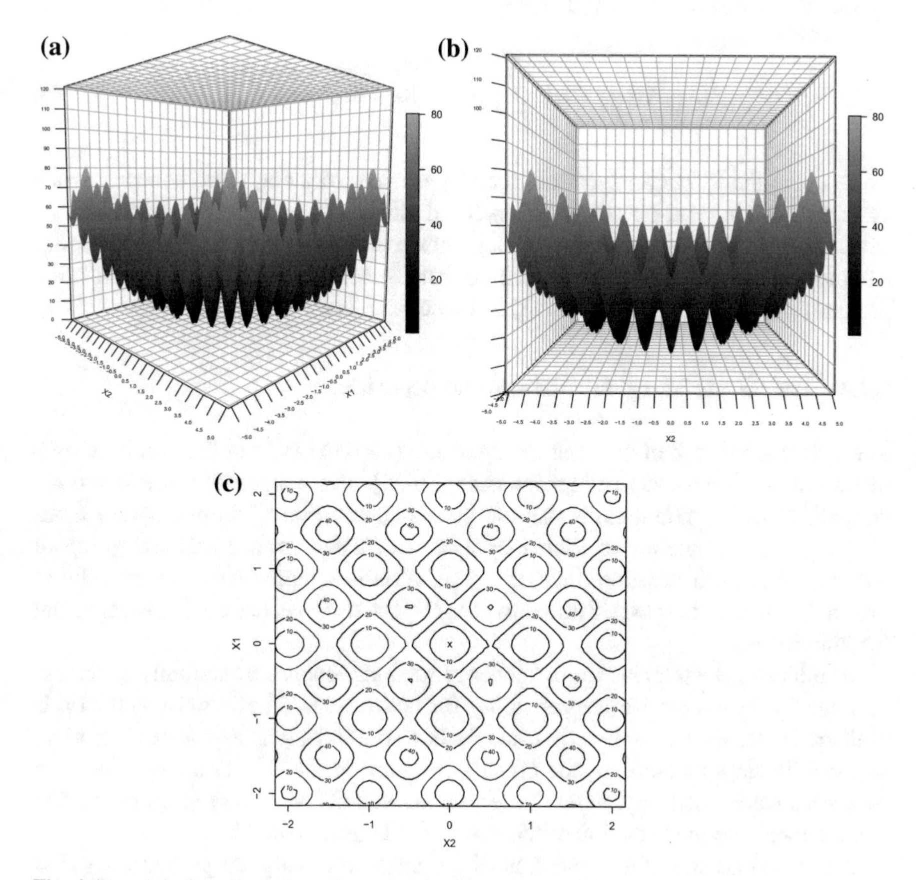

Fig. 4 Rastrigin function in two dimensions. **a** Isometric view in X_1 and X_2, **b** lateral view, and **c** level plot of the function, where the optimal value is located at $X_1 = 0$ and $X_2 = 0$

Table 1 Features coding for X_1 and X_2 using 3 bits and their corresponding values over the search space

Genes code	Variable value
000	−5.1200
001	−3.4133
010	−1.7066
011	0.0000
100	1.7066
101	3.4133
110	5.1200
111	6.8266

is defined in Eq. (3) and illustrated in Fig. 4. It is important to point out that the range for each variable of the function is $X_1 \in [-5.12, 6.8266]$, $X_2 \in [-5.12, 6.8266]$, and optimal is located on $(X_1 = 0.0, X_2 = 0.0)$.

$$f(X_1, X_2) = 20 + \sum_{i=1}^{2} \left(X_i^2 - 10 \cdot cos(2 \cdot \pi \cdot X_i)\right) . \quad (3)$$

To solve the Rastrigin function, the GA is encoded using three genes for each variable, building a 6-bit string for each individual. The genes are mapped to the search space uniformly distributed as it is shown in Table 1. Finally, Fig. 5 illustrates a numerical example about the procedure that GA follows to solve the optimization problem using Rastrigin equation (3) as objective function.

2.2.2 Univariate Marginal Distribution Algorithm

The univariate marginal distribution algorithm (UMDA) is from the family of estimation of distribution algorithms (EDAs) [23–25]. These algorithms are stochastic methods based on populations that are generated from marginal probability models [29]. The parameters of each model are calculated from a selected group of individuals in each iteration, then, the population is sampled from the probability model for the next iteration. Similar to GA, UMDA uses a genes coding to represent the individuals.

To initialize the population, the genes of the individuals are randomly generated between $\{0, 1\}$ with uniform probability for both values. The initial population is evaluated using the fitness function, and the best solution is kept as potential solution. In the following generations, the process will start in the selection step. This step uses a truncation strategy, which order the individuals according to their solution quality, then a defined fraction of the best individuals is selected.

In the second step, the estimation of the univariate marginal probabilities P is computed. In UMDA, the variables are considered as independent between them. The marginal probability model for independent variables can be defined as follows:

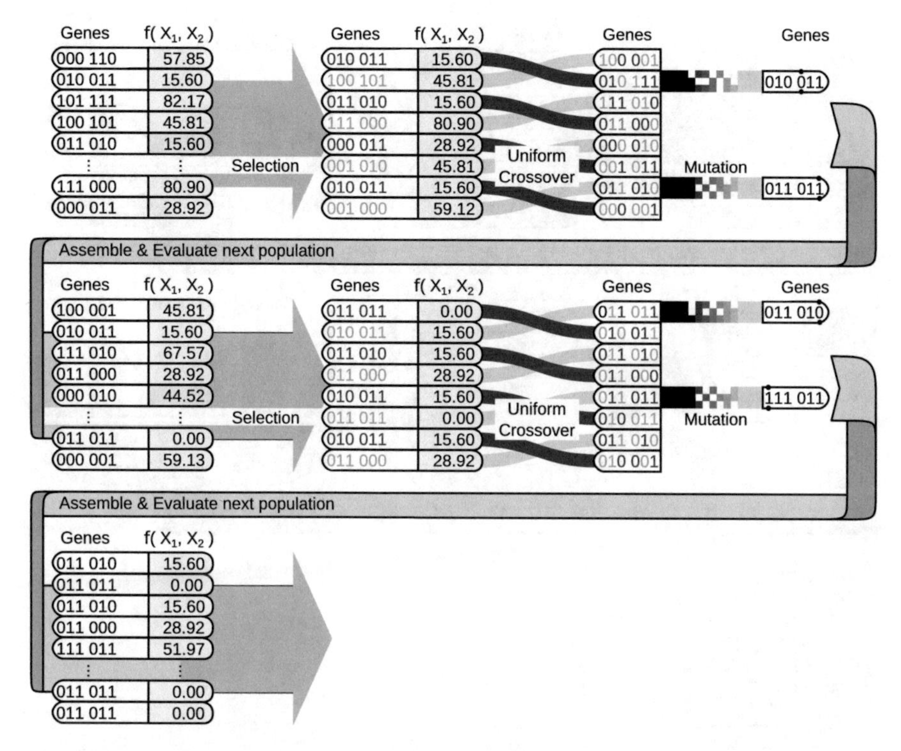

Fig. 5 Numerical example for solving the 2-D Rastrigin function using a GA as optimization strategy

$$P(x) = \prod_{i=1}^{n} P(X_i = x_i) \ , \tag{4}$$

where $x = (x_1, x_2, \ldots, x_n)^T$ represents the binary value of the ith bit in the individual, and X_i is the ith random value (from a uniform distribution) of the vector X.

UMDA samples a new population in the third step using the estimated marginal probability model to generate the new individuals genes. The best individual of each generation is compared to the current potential solution. If the new individual is better, the potential solution is updated. The process is iteratively performed until a convergence criterion is satisfied.

According to the above description, UMDA can be implemented as follows:

1. Initialize number of individuals n and generations t.
2. Initialize the individuals into the search space.
3. Select a subset of individuals S of $m \leq n$ according to the selection operator.
4. Compute the univariate marginal probabilities $p_i^s(x_i, t)$ of S.
5. Generate n new individuals by using $p(x, t+1) = \prod_{i=1}^{n} p_i^s(x_i, t)$.
6. Stop if convergence criterion is satisfied (e.g., number of generations), otherwise, repeat steps (3)–(5).

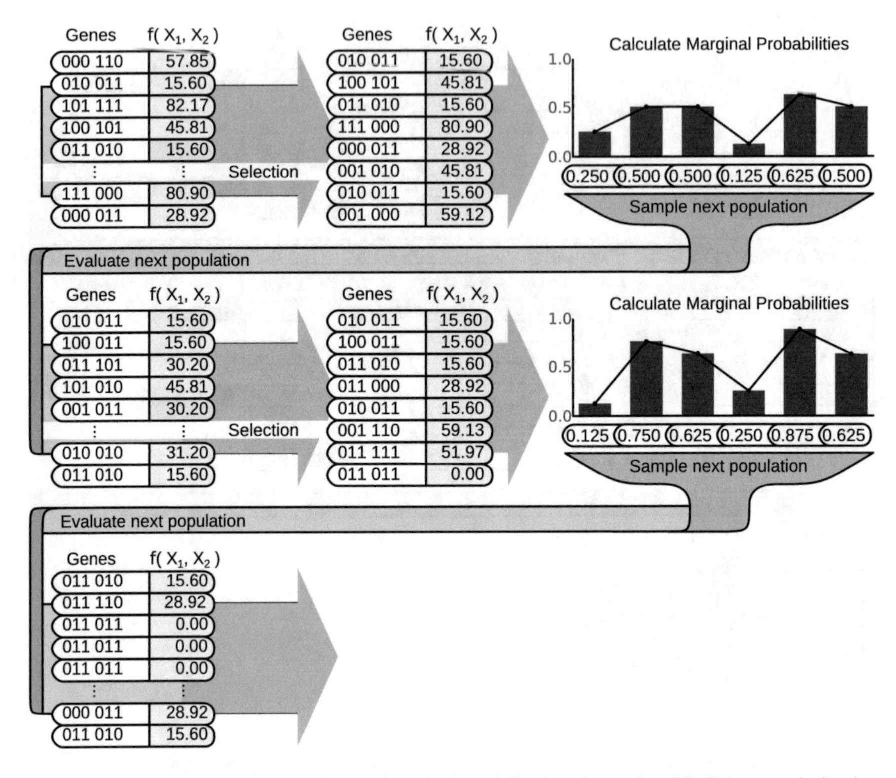

Fig. 6 Numerical example for solving the 2-D Rastrigin function using UMDA as optimization strategy

Implementation Details

Using the Rastrigin function discussed above (see Eq. 3), numerical example of the optimization process using UMDA is illustrated in Fig. 6. In this example, the genes mapping is the same used in the GA case (see Table 1).

2.2.3 Boltzmann Univariate Marginal Distribution Algorithm

Boltzmann univariate marginal distribution algorithm (BUMDA) [26], as well as UMDA, is an EDA, but in contrast, BUMDA uses an approximation of the Boltzmann distribution, built with a normal-Gaussian model, and the current search space instead of a gene-based mapping. Formulae to estimate mean and variance of the normal-Gaussian model are given as part of the algorithm, and the only parameter to be tuned is the number of individuals (population size). This method ensures the exploration of promising regions due to the sampling process. When the algorithm approximates to convergence, variance of the normal-Gaussian model tends to 0 and its mean tends to the best approximation of the optimum value. This algorithm has shown to be better for convex functions than other state-of-the-art EDAs based on multivariate Gaussian models. To estimate the Normal-Gaussian model, BUMDA computes a mean vector over a subset of selected individuals as follows:

$$\mu_t = \frac{\sum_{i=1}^{N_S} \bar{g}(x_i)x_i}{\sum_{i=1}^{N_S} \bar{g}(x_i)}. \quad (5)$$

where vector x_i are the features values of the individual i, N_S is the number of individuals in the selected set S. The $\bar{g}(x_i)$ value is the difference between the current and the worst individuals in S. Subsequently, a variance vector from the Normal-Gaussian model is calculated as follows:

$$v_t = \frac{\sum_{i=1}^{N_S} \bar{g}(x_i)(x_i - \mu_t)^2}{1 + \sum_{i=1}^{N_S} \bar{g}(x_i)}. \quad (6)$$

Finally, a new population is sampled from the normal-Gaussian model with the parameters μ_t and v_t for each feature by separate, and the process is iteratively performed until the variance vector of the Normal-Gaussian model is lower than a predefined value.

According to previous description, BUMDA can be implemented as follows[1]:

1. Initialize minimum variance v_{min} as convergence criterion.
2. Initialize N *individuals*.
3. Select the *individuals* with function evaluation above of φ and form subset S.
4. Calculate μ_t using Eq. (5) from S.
5. Compute v_t using Eq. (6) from S.
6. Generate N new *individuals* with attributes sampled from the marginal Normal distribution (mean μ_t, variance v_t).
7. Insert the individual with the best fitness to the new population.
8. If $v_t \preceq v_{min}$ then stop, otherwise, repeat from step 3.

Implementation Details

Following with the Rastrigin function discussed above (see Eq. 3), a numerical example of the optimization process using the real-coded BUMDA is illustrated in Fig. 7. In this numerical example, the range for each variable of the Rastrigin function is set as $X_1 \in [-5.12, 6.8266]$, $X_2 \in [-5.12, 6.8266]$, where the optimal value is located on $(X_1 = 0.0, X_2 = 0.0)$.

[1]Source code of BUMDA available at http://www.cimat.mx/~ivvan/public/bumda.html.

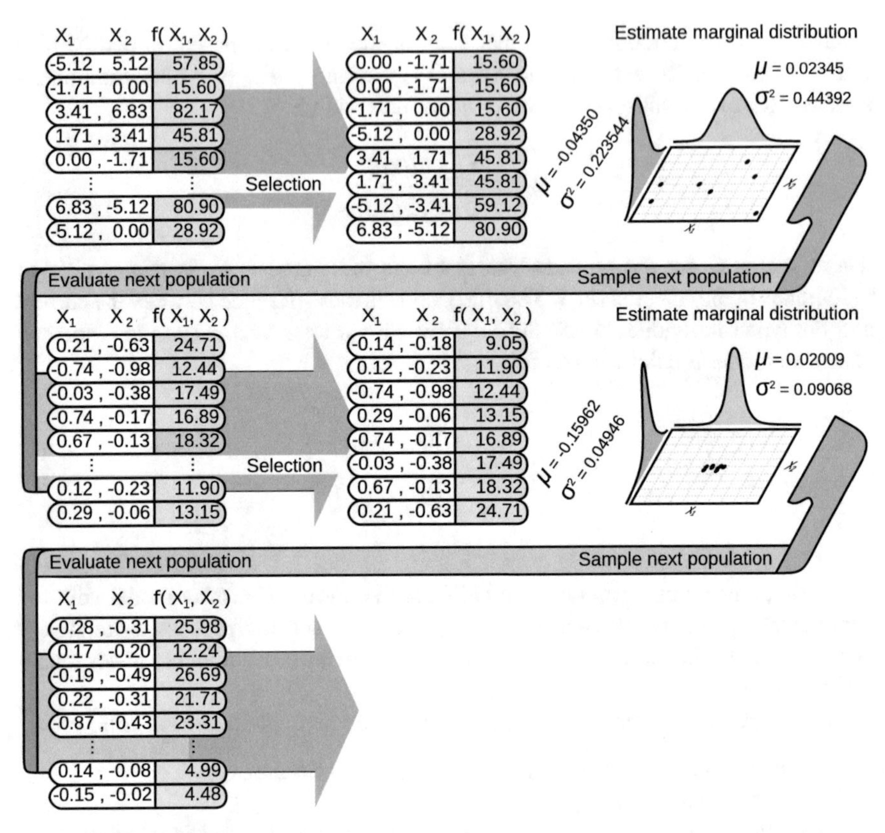

Fig. 7 Numerical example for solving the 2-D Rastrigin function using real-coded BUMDA as optimization strategy

3 Proposed Method

In the present section, the proposed method consisting of three steps is described. The first stage is the vessel enhancement through single-scale Gabor filters, which are tuned using the population-based method called BUMDA. In the second step, the Gabor filter response is segmented by comparing five thresholding methods and evaluated in terms of accuracy measure. The last step is carried out by mathematical morphology operators in order to estimate the vessel width of the processed X-ray angiogram.

3.1 Optimal Parameter Selection of SSG

Since the single-scale Gabor filters are governed by the three parameters of elongation (l), thickness (τ), and number of directional filters (κ), an optimization process to determine the most appropriate values for these parameters is required.

In general, this optimization process has been performed using an exhaustive global search over varying sets of parameters as applied over a training set of images.

Rangayyan et al. [14] proposed a search space as $l = \{1.7, 2.1, \ldots, 4.1\}$ and $\tau = \{7, 8, 9\}$, keeping constant the number of oriented filters as $\kappa = 180$ to be applied in the detection of blood vessels in retinal fundus images. Subsequently, Rangayyan et al. [15] extended the range for the elongation and average thickness parameters as it is shown in Table 2.

In these works, an exhaustive global search was applied for each combination of parameters (κ, τ, l) and evaluated by the area A_z under the ROC curve. The set of parameters with the best A_z value over the training set was directly applied on the test set of images.

The main disadvantages of the above-mentioned strategies for the training stage is the fact that the exhaustive search is computationally expensive, and also that the search space for the elongation parameter is discrete, which implies that the proposed space cannot be explored properly. Although the search space proposed by Rangayyan et al. [15] for the elongation and average thickness was originally defined for blood vessels in retinal images, it is suitable to be applied for the detection of coronary arteries in X-ray angiograms.

To avoid an exhaustive search in the training stage of the single-scale Gabor filters, three evolutionary computation techniques are analyzed in terms of efficiency and computational time to perform the optimization process. Since these evolutionary techniques require an objective function to be maximized, the area A_z under the ROC curve is used to evaluate the performance of the vessel detection results. The optimization process is carried out using the discrete space for the τ parameter, and with continuous values over the defined space for the l parameter as it was previously illustrated in Table 2.

Table 2 Search space for the single-scale Gabor filters proposed by Rangayyan et al. [15]

Parameter	Range of values
l	$\{1.3, 1.7, 2.1, \ldots, 17.7, 18.1\}$
τ	$\{1, 2, 3, \ldots, 15, 16\}$
κ	180

3.2 Thresholding of the SSG Filter Response

In order to discriminate vessel and nonvessel pixels from the single-scale Gabor filter response, the interclass variance thresholding method has been applied, which is described below.

The interclass variance method introduced by Otsu [30], assumes that the image to be processed has a bimodal distribution of the background and foreground pixels. This method can be described by three main steps. First, the mean intensity of the entire image can computed as follows:

$$\mu = \sum_{i=1}^{L} iP_i, \tag{7}$$

where L represents the intensity levels $\{0, 1, 2, \ldots, L-1\}$, and P_i the probability distribution of the number of pixels n_i with intensity i of the total number of pixels N in the image, which is calculated as follows

$$P_i = \frac{n_i}{N}. \tag{8}$$

In the second step, the average intensity for each class of pixels is computed as follows:

$$\mu_j = \sum_{i=t_{j-1}+1}^{t_j} \frac{iP_i}{w_j} \tag{9}$$

where w_j is the probability distribution for each class. Moreover, in the last step of the method, the interclass variance is computed as follows:

$$\sigma^2 = \sum_{j=1}^{n} w_j(\mu_j - \mu)^2, \tag{10}$$

where the threshold value is the intensity level with the maximum interclass variance. This step can be expressed as the maximization of the interclass variance criterion as follows:

$$\phi = \max_{1<t_1<\cdots<t_{n-1}<L} \{\sigma^2(t)\}. \tag{11}$$

On the other hand, to evaluate the segmentation results obtained from the thresholding methods, the accuracy measure has been adopted. This measure has been commonly used for the evaluation of binary classification systems, which can be calculated as follows:

$$Accuracy = \frac{TP + TN}{TP + FP + TN + FN}, \tag{12}$$

where *TP* and *TN* represent the fractions of vessel and nonvessel pixels correctly classified as such by the method, respectively, and *FN* and *FP* the fractions of vessel and nonvessel pixels incorrectly classified by the method.

In the accuracy measure, when the vessels and nonvessel pixels obtained from the method are completely superimposed with the ground-truth image, the obtained result is one, and zero otherwise.

3.3 Postprocessing of Segmented Vessels

The last step of the proposed method corresponds to the postprocessing of the segmented vessels, which is useful to work with the main coronary tree structure. The steps of length filtering and the process to estimate the vessel width along the coronary artery are described below.

3.3.1 Length Filtering

After the segmentation step, several isolated regions or misclassified pixels can appear in the resulting image. In order to remove this type of pixels, a length filtering is introduced. The length filtering removes isolated regions or pixels by using the concept of connected components. These regions represent individual objects, which are labeled to identify separate connected areas. The number of pixels for each connected region to be removed represents the main parameter of the filter, which needs to be experimentally determined. In Fig. 8, the length filter is applied with different parameter values using segmented angiograms with the proposed detection (SSG) and segmentation (Otsu's method) steps.

3.3.2 Measurement of Vessel Width

The process to estimate the vessel width in the segmented coronary angiograms is carried out by different steps, which are illustrated in Fig. 9.

First, edge detection over the segmented image (Fig. 9a) is performed using the Sobel operator, as it is shown in Fig. 9b. Second, the morphological skeleton is computed over the segmented image in order to obtain the medial axis (centerline) of the vessel-like structures as it is presented in Fig. 9c. Third, union operation is performed between the vessel boundary pixels and the vessel centerline as it is illustrated in Fig. 9d. Subsequently, for each pixel of the skeleton, Euclidean distance is computed from the boundary pixels of the coronary angiogram. The closer pixels to the skeleton (minimal distance) represents the vessel width for each skeleton pixel as it is shown in Fig. 9e. Finally, the distances (in pixels) are labeled along the skeleton of the coronary artery. The histogram of the vessel width estimation is presented in Fig. 9f.

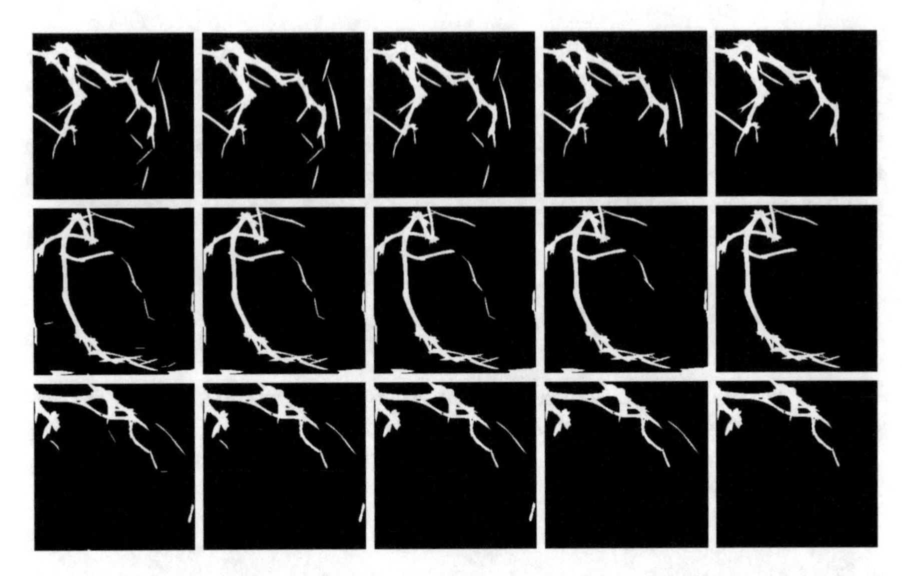

Fig. 8 *First colum* segmented angiograms by Otsu's method. The remaining *four columns* present the results of the lengthfiltering with 50, 100, 200, 500 pixels, respectively

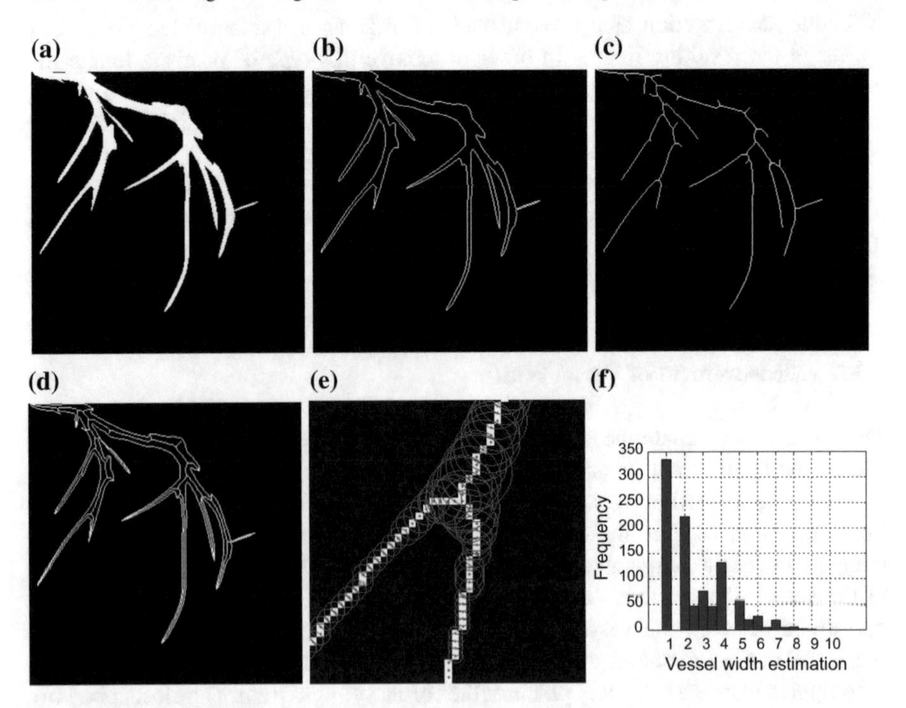

Fig. 9 **a** Segmented coronary angiogram. **b** Border detection of angiogram in (**a**). **c** Skeleton of angiogram. **d** Union between border detection and skeleton of angiogram in (**a**). **e** Process to estimate the vessel width of the angiogram. **f** Histogram of the vessel width estimation

4 Computational Experiments

The computational experiments presented in this section, were performed on a computer with an Intel Core i3, 2.13 GHz processor, and 4 GB of RAM using Matlab version 2012a. The dataset consists of 80 X-ray coronary angiograms of size 300 × 300 pixels from different patients, which were provided by the Mexican Social Security Institute, T1 León. To evaluate the performance of the proposed and comparative methods, the dataset was divided in the training and testing sets with 40 angiograms each one.

4.1 Results of Coronary Artery Detection

To analyze the performance of the single-scale Gabor filters, four state-of-the-art vessel detection methods are compared against each other using the A_z value and the test set of angiograms. In Fig. 10, the ROC curves acquired by the detection methods are presented. This comparative analysis shows that the single-scale Gabor filters obtain a higher coronary artery detection performance than the four comparative methods; therefore, single-scale Gabor filters are employed for further analysis.

Additionally, Fig. 11 shows a subset of angiograms along with the ground-truth images and detection results of the previously mentioned comparative methods. By visual inspection, it can be observed that the single-scale Gabor filter response shows

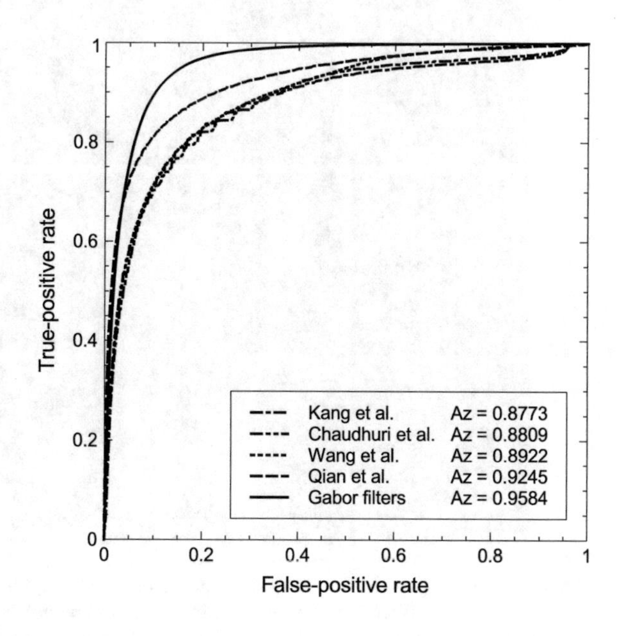

Fig. 10 Comparative analysis of ROC curves with the training set, using the single-scale Gabor filters and four vessel detection methods of the state-of-the-art

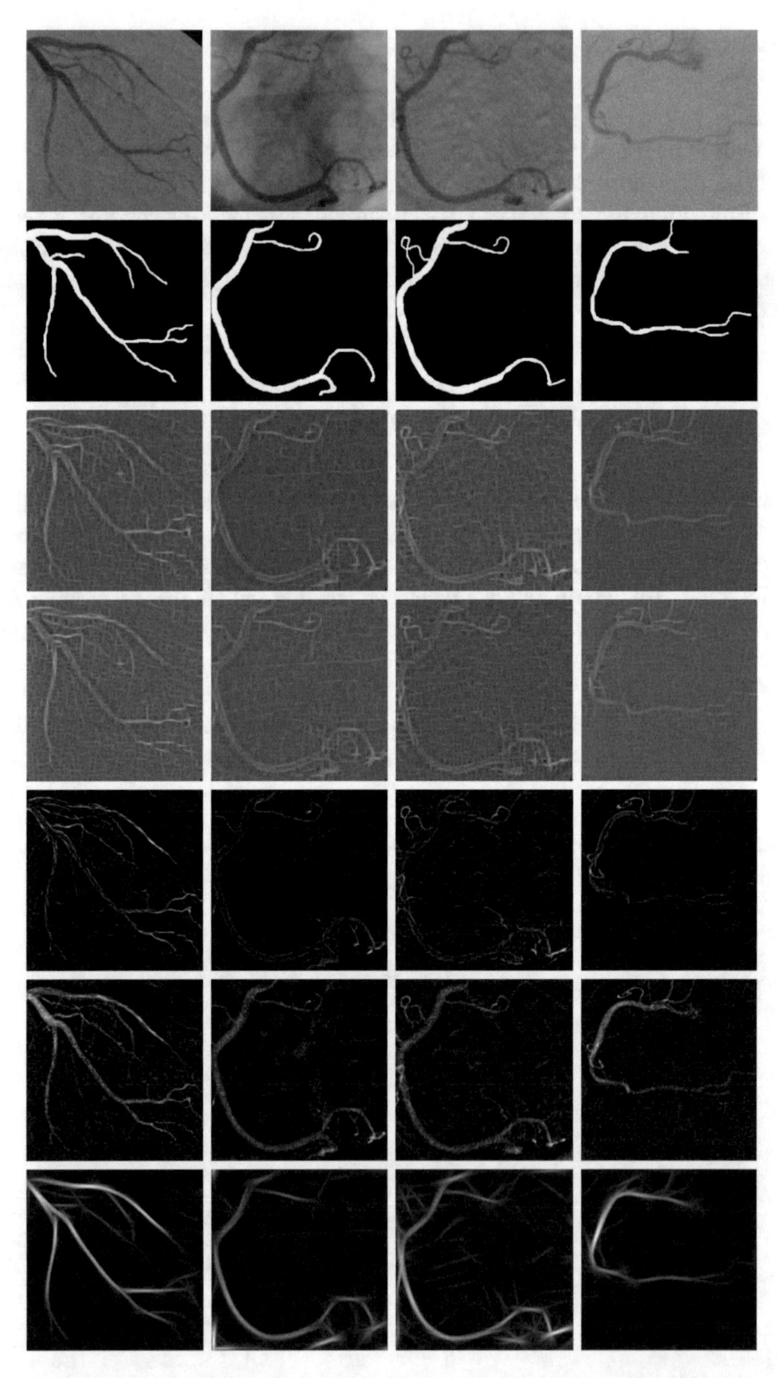

Fig. 11 *First row* subset of coronary angiograms. *Second row* ground-truth images. The remaining *five rows* show the filter response of the methods of Chaudhuri et al. [7], Kang et al. (Gaussian filter) [10], Wang et al. [13], Qian et al. [3] and single-scale Gabor filters, respectively

Table 3 Statistical analysis of GA, UMDA, and BUMDA, in terms of area A_z with 30 runs over the training set of angiograms

Measure	GA	UMDA	BUMDA
Maximum	0.9581	0.9512	0.9584
Minimum	0.9197	0.9188	0.9213
Median	0.9329	0.9258	0.9459
Mean	0.9349	0.9294	0.9442
Std. Dev.	0.0122	0.0098	0.0106

Table 4 Training time (in seconds) of evolutionary methods using the training set of angiograms

Method	Training time (s)
Exhaustive global search	15498.56
GA	7195.23
UMDA	6342.72
BUMDA	6511.87

a better level of enhancement between vessel and background pixels, while reducing noise from the angiogram.

Since the evolutionary computation techniques can be used for optimal parameter selection instead of the exhaustive global search, these techniques have to be evaluated in terms of the area A_z and computational time. In Table 3, a statistical analysis of the detection performance obtained by the evolutionary techniques is illustrated. Due to the EC techniques represent strategies of stochastic global search, the experiment was performed with 30 runs over the training set. According to the measures of maximum, mean and median, the analysis suggests that BUMDA is more stable and robust to work with Gabor filters than the GA and UMDA techniques.

On the other hand, to analyze the EC techniques in terms of computational time, in Table 4 the GA, UMDA, and BUMDA are compared against each other and with respect to the exhaustive global search. This performance analysis shows that UMDA obtains the lowest computational time using the training set. Taking into account the performance of the EC techniques in terms of vessel detection (A_z), and computational time, BUMDA was used for further analysis.

4.2 *Results of Coronary Artery Segmentation*

To assess the classification of vessel and nonvessel pixels from the single-scale Gabor filter response, a comparative analysis of five automatic thresholding techniques in

Table 5 Comparative analysis of five thresholding methods to discriminate vessel and nonvessel pixels from the single-scale Gabor filter response using the test set

Thresholding method	Accuracy
Rosenfeld and De La Torre [31]	0.7131
Pal and Pal [32]	0.8117
Kapur et al. [33]	0.9021
Ridler and Calvard [34]	0.9123
Otsu [30]	**0.9410**

terms of accuracy using the test set is presented in Table 5. In this performance analysis, the inter-class variance thresholding method proposed by Otsu [30] provides the highest segmentation accuracy when compared with the other thresholding strategies.

Moreover, Fig. 12 illustrates the results of the vessel detection and segmentation obtained from the single-scale Gabor filters and Otsu's method, respectively. The proposed method obtains a high rate of true-positive pixels, which leads to high detection of blood vessels of different diameters over the main vascular structure.

4.3 Vessel Width Estimation

The last step of the proposed method is the vessel width estimation, which requires the vessel detection and segmentation steps of the coronary angiogram to be performed. Figure 13, illustrates segmented X-ray angiograms from the test set. From these segmented angiograms, the first step to estimate the vessel width of the vascular structure is shown, which computes the boundary (edge pixels), and the vessel centerline (skeleton operator) of the coronary artery. Finally, the histogram of the vessel width estimation of the entire vascular structure is presented.

The considered vessel detection and thresholding methods provide appropriate performance according to the A_z value and accuracy measure, respectively. However, the proposed method consisting of the application of the single-scale Gabor filters trained by the evolutionary technique called BUMDA to avoid a global search for vessel detection and the Otsu's method for segmentation obtains superior performance in terms of the accuracy measure, as well as, computational time. The obtained results have also shown that the proposed method can be successfully applied for vessel width estimation, which can be useful for systems that perform computer-aided diagnosis in cardiology for the detection of potential cases of coronary artery stenosis.

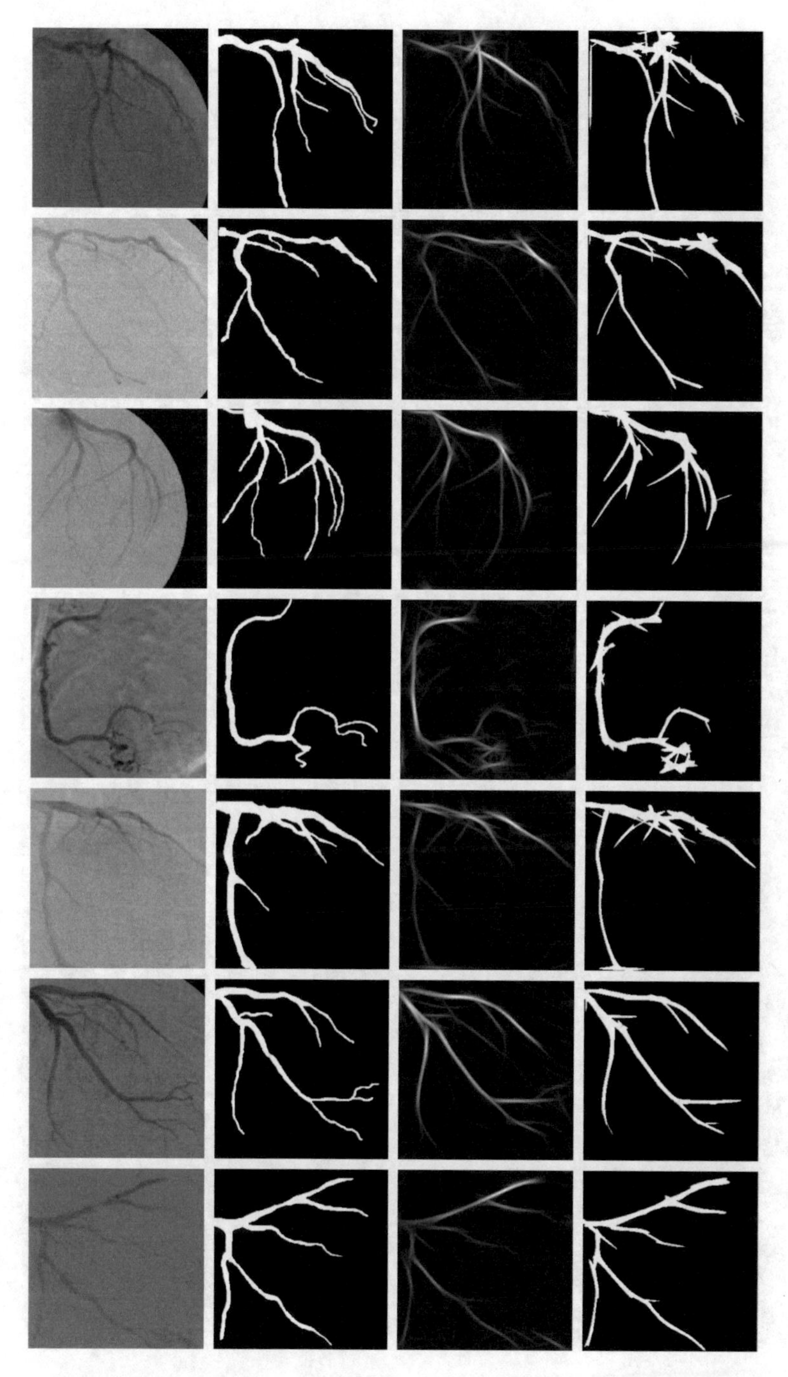

Fig. 12 *First column* angiographic images from the test set. *Second column* ground-truth images. *Third column* Gabor filter response of the angiograms in the *first column*. *Last column* Segmentation results of the Gabor filter response using Otsu's thresholding method

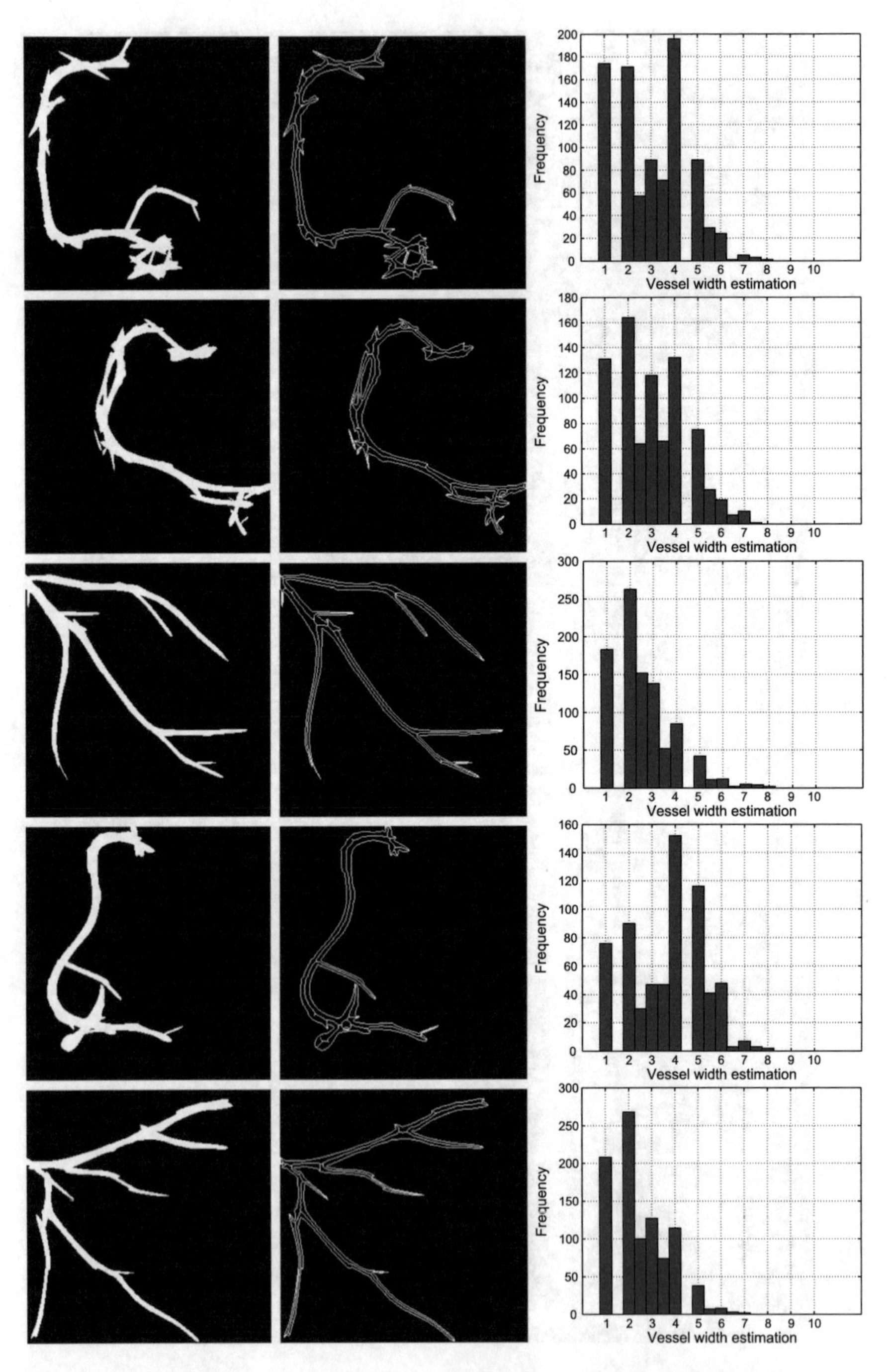

Fig. 13 *First column* subset of segmented angiograms with the proposed method. *Second column* vessel boundary and centerline of the angiograms in the *first column*. *Last column* Histogram of the vessel width estimation for the images corresponding to the *first column*

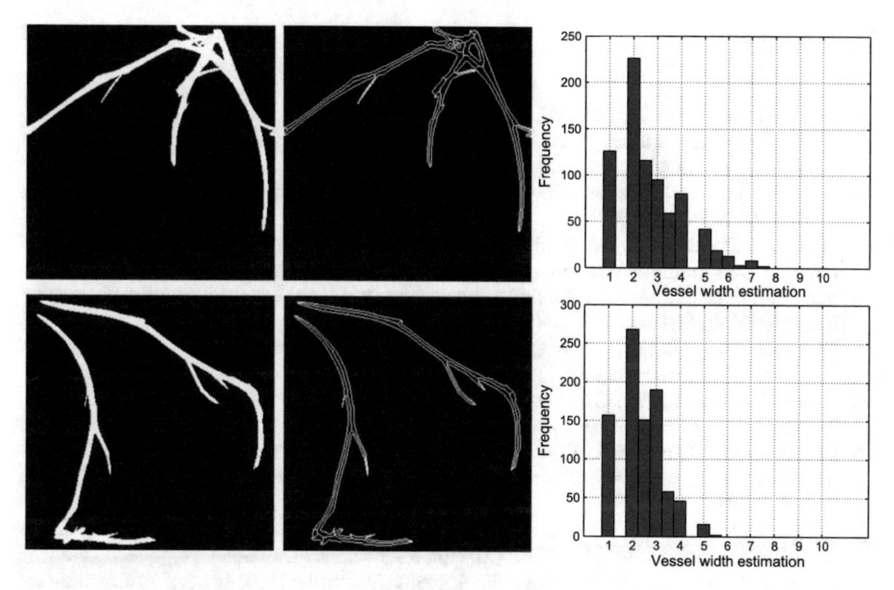

Fig. 13 (continued)

5 Concluding Remarks

In this chapter, a novel method based on single-scale Gabor filters (SSG) for vessel segmentation and vessel width estimation of X-ray coronary angiograms has been introduced. In the first step, a comparative analysis of three evolutionary computation (EC) techniques to improve the performance and reduce the computational time of the training stage of the SSG is performed. The analysis suggests that the Boltzmann univariate marginal distribution algorithm (BUMDA) is suitable for the selection of the optimal parameter values outperforming four detection techniques of the state-of-the-art achieving $A_z = 0.9584$ with a training set. In the second step, five automatic thresholding methods are compared to segment the Gabor filter response, where the inter-class variance method obtained the highest performance in terms of segmentation accuracy achieving a rate of 0.941 with a set of 40 angiograms. Finally, in the last step, a procedure to estimate the vessel width of the segmented vascular structure has been introduced. According to the experimental results, the proposed method can lead to higher accuracy because of the use of the Gabor filters than different state-of-the-art vessel detection methods.

Acknowledgments This research has been supported by the National Council of Science and Technology of México under the project: Cátedras-CONACYT 3150-3097. The authors would like to thank the cardiology department of the Mexican Social Security Institute, for the valuable collaboration and clinical advice.

References

1. Bouraoui, B., Ronse, C., Baruthio, J., Passat, N., Germain, P.L.: Fully automatic 3D segmentation of coronary arteries based on mathematical morphology. In: 5th IEEE International Symposium on Biomedical Imaging (ISBI): From Nano to Macro, pp. 1059–1062 (2008)
2. Eiho, S., Qian, Y.: Detection of coronary artery tree using morphological operator. Comput. Cardiol. **24**, 525–528 (1997)
3. Qian, Y., Eiho, S., Sugimoto, N., Fujita, M.: Automatic extraction of coronary artery tree on coronary angiograms by morphological operators. Comput. Cardiol. **25**, 765–768 (1998)
4. Sun, K., Sang, N.: Morphological enhancement of vascular angiogram with multiscale detected by gabor filters. Electron. Lett. **44**(2) (2008)
5. Chanwimaluang, T., Fan, G.: An efficient blood vessel detection algorithm for retinal images using local entropy thresholding. Proc. IEEE Int. Symp. Circuits Syst. **5**, 21–24 (2003)
6. Chanwimaluang, T., Fan, G., Fransen, S.: Hybrid retinal image registration. IEEE Trans. Inf. Technol. Biomed. **10**(1), 129–142 (2006)
7. Chaudhuri, S., Chatterjee, S., Katz, N., Nelson, M., Goldbaum, M.: Detection of blood vessels in retinal images using two-dimensional matched filters. IEEE Trans. Med. Imaging **8**(3), 263–269 (1989)
8. Cinsdikici, M., Aydin, D.: Detection of blood vessels in ophthalmoscope images using MF/ant (matched filter/ant colony) algorithm. Comput. Methods Progr. Biomed. **96**, 85–95 (2009)
9. Kang, W., Wang, K., Chen, W., Kang, W.: Segmentation method based on fusion algorithm for coronary angiograms. In: 2nd International Congress on Image and Signal Processing (CISP), pp. 1–4 (2009)
10. Kang, W., Kang, W., Li, Y., Wang, Q.: The segmentation method of degree-based fusion algorithm for coronary angiograms. In: 2nd International Conference on Measurement, Information and Control, pp. 696–699 (2013)
11. Frangi, A., Niessen, W., Vincken, K., Viergever, M.: Multiscale vessel enhancement filtering. In: Medical Image Computing and Computer-Assisted Intervention (MICCAI'98), vol. 1496, pp. 130–137 (1998)
12. Salem, N., Nandi, A.: Unsupervised segmentation of retinal blood vessels using a single parameter vesselness measure. Sixth Indian Conference on Computer Vision, Graphics and Image Processing. IEEE, vol. 34, pp. 528–534 (2008)
13. Wang, S., Li, B., Zhou, S.: A segmentation method of coronary angiograms based on multiscale filtering and region-growing. In: International Conference on Biomedical Engineering and Biotechnology, pp. 678–681 (2012)
14. Rangayyan, R., Oloumi, F., Oloumi, F., Eshghzadeh-Zanjani, P., Ayres, F.: Detection of blood vessels in the retina using Gabor filters. In: Proceedings of 20th Canadian Conf Electrical and Computer Engineering (CCECE 2007). IEEE, pp. 717–720 (2007)
15. Rangayyan, R., Ayres, F., Oloumi, F., Oloumi, F., Eshghzadeh-Zanjani, P.: Detection of blood vessels in the retina with multiscale Gabor filters. J. Electron. Imaging **17**(2), 023018 (2008)
16. Sang, N., Tang, Q., Liu, X., Weng, W.: Multiscale centerline extraction of angiogram vessels using Gabor filters. Computational and Information Science **3314**, 570–575 (2004)
17. Al-Rawi, M., Qutaishat, M., Arrar, M.: An improved matched filter for blood vessel detection of digital retinal images. Comput. Biol. Med. **37**, 262–267 (2007)
18. Al-Rawi, M., Karajeh, H.: Genetic algorithm matched filter optimization for automated detection of blood vessels from digital retinal images. Comput. Methods Prog. Biomed. **87**, 248–253 (2007)
19. Cruz-Aceves, I., Hernandez-Aguirre, A., Valdez-Pena, I.: Automatic coronary artery segmentation based on matched filters and estimation of distribution algorithms. In: Proceedings of the 2015 International Conference on Image Processing, Computer Vision, & Pattern Recognition (IPCV'2015), pp. 405–410 (2015)
20. Cruz-Aceves, I., Hernandez-Aguirre, A., Ivvan-Valdez, S.: On the performance of nature inspired algorithms for the automatic segmentation of coronary arteries using Gaussian matched filters. Appl. Soft Comput. p. 12 (2016)

21. Gabor, D.: Theory of communication. J. Inst. Electr. Eng. **93**, 429–457 (1946)
22. Ayres, F.J., Rangayyan, R.M.: Design and performance analysis of oriented feature detectors. J. Electr. Imaging **16**(2), 023007:1–023007:12 (2007)
23. Hauschild, M., Pelikan, M.: An introduction and survey of estimation of distribution algorithms. Swarm Evol. Comput. **1**(3), 111–128 (2011)
24. Larrañaga, P., Lozano, J.: Estimation of Distribution Algorithms: A New Tool for Evolutionary Computation. Kluwer, Boston (2002)
25. Pelikan, M., Goldberg, D., Lobo, F.: A survey of optimization by building and using probabilistic models. Comput. Optim. Appl. **21**, 5–20 (2002)
26. Ivvan-Valdez, S., Hernandez-Aguirre, A., Botello-Rionda, S.: A Boltzmann based estimation of distribution algorithm. Inf. Sci. **236**, 126–137 (2013)
27. Goldberg, D.: Genetic Algorithms in Search. Optimization and Machine Learning. Addison Wesley, New York (1989)
28. Mitchell, M.: An Introduction to Genetic Algorithms. The MIT Press, Cambridge (1997)
29. Lozada-Chang, L., Santana, R.: Univariate marginal distribution algorithm dynamics for a class of parametric functions with unitation constraints. Inf. Sci. **181**, 2340–2355 (2011)
30. Otsu, N.: A threshold selection method from gray-level histograms. IEEE Trans. Syst. Man Cybern. **9**(1), 62–66 (1979)
31. Rosenfeld, A., De la Torre, P.: Histogram concavity analysis as an aid in threshold selection. IEEE Trans. Syst. Man Cybern. **13**, 231–235 (1983)
32. Pal, N.R., Pal, S.K.: Entropic thresholding. Sig. Process. **16**, 97–108 (1989)
33. Kapur, J., Sahoo, P., Wong, A.: A new method for gray-level picture thresholding using the entropy of the histogram. Comput. Vis. Graph. Image Process. **29**, 273–285 (1985)
34. Ridler, T., Calvard, S.: Picture thresholding using an iterative selection method. IEEE Trans. Syst. Man Cybern. **8**, 630–632 (1978)

Hybrid Intelligent Techniques for Segmentation of Breast Thermograms

Sourav Pramanik, Mrinal Kanti Bhowmik, Debotosh Bhattacharjee and Mita Nasipuri

Abstract The incidence of breast cancer has rapidly increased over the past few decades in India and the mortality rate is more than other countries across the entire world. These facts have motivated the development of new technologies or modification of the existing technologies for the identification of breast cancer before it metastasizes to the neighboring tissues. Breast thermography is a promising front-line breast screening method, which is noncontact, cheap, quick, economic, and painless. The use of thermal imaging for the identification of breast abnormality is based on the principle that the temperature distribution in precancerous tissue and its surrounding area are always higher than that in normal breast tissue. However, the accurate interpretation and classification of the breast thermograms for proper diagnostic decision-making is a major problem. Proper segmentation of hottest region from the segmented breast region plays a key part in the diagnosis of breast cancer that calls for the application of hybrid intelligent methods in the segmentation of hottest region. The shape and size of the hottest regions are used to determine the degree of malignancy of the tumor and classify its type. Hybrid intelligent systems have been successfully applied in the classification of breast thermal images over the last few years. In this chapter, we have proposed a sequential hybrid intelligent technique for the segmentation of the hottest region and also shown the significance of hybrid intelligence systems over the conventional methods for the segmentation of hottest region. A detailed review related to the segmentation of breast region and the segmentation ofhottest region is included in this chapter. In addition, this chapter

S. Pramanik (✉) · D. Bhattacharjee · M. Nasipuri
Department of Computer Science and Engineering, Jadavpur University, Kolkata, India
e-mail: srv.pramanik03327@gmail.com

D. Bhattacharjee
e-mail: debotosh@cse.jdvu.ac.in

M. Nasipuri
e-mail: mnasipuri@cse.jdvu.ac.in

M.K. Bhowmik
Department of Computer Science and Engineering, Tripura University
(A Central University), Suryamaninagar 799022, Tripura, India
e-mail: mkb_cse@yahoo.co.in

S. Bhattacharyya et al. (eds.), *Hybrid Soft Computing for Image Segmentation*, DOI 10.1007/978-3-319-47223-2_11

also contains the detailed overview of the principles, reliability, and predictive ability of the breast thermogram in early diagnosis of breast cancer.

Keywords Segmentation · Fuzzy c-means · Devies–Bouldin Index · Breast region · Hottest region · Color segmentation

1 Introduction

Breast cancer is the second most frequent cancer type among the women in the world. According to the report presented by the International Agency for Research on Cancer (IARC) in 2012, around 1.67 million new breast cancer patients were diagnosed globally, which is nearly 25 % of all cancer types in women. Over the time, incidence rate of breast cancer has been increasing in most of the countries across the world. Since 2008, it has escalated by 20 % in 2012, while mortality rate has escalated by 14 % [1]. However, there are immense inequalities between the developed and developing countries with respect to the incidence and mortality rates. In comparison to the developed countries, the mortality is considerably greater in developing countries but the incidence rates are still considerably greater in the developed countries. For example, in Western Europe, more than 90 new breast cancer cases per 100000 women were diagnosed each year, while there are only 30 cases per 100000 in Eastern Africa [1] but the mortality number is almost identical in these two regions accounting nearly 15 per 100000 women. According to Dr. David Forman, head of the IARC, the increase in incidence rates and mortality rates in the developing countries because the clinical advances, treatment facilities and awareness program to fight against the disease are not reaching to the women living in these countries. Figure 1 illustrating the incidence and mortality rates across the different world regions with rates per 100000 women [1].

According to the National Cancer Registry Program (2007–2011), the occurrence of breast cancer has rapidly increased over the past few decades in India that accounts for approximately 25–32 % of all types of cancers in women, which is the major concern nowadays. As per the data obtained from Globocan-2008 and Globocan-2012, over just 4 years, the breast cancer has overtaken cervical cancer in India [1]. It is the most frequent occurring disease among the women in maximum cities in India and the second most occurring disease in rural areas [2]. Because breast cancer is still a nonexisting entity and screening is an alien word for most of the women in India until and unless a near and dear one suffers from it. However, the major concern in India regarding the incidence of breast cancer is the significant shift of average age (20–50 years) that means more young ladies are getting this ailment. Due to the lack of treatment facilities and awareness programs, the mortality rate of women with breast cancer is very high in India accounting about 50 %, while 16.6 % in United States and 25 % in China.

Owing to this high incidence and mortality rates, research on early breast cancer detection has been getting importance these days. Some studies have suggested that

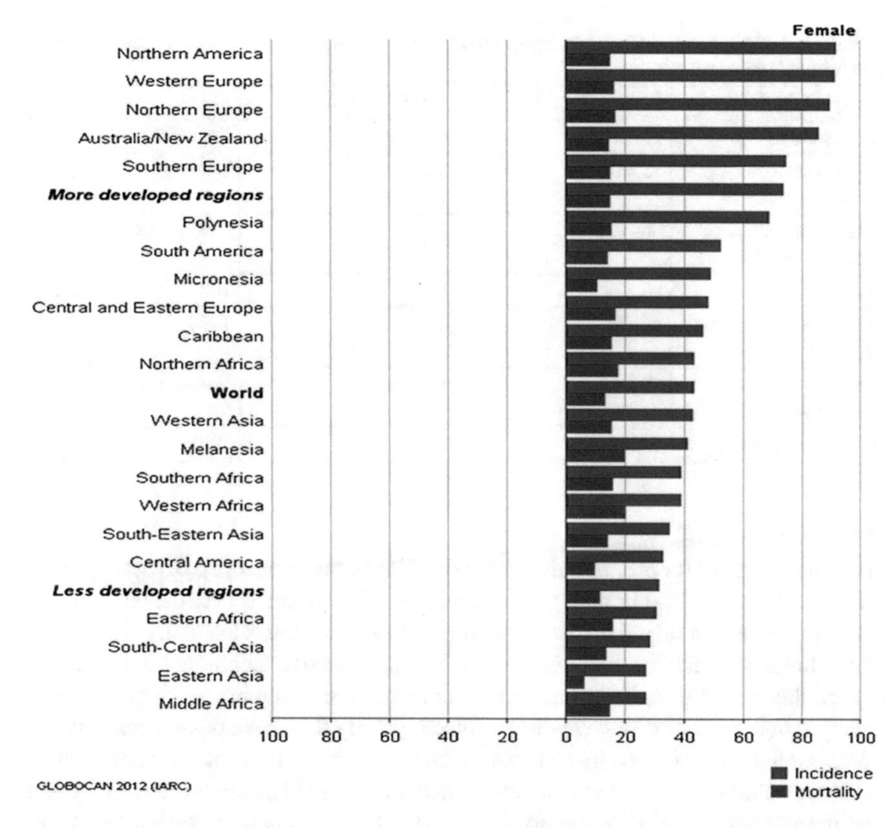

Fig. 1 Estimated incidence and mortality age-standardized rates (world) per 100000 women [1]

the breast cancer has 95 % survival rate if it is detected at the very early stage [3]. However, there is no such effective screening method available which can ensure early prevention of breast cancer. Over the last few decades, researchers have put many endeavors to develop new breast imaging modality or ameliorate existing modality for early identification and effectual therapy of the disease. As a result, a number of diagnostic imaging modalities such as ultrasound, mammography, MRI, CT scan, and breast thermography have been developed for the early detection of breast cancer. Among them, mammography has been considered as the gold standard since 1976 and still is mammography which has sensitivity between 13–90 % [4]. However, mammography-based screening has lots of risk factors. The ionizing radiation which is associated with the mammography screening can increase the chances of developing breast cancer, specially in young women. Also, it is very hard to obtain acceptable mammogram image for an implant breast, dense breast, and fibrocystic breast. Since the mammography is an invasive technique, rupture risk of cancerous tumor is associated with it. For these reasons, thermography has received lot of attention by the medical practitioners and considered as the most promising and

Table 1 FDA approved different imaging modalities and their sensitivity and specificity for the diagnosis of breast cancer

Imaging modalities	Sensitivity(%)	Specificity(%)	Year
Ultrasound	13–98.4	67.8–94	Late 1970s
Film-screen mammography	13-90	14–90	1976
Magnetic resonance imaging (MRI)	86–100	21-97	Late 1980s
Thermography	90	90	1982
Electrical impedance imaging	62–93	52–69	1984
Positron emission tomography (PET)	80-86	91–100	Mid–1990s
Digital mammography	63.3	88.2	1997

acceptable options for preliminary diagnosis of the breast cancer [5]. Breast thermography is a functional examination with sensitivity 90 % which senses the temperature distribution pattern on the breast using the infrared radiation released by the surface of the breast. Unlike the mammogram, it is a noninvasive technique that does not expose the patient to radiation or involve compression of the breast. Specifically, it is very much useful for young women since typically they have dense breast tissue. Some studies have showed that a thermography has ability to locate precancerous or cancerous tumor much earlier than others methods [6]. In [4], authors have suggested that thermography has potential to alert women 8 to 10 years earlier than a cancer is detected in the mammogram. Table 1 shows the FDA approved different imaging modalities used for early breast cancer detection and their sensitivity and specificity.

Thus, in comparison to other techniques, the thermography is the only method which provides physiological information of the breast and its result is independent of age, hormone replacement therapy, and tumor location that justify a pivotal role of thermography in early identification of breast cancer. However, interpretation of thermal breast images still relies on the visual analysis made by the trained radiologists. But, the human perceptibility for the interpretation of breast thermal images is often affected by the carelessness, absent-mindedness, and fatigue of the radiologist. Hence, computer-aided analysis of breast thermogram has been playing a significant role from the past few years. Accurate analysis of breast thermal image in search of abnormality using computer-aided analyzing tool comprises of mainly three steps, such as segmentation of interested region, extraction of features, and classification. Among them, segmentation of the interested region is the kernel phase for the analysis of thermal breast image. Segmentation of the region of interests with regard to the breast thermal image processing is mainly two types, namely, breast region segmentation and hottest region segmentation. Usually, the breast thermal images are captured in a larger area that includes upper body area, breast region, and lower body area. Thus, it is required to separate the breast region area from the rest of the body to further process it. Once the breast region is segmented, another most important phase is hottest region segmentation. The hottest region in the breast thermal image describes the most useful information regarding the goodness of the breast. Accurate segmentation of this region leads to proper identification of abnormal case and the degree of malignancy. However, segmentation of the hottest region is a very challenging problem because the boundaries of the hottest regions are not clearly visible and also the size and shape vary in these images. Over the past few years, some techniques have been delineated in the literature by various researchers for the segmentation of breast region and to detect potentially hottest regions from the segmented breast region [7–10]. In contrast to the conventional segmentation techniques, the segmentation technique based on hybrid intelligent method provides relatively better results for the segmentation of hottest region that we have showed in the this chapter. The experimental results justify the use of hybrid intelligent method is proficient enough for the segmentation of the hottest region.

The organization of this chapter is as follows. Section 2 illustrates a summary on breast cancer types, possible symptoms and risk factors, and importance of breast cancer awareness program. A brief overview on predictive ability of breast thermography and the interpretation of breast thermography are presented in Sect. 3. Section 3 also includes the interpretation techniques of breast thermography. The necessary breast thermal image acquisition protocol and the databases available for this research are described in Sect. 4. Section 5 presents a brief outline on importance of region of interest segmentation for the analysis of breast thermogram along with a review work on region of interest segmentation. The proficiency of hybrid intelligent system for the segmentation of hottest region is also included in this section. Finally, the concluding remarks of this chapter are reported in Sect. 6.

2 Breast Cancer

Breast cancer is an uncontrolled division of abnormal cells that starts in the tissue of the breast which include either the milk producing glands called lobules glands or in the ducts that convey milk to the nipples. This type of growth of the cells in the breast forms a mass or lump, called a tumor. Based on the growth and origin, the tumor can be classified as either benign which is not cancerous or malignant which is cancerous. However, breast cancer occurs exclusively in women, but it also occurs in men which is very rare. To understand the breast cancer properly, it is necessary to understand the anatomical structure of the normal breast. Figure 2 illustrates the anatomical structure of the breast. Anatomically, a woman breast is composed mainly of lobules glands that produce milk, ducts tubes that convey milk from the lobules glands to the nipple, nipple, areola surrounding the nipple, fibrous or connective tissue that surrounds the lobules and ducts, fatty tissue, blood vessels, and lymphatic vessels and nodes. Each breast of a woman sits over a chest muscle called the pectoral muscle. The breast tissue expands horizontally from the sternum edge to the midaxillary line and enclosed by a thin surface of connecting tissue called

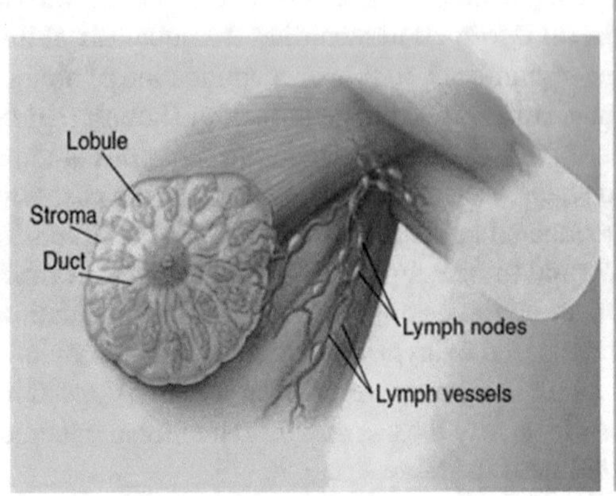

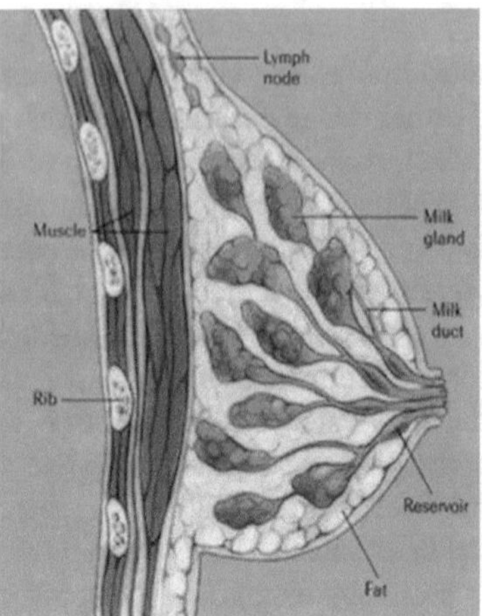

Fig. 2 Anatomical structure of the female breast [11]

fascia. The deep surface of the fascia typically lies immediately on top of the pectoral muscle while the superficial layer lies just below the skin. The breast consists of a large number of blood and lymph vessels. The lymph vessels flow just opposite direction of blood vessels whose main task is to collect and passage lymph fluid away from the breast tissue into lymph nodes. The lymph nodes and lymph vessels together called the lymphatic system whose main purpose is to fight against infection. Understanding of the lymphatic system is important as breast cancer metastasizes through this system. It is also essential to note that an axillary tail of Spence or simply a tail of the breast tissue can expand into the axilla. This is important because the breast cancer can also develop in this axillary tail.

The most frequent types of cancer occur in the breast are ductal carcinoma that start in the milk duct and lobular carcinoma that start in the lobules gland. Based on the level of spreading they are also classified as invasive and noninvasive breast cancer. The term invasive refers to as the cancer cells that spread to the neighboring breast tissue while noninvasive cancers cells do not spread to the neighboring cells. Invasive ductal carcinoma (IDC) is the most occurring breast cancer among the women accounting about 80 % of all types of breast cancer. Typically, this type of cancer begins in the duct tube and then over time it invades to the neighboring breast tissue by breaking through the wall of the duct. Another type of most frequent breast cancer after the IDC is invasive lobular carcinoma (ILC) accounting about 10 % of all breast cancer types. It begins in the milk producing gland called lobules and over time it metastasizes to the neighboring breast tissue, lymph nodes, and other regions of the body. However, IDC and ILC are more frequently occurring disease as women grow older but it can also occur at any age. The most frequent noninvasive breast cancers are ductal carcinoma in situ (DCIS) that starts inside the duct tube and lobular carcinoma in situ (LCIS) that begin inside the lobules, which remain in the same location over the time. DCIS and LCIS are not life-threatening, but

Table 2 Breast cancer stages and their five years survival rate [12]

Stage	Type	5-years survival rate (%)
0	DCIS and LCIS	92
I	Invasive ductal or lobular carcinoma less than or equalto 2 cm in size without nodal involvement and not metastasize to the neighboring tissue	87
II	Invasive ductal or lobular carcinoma less than 5 cm in size without nodal involvement but with movable axillary nodes and not metastasize to the neighboring tissue	75
III	Invasive ductal or lobular carcinoma less than 5 cm in size with nodal involvement and fixed axillary nodes	46
IV	Any form of breast cancer with distance metastasis	13

having DCIS or LCIS can increase the possibility of getting an invasive breast cancer afterwards in life. Other very uncommon types of breast cancers are inflammatory breast cancer which is invasive and paget disease of the nipple which arises in the ducts and metastasizes to the skin of the nipple and then move to the areola Breast cancer can also be categorized into different stages based on the survival rate [12]. Table 2 illustrates different stages of breast cancer and their five years survival rate.

2.1 *Different Risk Factors of Breast Cancer*

The most fearing disease for the women is breast cancer. Because the mortality rate of breast cancer is the second highest among all type of cancer in women. However, most women are unaware about the actual cause of developing their cancer. There are lots of risk factors connected with the growth of breast cancer. Among them some risk factors, such as being a woman, family history, age, genetic factors, personal history, and menstrual history could not be altered. On the other hand, environmental and lifestyle risk factors like overweight, lack of physical activity, consumption of alcohol, hormone replacement therapy, smoking, etc., could be avoid. But a woman having single or multiple risk factors do not necessarily signify will develop breast cancer. Similarly, if she does not have any risk factors, not justify she would not get breast cancer. However, knowing the possible risk factors for the development of breast cancer may help the women to take preventative measures to decrease the likelihood of developing the breast cancer. The certain established risk factors those are linked to the growth of breast cancer are described below:

- Gender: The fundamental risk factor for getting breast cancer is just being a woman. Although it also occurs in men but it is a very rare disease in comparison to women. Typically, the breast cancer incidence rate for a woman is hundred times more usual than a man [13].
- Age: Another strongest risk factor associated with the development of breast cancer is ageing. Typically women with age of 50 or more have higher chances of developing breast cancer. As stated by the American Cancer Society, two out of three women have been found of getting invasive breast cancer at the age of 55 or older. Furthermore, less than five percent women of under age 45 have been found developing invasive breast cancer. Thus, every women of above age 20 should undergo regular breast screening program [14].
- Family History and Genetic Factors: A woman, whose first or second-degree female or male relatives diagnosed with breast cancer has higher chances of getting the disease compared to a woman without such a family history. But a woman without a family history of the disease not necessarily means she will not get it. If she has one first-degree relative who was diagnosed at the younger age or has more than one relative with breast cancer history then her risk of getting the disease is three to four times higher [15]. However, breast cancer is also caused by some inherited gene mutations accounting for only 5–10 % of all cancer cases. Usually, two genes such as BRCA1 and BRCA2 are the most common genes that increases the possibility of getting breast cancer [15]. Estimated risk of these two genes are different. Typically, BRCA1 gene has about 55–65 % possibility while BRCA2 has about 45 % possibility of developing breast cancer by age 70 [15].
- Personal history: A women diagnosed with breast cancer or with some benign breast conditions such as hyperplasia, lobular carcinoma in one breast have 3 to 4 times more likelihood of developing a new cancer in a different part of the same breast again or in the other breast.
- History of breast biopsy: Another important risk factor that is linked with the growth of breast cancer is higher number of breast biopsies.
- Having dense breast tissue: High breast density is associated to an increased possibility of developing breast cancer. A woman breasts are made up mostly of fatty and fibrous tissues and epithelial components which includes lobules and ducts tubes. A dense breast means that it consists of more lobules and fibrous tissue and considerably less fatty tissue. Compared to a woman with low breasts density, very dense breasts woman has about four to five times greater chances to develop the disease [16].
- Menstrual history: A woman who had started menstruation at younger than age 12 have higher chances of getting breast cancer in subsequent life. Similarly, if she went through menopause at an older age of 55 have greater chances of getting breast cancer. Because more menstruation cycle causes longer exposure to the oestrogen and progesterone hormones which stimulate the growth of breast cancer cells [17].
- Overweight: A woman ovaries usually produce maximum amount of oestrogen hormone while fatty tissue produce very less amount of oestrogen hormone to keep balance of oestrogen levels before menopause. But after menopause ovaries stop producing oestrogen hormone and thus maximum oestrogen hormones comes

from fatty tissue. Overweight or obese women have extra fat cells which typically increase the oestrogen levels in the body, as a result have a greater chance of getting breast cancer [13].

- Radiation therapy to the chest in early life: Radiation therapy is always associated with some side effects when given to the chest area. A women who had radiation therapy to the chest at the very early age, specifically during adolescence, for the treatment of other disease such as Hodgkin or non-Hodgkin lymphoma, have 20 times higher risk factor than average [17].
- Having children in later life or no child: The development of breast cancer risk to a woman is usually related to exposure to the endogenous oestrogen and progesterone hormone which are produced by woman ovaries. Breast cells of a woman are immature and very sensitive before first full-term pregnancy and respond to the oestrogen and progesterone hormone. Pregnancy helps breast cells to grow in a regular way and make it fully mature which is the main reason to get protecting benefit against breast cancer. Some studies have reported that a woman who gives birth to her first baby after age 30 have increase risk of developing the disease compared to a woman who gives birth of her first child before age 30 [17].
- Breast feeding: Breastfeeding for a longer period of time, i.e., two years or more can reduce breast cancer risk for a woman. Some related study have shown that a woman who breastfed at least two years for her lifetime got twice the benefit compared to those who breastfed one year throughout her lifetime [18].
- Use of birth control pills for long time: In addition to effectively stopping undesired pregnancy, it is also very much important that birth control pills be safe. Women who use birth control pills for a longer time have increase risk of developing breast cancer. Typically, birth control pills apply hormones to block pregnancy that also over stimulate breast cells escalates the chance of getting breast cancer [17].
- Using hormone replacement therapy (HRT): Use of HRT by the women for a longer period of time after menopause causes higher risk of being develop breast cancer. There are mainly two types of HRT such as oestrogen only-HRT and combined HRT that contains oestrogen and progesterone hormone. The use of combined HRT, even for a short time, increases possibility of getting breast cancer by around 75 %, while oestrogen only-HRT increases the chance of contracting breast cancer if it used for more than 10 years [17].
- Consume alcohol: Regular consumption of alcoholic beverages increase breast cancer risk for the women compared to the women who consumed alcohol only occasionally. Typically, alcohol increases amount of oestrogen and other hormones linked with hormone-receptor-positive breast cancer. Experts estimated that the women have 25 % higher risk of developing breast cancer if she consumed one drink per day compared to the women who never drink at all [17].
- Lack of physical activity: A sedentary lifestyle of a woman can increase the possibility of breast cancer. According to the American Cancer Society reports, regular exercise of about 4–7 h per week can reduce the possibility of breast cancer by 18 % [13].
- Smoking: Cigarette smoking has been linked to a number of diseases. Recent research has been suggested that smoking is highly connected with the growth

of breast cancer for the women who smoke in younger age and premenopausal stage [13].

- Night work: Lots of studies have been reported that the women have a greater chance risk of developing breast cancer if they work at night or lives in areas where high levels of external light available at night [17]. This increase in risk of breast cancer is associated with the level of Melatonin hormone in the body. Typically, the Melatonin hormone maintained the body's sleep cycle. The production Melatonin hormone is high during the night time, while low during the day time or when eyes register light exposure. But still it is not known how much darkness is required to begin Melatonin production.

2.2 Importance of Breast Cancer Awareness

The incidence of breast cancer and mortality due to the breast cancer is very common among the women in worldwide. According to the World Cancer Research Fund report about 1.67 million (around 12 %) new breast cancer cases are diagnosed in 2012, which is the second most frequent cancer overall. Incidence rates of breast cancer around the world are increasing day-by-day which is a major concern for the health professional. However, detecting and treating the warning signs of the disease early can significantly increase the chances of survival. Breast cancer awareness program has become a very prevalent term these days, whose main aim is to raise awareness about the disease to the people through education about the symptoms, risk factors, treatment options, preventive measure, survivorship, and importance of getting tested early as cancer is highly curable disease if detected early, and assimilate about protecting yourself. There are different things such as self-breast examinations, taking healthy diet, regular exercise, clinical breast exam (CBE) in every three years, and not smoking, etc., women can do to aid prevent breast cancer. There are various organizations available in the globe whose main purpose is to raise awareness about the breast cancer. Among them some of the most active organizations include the Living Beyond Breast Cancer, Susan G. Komen breast cancer foundation, Indian Cancer Society, American cancer society, Breast Cancer Research Foundation, Breastcancer.org, The Rose, Bay Area Cancer Connections, Dana-Farber Cancer Institute, God's Love We Deliver, and It's The Journey Inc. Compared to the developed country, breast cancer is still a nonexistent entity for the women in the developing country until and unless it moves to the stage 2B and beyond. Specifically, the situation is horrific in the rural areas of the developing country where breast screening is an alien word for majority of the women. In India, Indian Cancer Society is working hard to educate people about the usefulness of regular breast screening and possible risk factors of getting breast cancer since 1951. Recently, a good initiative has been taken by Tripura University, Tripura, India, along with Regional Cancer Centre, Agartala Government Medical College, Tripura, India and Jadavpur University, Kolkata, India to publish a breast cancer awareness booklet in Bengali language which will include breast cancer symptoms, breast cancer staging, methods of performing breast self-examinations, presence of different breast imaging modalities, and different treatment options to educate people in rural areas.

3 The Role of Thermal Breast Imaging in the Identification of Breast Cancer

Surface temperature of the human body has been a long established criterion about the goodness of the health since 400 B.C. The Greek physician Hippocrates wrote that "In whatever part of the body excess of heat or cold is felt, the disease is there to be discovered" [19]. He smeared the body with wet mud and identified the region that dried quickly as the existence of the diseased tissue. In [20], authors have investigated the relation between the breast cancer and breast skin temperature. They found and concluded that the difference between the rhythmic changes in breast surface temperature of a pathology breast and a healthy breast were measurable and real. Another study by Gautherie et al. [21] mentioned that the temperature information of the breast is a potential indicator of breast cancer in its early phase. The presence of a tumor, endocrine changes, and inflammation in the breast usually generates more heat than the normal breast tissue that influences the changes in the breast skin surface temperature which is recorded by the thermal camera [22]. There are various explanations available for this temperature changes that includes angiogenesis, presence of nitric oxide, inflammation, and estrogens [22]. To support the growth of cancerous tumor, it creates new blood vessels, and also opens the dormant vessels for the supply of necessary nutrients and oxygen. The process of creation of new blood vessels and opening of dormant vessels is called pathologic angiogenesis. Typically, cancerous cells used nitric oxide (NO) as a local vasodilator to increase the oxygen and nutrients to the cancerous cells, which causes the localized increase of temperature. On the other hand, inflammation is another process by which localized increase of temperature in the breast may be generated. Furthermore, imbalance of estrogens may cause localized temperature changes, since it also increases the production of nitric oxide.

3.1 *Predictive Ability of Breast Thermography*

Since late 1960s, several researches have been carried out to detect breast abnormality using thermal imaging. The Congressionally Directed Medical Research Program has defined some ideal features for the breast cancer screening method, such as identification of tumors in its early stage, high sensitivity and specificity, noninvasive, cheap, and decrease mortality [23]. Breast cancer screening using thermal imaging met all the mentioned requirements [24]. Studies have suggested that an abnormal thermal image of the breast is probably the single most reliable indicator for the identifying of breast cancer in its very early phase [3]. In 1965, a researcher and radiologist from the Albert Einstein Medical center, introduced the potential of thermal infrared imaging in breast cancer detection in United States [25]. He reported a total of 4000 cases by using a Barnes thermography with sensitivity of 94 %. In prospective studies, the thermography was first used in a gynaecologic practice by Hoffman [26]. He identified 23 carcinomas in 1924 patients with a sensitivity of 91.6 % and a specificity of 92.6 %. In a study, Amalric et al. [27] used total 25000 subjects breast thermal images in which 1878 were confirmed breast cancers. They found sensitivity and specificity of about 91 %. In [28], authors diagnosed 37506 patients using thermal imaging. They reported 5.7 cancers per 1000 patients with 14 % false positive and 12 % false negative rate. Study by Spitalier et al. [29], investigated 61000 patients using thermal imaging for more than 10 years. They reported sensitivity and specificity of 89 % and concluded that the thermal image alone has an ability to detect the first signs of tumor development about 60 % cases. A detail comparison among thermography, mammography, and physical examination in search of breast abnormality has been made by Nyirjesy et al. [30]. They used three different groups of patients that include 4716 patients with proven cancer, 8,757 normal, and 3,305 patients with histologically confirmed benign breast disease. They reported that physical examination had an average sensitivity of 75 % for detecting all types of tumors. The average sensitivity and specificity of mammogram was achieved 80 % and 73 %, respectively, while thermography achieved 88 % average sensitivity and 85 % specificity. In [31], authors have used thermography to the mammographically questionable cases for a period of 4 years. In the recent years, due to the enhancement of infrared thermal technology, advancement in image processing technique, and better understanding of the pathophysiology of heat generation researchers have even achieved better sensitivity, specificity, and accuracy [32] in breast cancer detection. However, an examination over 15 large scale projects for a period of 1967–1998 indicated that the average sensitivity and specificity of abnormality detection from breast thermal image has achieved 90 %, which justify the use of thermal imaging in search of breast abnormality detection.

3.2 Interpretation of Breast Thermography

The presence of infection, tissue inflammation, or hormone imbalances increase the false positive findings of breast thermography, especially on initial studies of breast cancer of an individual. Thus, proper interpretation of breast thermography by experienced interpreters with standardized interpretation rules may diminish the false positive rate of breast thermography during the diagnosis of breast cancer. Early methods of breast thermal image interpretation was entirely based on subjective criteria and used to read only variation of vascular patterning of each breast without considering temperature difference of two breasts [33]. As a result huge variations in the outcomes are generated by the interpreters. In 1970, researchers have tried proving that the use of subjective information along with quantitative data can ameliorate the accuracy, sensitivity, and specificity in the interpretation of breast thermography. A standardized method for breast thermal image interpretation was described in 1980s, which is based on large-scale studies and previous research [28, 34]. Using this methodology, the internationally standardized breast thermal image classification, known as Marseille System, describes five standard and recognizable reporting classes, ranging from TH-1 (normal) to TH-5 (severely abnormal) and two specialized classes: TH-0 (incomplete or technically imperfect) and TH-6 (evaluating thermography features for already proven cancer using biopsy) [34]. The use of these standardized explanation methods notably increased the breast thermograms positive and negative predictive value, sensitivity, specificity, and intra or interexaminer interpretation reliability [35]. Figure 3, shows the examples of six classes. The classes of TH1 to TH6 are explained as follows:

- TH1: This category of breast thermal image defines a temperature profile of the breasts that do not exhibit any of the thermology features linked with risk for the breast cancer.
- TH2: Defines the thermal profiles of the breast, which are uniform, regularly patterned, and comparatively large blood vessels. These types of thermal features are usually associated with functional benign changes, such as hormone imbalances or lactation, and pregnancy.
- TH3: Defines unusual thermal profile of breast tissue or blood vessels that indicates a minor or equivocal risk for the diagnosis of breast cancer. Typically these types of features indicate benign changes, such as acute cysts or fibro-adenoma development, inflammation, infection, etc.
- TH4: Represent abnormal tissue function or blood vessels that are probable indication of risk for breast cancer.

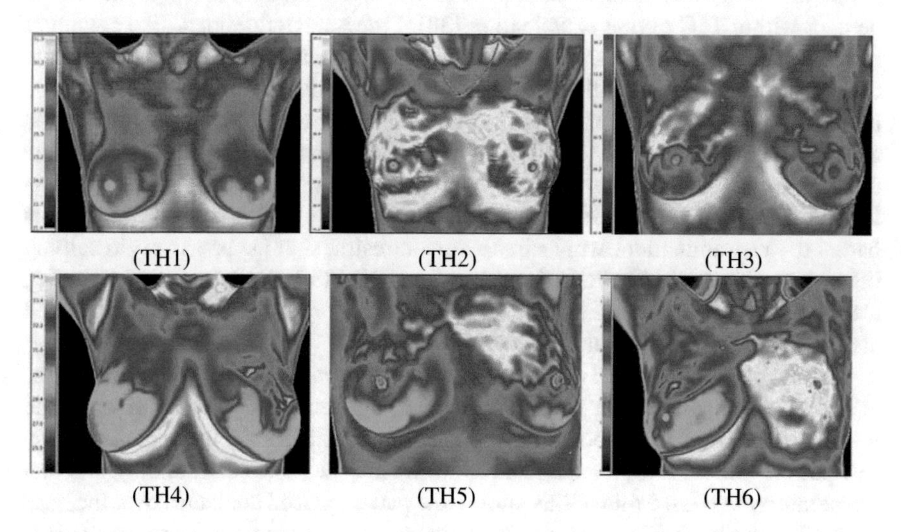

Fig. 3 Example of classes TH1 to TH6 [36]

- TH5: Defines severely abnormal tissue or blood vessels function that is linked with a very high risk of breast cancer.
- TH6: This category is a specialized classification applied to breast thermography when the presence of breast cancer has already been identified using tissue biopsy and no traditional therapy has been applied.

4 Breast Thermal Image Acquisition

4.1 Laboratory and Patient Preparation

Acquisition of diagnostic quality breast-thermal-image entirely depends on the proper laboratory setup and preparation of patient. Thus, it is necessary to follow some strict protocol for laboratory setup and patient preparation prior to image acquisition. The skin in the human body typically helps to maintain the body temperature in a very small scale by altering heat transfer procedures from the body to the surroundings and vice versa. However, the physiological activities and thermal properties of the skin may get affected by the several factors such as pressure, temperature, damage, infection, inflammation, hormone imbalance, and age. Thus the thermal image acquired in an uncontrolled environment and from an unprepared patient may increase the false positive findings in the thermography. Different factors such as room size, environmental control system, computer equipment, patient cubicle, and minimized external infrared interferences are associated with the preparation of examination room [37]. Some studies suggested that the room size should be 2×3 m to 3×4 m and ambient temperature must be maintained between 18–25 °C and changes of temperature must be kept within 1 °C during examination [38]. Processing equipment like computer must be kept sufficiently away from the patient. A patient cubical must be present in the room, where patient will take rest and maintain privacy during acclimatization period. Any external IR radiation coming into the room must be prevented by shielded all the windows in the room [39]. The patient must be instructed prior to examination to avoid the sun bath five days prior to the image acquisition, avoid using of cosmetics and deodorants on the day of the examination, no physical exercise four hours before to the examination, avoid stimulation or treatment of the breast, avoid bathing before examination and bathing should not be greater than one hour, avoid smoking, alcohol, heavy meals, above average intake of coffee or tea before examination, and also tight fitted clothing should be avoided [39].

Finally, patients upon arrival in the examination room must be informed about the examination process, instructed to remove clothing from the upper part of waist and jewelry in the cubicle, and be informed to sit for about 15 min in the cubicle. During this period, patient body temperature is acclimated with the surrounding ambient temperature. The last 5 min of this period, the patient placed her hands over the head to ameliorate anatomic presentation of the breasts for imaging. Furthermore, patients are advised to not fold or cross their arms or legs and not to place their feet on a cold surface [37].

4.2 Acquisition System

Accuracy of the breast thermal image is entirely dependant on the sensitivity and resolution of the thermal camera [23]. Studies suggested that a thermal camera with good sensitivity can identify a very slender temperature difference and can present useful thermal details with a resolution as high as 640 × 480 pixels. Qi et al. [40] acquired breast thermal images using Inframetrics 600M infrared camera, which have sensitivity of 0.05 °K. All the patient's images were captured at Elliott Mastology Centre (EMC). Arena et al. [41] have captured breast thermal images of 517 women ranging in age from 35 to 80. For their study, all the breast thermal images are divided into three groups: 343 women with no cancer, 110 women with newly detected cancer, and 63 women with malignancy detected 0 to 10 years previously. A state-of-the-art thermal camera with a resolution 320 × 240 pixels and sensitivity of 0.05 °C had been used to capture those breast thermal images. For the acquisition, they have a well equipped suite with an ergonomically designed height adjustable chair along with infrared reflecting side mirrors and ask the woman to site about 5 feet away from the camera. Arora et al. [42] have used the Sentinel BreastScan thermal camera having a resolution 320 × 240 and sensitivity of 0.08 °C to detect breast abnormality in a group of 92 patients. Before examination, each patient was signed consent term and they obtained approval for capturing breast thermal images from their Institutional Review Board. Tang et al. [43] used TSI-21 thermal camera from BIOYEAR Inc. to capture the breast thermal images of 117 female patients. The sensitivity of the camera is 0.05 °C and captured image resolution is 256 × 256. All the images were captured at People's Liberation Army General Hospital, China. Among 117 patients, 47 patients were found malignant tumor cases and remaining with benign cases. Avio TVS-2000 MkII ST (Tokyo, Japan) thermal infrared camera of thermal sensitivity 0.1 °C had been used by Ng et al. [44] to capture breast thermal images of 90 patients in Department of Diagnostic Radiology (DR), Singapore General Hospital (SGH). They have instructed the patients to come for images acquisition within the period of the 5th to 12th and 21st days after the onset of menstrual cycle. Agostini et al. [45] used AIM256Q from AEG infrarot-Module GmbH thermal camera which has a noise equivalent temperature difference (NETD) equal to 17.3 mK at 300 K, with an integration time equal to 20 ms to acquire breast thermal images. They asked the patient to recline on the examination table which is approximately 220 cm away from the camera. The resolution of each breast thermal image is 256 × 256. Wishart et al. [46] used Sentinel BreastScan to capture breast thermal image of 117 patients. They instructed the patient to remove clothing from the upper part of waist and sit in an ergonomic chair with arms supported at eye level. Then temperature controlled air flow was directed at the breasts for about 5 min. After that the thermal camera is used campture a series of thermal breast images. Acharya et al. [47] captured thermal breast images of 50 patients ranging in age from 43 to 59 and 36 to 46 in Department of DR, SGH using NEC-Avio Thermo TVS2000 MkIIST. The sensitivity of the camera is 0.1 °C and image resolution is 256 × 200 pixels. In [48], authors performed the experiment at Dr. Sarjito Hospital Yogyakarta. Total 150 women were examined and their breast thermal images were captured using Fluke digital thermal camera. Kontos et al. [49] had used the Meditherm med2000 thermal camera for collecting the breast thermal images of 63 patients, where 20 patients were diagnosed with breast cancer. For the image acquisition, camera was positioned 1 m away from the chair. Zadeh et al. [50] collected the breast thermal images of 200 women ranging in age from 18 to 35 using SDS Dseries camera at Hakim Sabzevari University in Sabzevar with the cooperation of Sabzevar University of Medical Science. The sensitivity of the camera is 0.1 °C and a resolution of 160 × 120 pixels. The researchers of the PROENG project [51] used FLIR ThermaCAM S45 camera to capture the breast thermal images of 220 patients at the University Hospital of the UFPE, Brazil. The resolution of the camera is 320 × 240 pixels. L.F. Silva et al. [52] had used FLIR SC-620 thermal camera with thermal sensitivity less than 0.04 °C and resolution 640 × 480 pixels to capture breast thermal images of 287 patients. The experiment was carried out at Hospital Universitário Antônio Pedro (HUAP) of the Federal University Fluminense. Table 3 Summarizes the available acquisition system and their sensitivity and resolution.

4.3 Capturing Views and Number of Captured Images

The actual procedure of breast thermal imaging is undertaken with the purpose to adequately detect the infrared emissions from the relevant surface areas of the breasts since the emissivity values changes with different angles of infrared measurements. Till now there is no such universally accepted thermal image acquisition protocol available. Hence, each hospital or clinic acquires arbitrary number of images by assuming different angles of the patients with respect to the camera. However, the accuracy of breast abnormality detection using breast thermal images is completely rely on some factors associated with the breast thermogram such as thermal sensitivity, image resolution, and number of views of the breast. Ng et al. [44] captures three thermograms of the breast, one frontal and two laterals. The patients of Agostini et al. [45] are required to lie-down on an examination table with arm up and resting the hands over the head and only frontal image is taken per patient. Tejerina [53] have used total three positions of each patient to capture the breast thermal image: including frontal view with hands resting over the head, internal lateral view of the left and right breasts, and external lateral view of the left and right breasts. Delgado et al. [54] and Kontos et al. [49] have used static protocol to capture three different views breast thermograms: frontal view, left oblique view, and right oblique view. Antonini et al. [55] and Kolaric et al. [56] have acquired total five views breast thermograms of each patient: right semi oblique, left semi oblique, left oblique, right oblique, and frontal. Motta et al. [57] have used total eight different views thermal

Table 3 Summary of the breast thermal image acquisition system

Paper name	Camera name	Sensitivity	Resolution	Image acquisition location	No. of patients	No. of abnormal patient	Age
Zadeh et al. [50]	SDS Dseries camera	0.1 °C at 30 °C	160 × 120	HSU and SUMS, Sabzevar	200 women	15 patients with abnormal lesions in their breasts	18 to 35
Arena et al. [41]	State of the art camera	0.05 °C	320 × 240	…	517	110 newly detected, biopsy proven cancer. 63 patients who underwent treatment for proven malignancy 0 to 10 years previously	35 to 80
Acharya et al. [47]	NEC-Avio Thermo TVS2000 MkIIST	0.1 °C at 30 °C black body	256 × 200	Dept. of DR, SGH	50	25	43 to 59 and 36 to 46
Qi et al. [40]	Inframetrics 600M camera	0.05 °K	…	EMC	…	…	…
Ng et al. [44]	Avio TVS-2000 MkII ST	0.1 °C at 30 °C	…	Dept. of DR, SGH	90	30 asymptomatic patients, 48 with benign breast, and 4 with cancerousbreast	43 to 59, 36 to 56, and 40 to 50
Wishart et al. [46]	Sentinel BreastScan	…	…	…	113	…	…
Arora et al. [42]	Sentinel BreastScan	0.08 °C	320 × 240	…	92	…	…

(continued)

Table 3 (continued)

Paper name	Camera name	Sensitivity	Resolution	Image acquisition location	No. of patients	No. of abnormal patient	Age
Tang et al. [43]	TSI-21 from BIOYEAR Inc	0.05 °C	256 × 256	PLAGH, China	117	70 benign and 47 malignant	...
Agostini et al. [45]	AIM256Q from AEG infrarot-Module GmbH	...	256 × 256	...	...	...	...
Nurhayati et al. [48]	Fluke	...	256 × 256	Dr. Sarjito Hospital Yogyakarta	150	50 Healthy images, Chemotherapy Group with 50 images, and Advanced Group with 50 images	...
Kontos et al. [49]	The Meditherm med2000	...	...	...	63	20	26 to 82
PROENG [51]	FLIR ThermaCAM S45	...	320 × 240	University Hospital of the UFPE, Brazil	220	...	...
L.F. Silva et al. [52]	FLIR SC-620	<0.04 °C	640 × 480	Hospital Universitrio Antnio Pedro (HUAP) of the FUF	287	47	...

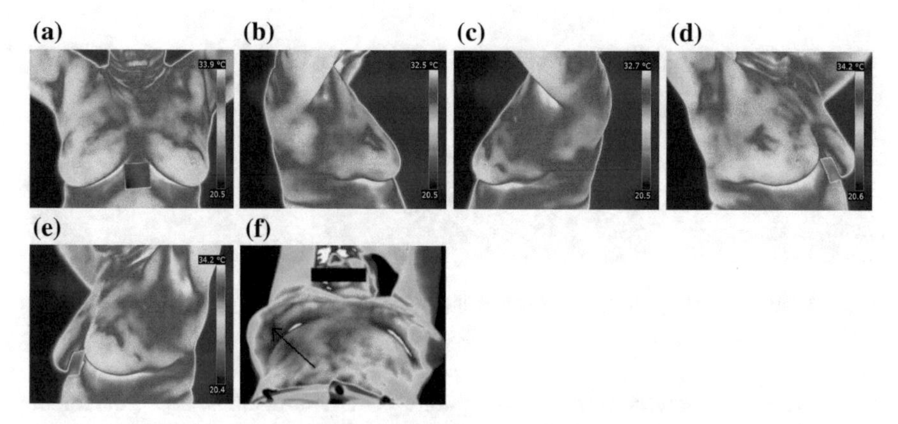

Fig. 4 **a** *Frontal* view breast thermal image, **b** 90° *right* lateral view, **c** 90° *left* lateral view, **d** *right* oblique view, **e** *left* oblique view, **f** abnormal *supine* view of the breast where arrow point out the abnormal sign in the underside of the *right* breast [52]

breast images of each patient: only left breast, only right breast, external lateral view of the left and right breasts, internal lateral view of the left and right breasts, frontal with hands over the head, and frontal view with hands on the hips. Figure 4, shows the example of frontal view, 90° right lateral view, 90° left lateral view, right oblique, and left oblique view breast thermograms [52]. Bharathi et al. [58] used twelve views of breast thermal images of each patient which was captured at an angular interval of 30. Importance of capturing the underside view of the breast has been proved by Campbell [59]. Figure 4f, shows the example of underside view of the breast [59]. However, among the different views, frontal view breast thermogram is considered as the most informative image by the medical specialists for the early breast cancer detection.

4.4 Available Databases for Research

Some studies have reported that the breast cancer is a highly treatable disease with 95 % possibility of survival if it is detected early [37]. Over the last few decades, breast thermography has been considered by the medical practitioners as the most promising and suitable alternatives for early detection and preliminary screening of the breast cancer [5]. Due to the limitation of human perceptibility, computer-aided analysis of breast thermal images has been playing a significant role from the past few years [60]. For accurate analysis and diagnosis of breast cancer using computer-aided analysis tool, it is necessary to improve the performance of different stages of computer-aided analysis tool like segmentation algorithm, feature extraction method and classification accuracy. But for the extensive research on it, it is equally important to get adequate number of breast thermal images. Although to stimulate the prospect

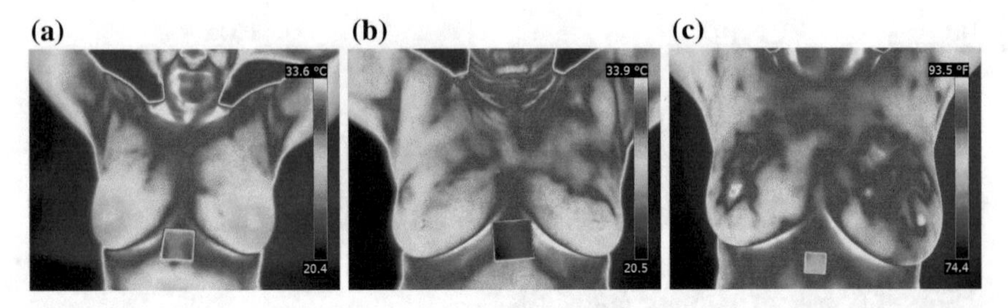

Fig. 5 **a** Normal breast thermogram, **b** benign breast thermogram, **c** malignant breast thermogram [52]

of research in this area, conservation of a publicly available database is of great importance to the researchers. But unfortunately compared to other modality images like mammography, MRI, and ultrasound, there is no such demonstrative publicly available databases in breast thermal images are available. However, in most of the cases breast thermal images are stored in private image database for internal diagnosis purpose only and the physician and the patient can access them [53, 61]. Till now there are only few publicly available databases available for this purpose [62, 63]. In [62], authors have used 19 breast thermal images from a set of 86 images collected from the Moncton Hospital to validate their proposed method. The images were captured in 1984 by the team of Moncton Hospital. A publicly available database [63] was first described by Araujo et al. [64] in their works that was later used by a number of researchers in their works [32, 57, 65–67]. The images of this database are collected from the University Hospital of UFPE, Brazil. L.F. Silva et al. [52] developed a database entitled "Database for Mastology Research with Infrared Images (DMR-IR)," which is the most expressive publicly available database for breast cancer research. The images can be accessed through a user friendly online interface (http://visual.ic.uff.br/dmi). Database contains breast thermal images of total 287 patients with three different conditions such as normal, benign, and cancerous for early detection of breast cancer, as shown in Fig. 5. Each image was captured using the FLIR SC-620 thermal infrared camera and digitized into 640 $\times$ 480 pixels with a resolution of 24-bit per pixel. In [68], authors have collected breast thermal images from the Perm Regional Oncological Dispensary [69] by going through the procedures approved by the Local Ethics Committee. Their database consist of total 47 women breast thermal images in which 33 women (between age 37–83) have confirmed malignant tumor and 14 women (between age 23–79) with intact mammary glands.

5 Region of Interest Segmentation from the Thermal Breast Image

In the analysis of thermal breast images in search of breast abnormality detection, segmentation of the required region plays a vital role prior to the feature extraction and classification phase and considered as one of the kernel phase. However, the accurate segmentation of the required region from the thermal breast images is a very difficult task due to the large variation in shape, size, and lack of clear edge definitions in the breast thermogram. Although, the region of interest segmentation can be categorized mainly in two types such as breast region segmentation and hottest region segmentation. Usually the breast thermal images are captured in a large area that includes upper body area, breast region, and lower body area. Thus, it is necessary to remove the unnecessary body parts in order to obtain the breast region only, since hottest region segmentation strongly depends on the precise segmentation of the breast region. After segmentation of the breast region from the acquired breast thermal image, hottest region segmentation from the segmented breast region is another significant step toward the analysis of breast thermogram. There are several benefits of hottest region segmentation from the segmented thermal breast image for identification of abnormal cases, and then computation of the degree of malignancy [7]. Typically, the temperature profile in the normal breast thermograms of the two breasts is almost symmetrical [70]. A small asymmetry signifies the abnormal cases. This asymmetry can be identified by comparing the segmented hottest regions of the left and right breast. After segmentation of the hottest regions, different features such as size, ratio of the area of the hottest region to the whole breast, and location could be extracted from those regions to determine the level of malignancy and the level of expansion of the tumor [7]. Thus, the segmentation of the hottest region is a very significant phase in the analysis of thermal breast images. This section is further subdivided into two subsections: breast region segmentation, and hottest region segmentation. A detailed review on the breast region segmentation is included in the breast region segmentation section. In the hottest region segmentation section, a detailed review on the hottest region segmentation is included and also shows the reliability of the hybrid intelligent techniques in the segmentation of the hottest region.

5.1 Breast Region Segmentation

Original acquired breast thermal image includes the breast region along with some unnecessary body regions such as upper body part, and lower body part, as shown in Fig. 5a. Hence, it is very important to segment the breast region only from the rest of the image for further processing. Accurate segmentation of the breast region can improves the performance of the hottest region segmentation algorithm, hence improves the overall accuracy of the breast cancer detection system. However, over the last one decade, some significant amounts of works on breast region segmentation have been reported in the literature by various researchers. In the breast thermal images, the boundaries of the breasts are not clearly visible and also the size and shape of the breasts vary in these images. Thus, due to the above-mentioned problems, most researchers prefer the manual segmentation to avoid erroneous analysis of the thermal breast image. But the automatic segmentation of the breast region is very much important for the computer-aided diagnosis of breast thermal images. In this section, we have thus highlighted only automatic systems related to the breast region segmentation. Qi et al. [70] had segmented the breast region based on edge detection followed by Hough transformation. In their algorithm, breast region is segmented based on four key points: two armpit points that are identified by computing the maximum curvature point in the left and right body boundary, lower boundary of the breast, and the point where parabolic curves of two breasts are intersected. Chaotic two-dimensional Otsu method-based genetic algorithm was used by Jin-Yu et al. for the segmentation of breast region from the breast thermal image [71]. The specialty of their algorithm is that it can be applied in generic thermal images. However, their method includes mainly four steps: the population of the genetic algorithm is initialized by using logistic mapping equations, the chaotic 2D Otsu method is applied to compute the fitness of the current population, the beast individual is returned when it reachs the end condition that leads to end of the algorithm, otherwise the process continues, and finally, new population is generated by selection, crossover, and mutation operations and then return to the previous step. Kapoor and Prasad used Canny edge detector and Hough transform to segment the breast region. Canny edge detector was used to identify the outward boundary of the breast and in the next step lower breast boundary was identified using the Hough transform [72]. Zadeh et al. [73] segmented the breast region by applying logarithmic edge detection method followed by parabolic Hough transform. To decrease the computational cost, a 6×6 Gaussian filter was first applied to remove the noise. Thereafter, Hough transform was applied for the segmentation breast region. Motta et al. [57] proposed a segmentation technique that is based on the selection of automatic threshold and border, and identification of inframammary folds by applying mathematical morphology and cubic-spline interpolation method. Pramanik et al. [9] proposed an automatic segmentation method consisting of four main steps: (i) removal of background by applying Otsu's thresholding followed by gray level reconstruction technique, (ii) horizontally partitioning of the image into halves, (iii) finding the breast lower limit by identifying inframammary fold region followed by the detection of maximum x-coordinate value, (iv) detection of the body boundary of the top half image and computation of the maximum curvature point as it is considered as the upper limit of the breast. Table 4 summarizes the different methods used for breast region segmentation.

Table 4 Summary of different method used for breast region segmentation

Authors	No. of breast thermograms used	Method of validation
Qi et al. [70]	...	Not to mention
Jin-Yu et al. [71]	10	Not to mention
Kapoor et al. [72]	...	Not to mention
Zadeh et al. [73]	...	Not to mention
Motta et al. [57]	150	Compared with manually segmented image
Pramanik et al. [9]	306	Not to mention

5.2 *Hottest Region Segmentation*

This section described the proposed hybrid intelligent technique for the segmentation of the hottest region. As discussed earlier, presence of tumor in the breast increased the local temperature that leads to the changes in the skin temperature of the breast. This temperature changes can be recorded with a fair degree of accuracy using thermal infrared camera. In the captured breast thermal image, different false colors are used to represent the different rates of temperature based on the camera and software used. In most of the captured thermal image, color scale usually emerged alongside the image, where temperature for each specific color is noted. In the images those are used in this study, lighter shades of red color demonstrating the first hotter color and red color representing the second hotter color designating it to be the top of the scale, while dark blue color signifying the cooler color and are at the bottom of the scale. The segmentation of hottest color region is one of the kernel phase in the analysis of thermal breast image. But unfortunately the contributions of researchers are very less in this domain [7, 8, 74]. Milosevic et al. [8] have segmented the hottest region using minimum variance quantization method followed by morphological erosion and dilation operations. They applied iteratively minimum variance quantization method on the breast thermal image to segment the hottest region. If the output contains hottest region along with the normal region, morphological erosion and dilation operations have been applied to eliminate the normal region. But they did not mention quantization level, number of iteration, and radius of the structuring element. K-means and fuzzy c-means (FCM) clustering methods have been used by EtehadTavakol et al. for the segmentation of different color regions [7]. The major problem of their work is that they did not provide any information regarding the number of clusters. For different images they used different number of clusters. In addition, they have not used any intelligent technique to identify the desired cluster. In [74], authors have used level set method (LSM) for the segmentation of hottest region and compared with the method proposed in [7]. They showed that the level set method provides better result over the k-means and FCM methods. But the problem with this technique is the initialization of the initial contour points. Table 5 Summarizes the different methods used for the segmentation of the color regions.

Segmentation of the hottest region from a color thermal breast image is a very significant step toward the analysis of the thermal breast image and relatively a new field of research in this arena. The earlier works presented in [7, 8, 74] are mainly focused on the segmentation of the color regions from the color thermal breast image. But up till there is no such method that exists in this respect, which can identify the desired region/regions from a set of segmented regions. However, it is very much important to identify the desired hottest regions during the analysis of the thermal breast image. From the above literature it can also be noted that only one soft computing method, fuzzy c-means (FCM), is used for color region segmentation. But in the segmentation of the medical images like digital mammography, X-rays, MRI, ultrasound, CT scan, etc., soft computing techniques, such as fuzzy c-means, neural network, particle swarm optimization (PSO), self-organizing map (SOM),

Table 5 Summary of different method used for color regions segmentation

Paper	Method used	Image type (gray/color)	Validation	Number of patients/Breast thermal images used
Milosevic et al. [8]	Minimum variance quantization	Color	Compared with original image	4
EtehadTavakol et al. [7]	K-means and FCM	Color	Compared with original image	6
Golestani et al. [74]	k-means, FCM, and LSM	Color	Compared with original image	30

genetic algorithm (GA), etc., are the most popular methods used in this respect [75, 76]. Although, our problem of hottest region segmentation involves mainly two steps: color regions segmentation, and identification of the desired region. Thus, in this study, we presented a sequential hybrid intelligent technique to segment the hottest region. The proposed method is based on hybridization of two computational intelligence techniques: fuzzy c-means (FCM) for the segmentation of different color components, and neural network to identify the desired hottest region clusters. A hybrid intelligent technique is a combination of different soft computing techniques that utilize the advantages of various soft computing techniques and overcomes their individual's limitations. Over the last one decade, hybrid intelligent techniques have been successfully applied by the various researchers in the classification of thermal breast images into normal and benign breast [77]. They have shown that the hybrid intelligent technique is a very efficient method to differentiate normal and abnormal breast thermogram. But there is no such hybrid intelligent technique that has been reported in the literature to segment the hottest region. Thus, in this study, we have tried to employ the hybrid system for the segmentation of the hottest region from the color thermal breast image. Our proposed method is inspired from the method presented by EtehadTavakol et al. [7]. Our method not only segments the different color regions, it also identifies the desired region/regions. The block diagram of the hottest region segmentation using the proposed hybrid intelligent technique is shown in Fig. 6. The method is described in detail in the following subsections.

5.2.1 Color Segmentation

At the first step of the color segmentation, we have transformed all the breast thermal images into uniform color space. In 1976, the International Commission on Illumination (CIE) suggested two uniform color spaces: CIELUV and CIELAB [78]. In this work, CIELUV is used to transform the breast thermal images into uniform color space. The LUV space was designed particularly for emissive colors that correspond

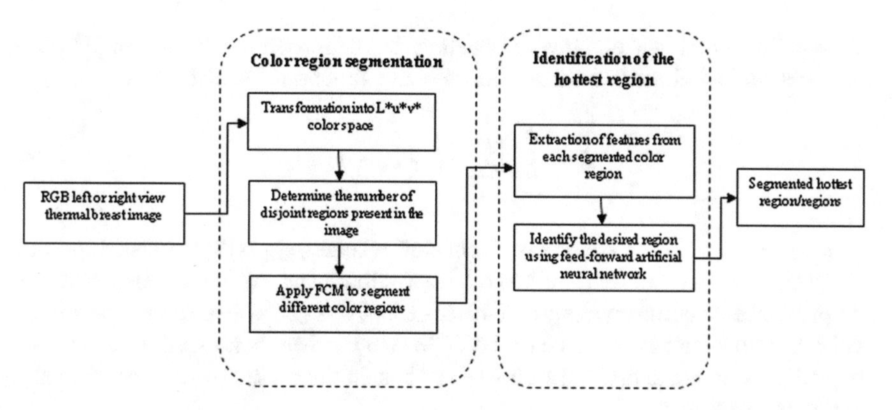

Fig. 6 Block diagram of the proposed hybrid technique for the segmentation of the hottest region

to images captured by a camera. In CIE $(L^*u^*v^*)$ color space, L^* is a luminosity layer, u^* and v^* are the chromaticity elements. The CIELUV color space for breast thermal image is computed as follows:

$$L^* = \begin{cases} 903.3(\frac{Y}{Y_p}), \frac{Y}{Y_p} \leq 0.008856 \\ 116(\frac{Y}{Y_p})^{\frac{1}{3}} - 16, \frac{Y}{Y_p} > 0.008856 \end{cases} \tag{1}$$

$$u^* = 13L^*(u' - u'_p) \tag{2}$$

$$v^* = 13L^*(v' - v'_p), \tag{3}$$

where, $X_p, Y_p, and\ Z_p$ are the nominally white objects color stimulus and the typical values for X_p, Y_p and Z_p are 95.047, 100, and 108.883, respectively.

The quantities u', v', u'_p, and v'_p are computed as follows:

$$u' = \frac{4X}{X+15Y+3Z}, u'_p = \frac{4X_p}{X_p+15Y_p+3Z_p}, v' = \frac{9Y}{X+15Y+3Z}, v'_p = \frac{4Y_p}{X_p+15Y_p+3Z_p},$$

where, the tristimulas X, Y, and Z are calculated as:

$X = 0.4303R + 0.3416G + 0.1784B$
$Y = 0.2219R + 0.7068G + 0.0713B$
$Z = 0.0202R + 0.1296G + 0.9393B$

After conversion into CIELUV space, FCM clustering technique is used for the segmentation of color in thermal breast images.The basic idea behind the FCM algorithm is that it allocates data points to each class based on the fuzzy membership values. In our case, data points are the image pixels, which need to be grouped into different classes. Let $I = \{a_1, a_2, \ldots, a_p\}$ be an image having p pixels that to be classified into n clusters. During the process of allocation, it iteratively minimizes an objective function J along with updating the membership value m_{ij} of jth pixel in i^{th} cluster and cluster center c_i. The objective function J is defined as

$$J = \sum_{i=1}^{n} \sum_{j=1}^{p} m_{ij}^{r} \parallel a_j - c_i \parallel^2, \tag{4}$$

where, $K = \{c_1, c_2, \ldots, c_n\}$ be a vector of cluster center, m_{ij} is the membership value of jth pixel a_j in i^{th} cluster, $\parallel . \parallel$ is a L^2 norm that measures the distance between j^{th} pixel and its corresponding i^{th} cluster center, $r \in [1, \propto]$ is a control parameter which controls the performance of the FCM. Up till there is no such well accepted method is available to optimally choose r. Thus, in this study, we have heuristically chosen the r value, i.e., 2.

The cluster center c_i and membership value m_{ij} are updated as follows:

$$c_i = \frac{\sum_{j=1}^{p} m_{ij}^{r} a_j}{\sum_{j=1}^{p} m_{ij}^{r}} \tag{5}$$

$$m_{ij} = \frac{1}{\sum_{k=1}^{n} \{\frac{\|a_j - c_i\|}{\|a_j - c_k\|}\}^{\frac{2}{r-1}}} \tag{6}$$

The FCM algorithm returns a matrix $M = [m_{ij}]_{n \times p}, \forall i, j, m_{ij} \in [0, 1]$, called partition matrix that contains the membership value between each pixel and clusters centers. The detailed algorithm for FCM clustering could be found in [79].

In [7], EtehadTavako et al. also used FCM clustering technique for the segmentation of color in breast thermal image. The major drawback of this algorithm is prior specification of the number of cluster. If the number of specified cluster is less than the number of colors present in the given image, then it can generate under segmentation results. Similarly, if the number of specified cluster is greater than the number of colors, then it can generate empty cluster or over segmentation results. Thus, to overcome this problem, we have first used The Davies–Bouldin Index [80] to compute the optimal number of clusters present in the given breast thermal image and then fuzzy c-means clustering algorithm is used to segment the different color regions. The Davies–Bouldin index is a metric that was introduced by D.L. Davies and D. W. Bouldin in 1979 for the evaluation of clustering algorithm. The algorithm is based on the similarity measure of the clusters. According to this method, the images used in this study consist of approximately five optimal clusters. Figure 8b–f shows the five clusters of the corresponding Fig. 7a.

5.2.2 Identification of the Hottest Regions

After segmentation of different color regions from the breast thermal image, it is necessary to identify the desired region or regions for further processing. In [7], authors have used FCM algorithm to segment the different color regions, which are

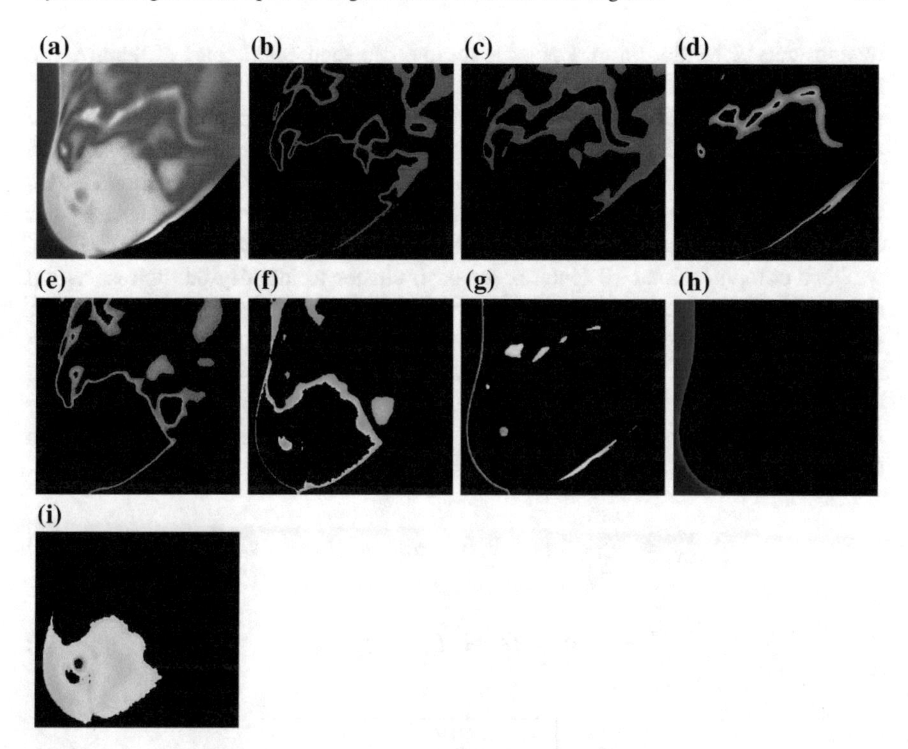

Fig. 7 **a** Manually segmented *right* view breast thermal image, **b–i** are the colors segmented by the method of EtehadTavakol et al. [7]

Algorithm:1
Input: RGB left or right view thermal breast image I^{LorR}
Output: Segmented different color regions
Step 1: Transform the original RGB left or right view thermal breast image into $L*u*v*$ color space.
Step 2: Apply Davies–Bouldin index to compute the optimal number of clusters present in the given image, say n.
Step 3: Select the weighting exponent r.
Step 4: Randomly initialize the membership matrix $M = [m_{ij}]$.
Step 5: Compute the centroids c_i for each cluster using equation-6.
Step 6: Compute new membership values m_{ij} using equation-7, for $i = 1, 2, \ldots, n$ and $j = 1, 2, \ldots, p$.
Step 7: Update the membership matrix $M^{k+1} \leftarrow M^k$.
Step 8: Repeat Step 5 to 7 until the difference of membership values is very less, $M^{k+1} - M^k < \epsilon$, where value of ϵ is very small, typically 0.01.
Step 9: Based on the final obtained membership matrix M (also called partition matrix), segment the different color regions.

then manually compared with color palette spectrum to identify the first and second hottest regions. In this study, we have used a feedforward Neural Network classifier to identify the first and second hottest regions from a set of five clusters obtained by applying FCM algorithm. Once the regions are segmented, a set of features are derived from each cluster to identify the desired cluster or clusters. At the first step of feature extraction, three color components such as red, green, and blue are extracted from each segmented cluster. Thereafter, three statistical features: standard deviation, skewness, and kurtosis are calculated for each color component. Then, features are normalized using the equation-7. In addition, ratio of the cluster center is also considered as the dominant information of the corresponding cluster. Thus, we have computed total 10 features for each cluster to identify the hottest region cluster. These features are then used in desired cluster identification stage for which we employ a three layer feed forward neural network. The gradient decent training rule is used to train the classifiers. The network consists of 10 input neurons, 20 hidden neurons, and one output neurons as it is a two class problem. Linear transfer function is used in the input layer, while sigmoid transfer function is used in both the hidden layer and output layer, respectively.

$$f_N^{(i)} = \frac{f_C^{(i)}}{S}, i = 1, 2, 3 \quad and \quad C = R, G, B \tag{7}$$

$$S = f_C^{(1)} + f_C^{(2)} + f_C^{(3)} \tag{8}$$

$$f_C^{(1)} = \sqrt{\frac{1}{K} \sum \sum_{i,j \in R}^{MN} (\mu - f_C(i, j))^2} \tag{9}$$

$$f_C^{(p)} = \frac{1}{K} \sum \sum_{i,j \in R}^{MN} (\mu - f_C(i, j))^q, p = 2, 3 \quad and \quad q = 3, 4 \tag{10}$$

where $f_C^{(1)}$, $f_C^{(2)}$, and $f_C^{(3)}$ are the standard deviation, skewness, and kurtosis of each color component C, K is the number of pixels having the color component value greater than zero, $M \times N$ is the size of the image, R is the region within the image having color component value greater than zero, and $\mu = \frac{1}{K} \sum \sum_{i,j \in R}^{MN} f_C(i, j)$.

5.2.3 Experimental Results and Discussion

The experiment has been carried out on 40 anterior view breast thermal images of DMR database [52]. Out of which 20 images are of malignant cases and remaining are of benign cases. However, it is very difficult to give the experimental results for all the images from the visual perception. Thus, we provide the experimental results for two cases, such as malignant and benign from the visual perception. Figure 7a shows the anterior view thermal breast image of a malignant right breast. Figure 9a

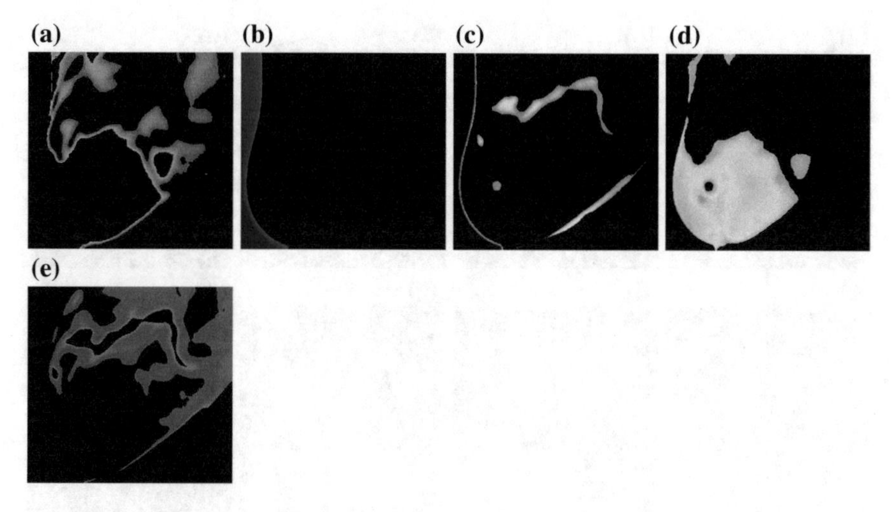

Fig. 8 **a–e** Shows the segmented clusters obtained using our proposed hybrid intelligence method of the Fig. 7a

shows the anterior view thermal breast image of a benign right breast. Figures 7b–i and 9b–i shows the color segmentation results using fuzzy c-means clustering technique proposed by EtehadTavakol et al. [7] as they have used eight clusters for cancerous breast. Figures 8a–e and 10a–e shows the color segmentation result using the combination of the Davies–Bouldin Index and Fuzzy c-means clustering methods for the Figs. 7a and 9a, respectively. By carefully observing these results for two cases, it can be concluded that the method of EtehadTavakol et al. provides over segmentation results.

During the identification of the desired region, 24 individuals images are chosen arbitrarily from a set of 40 individuals images to train the feedforward neural network and remaining samples are used to test and validate the network. Based on the outcomes of the network, we have computed the classification accuracy (using equation-11) for the identification of the desired clusters. The obtained accuracy is 99 %, which signify the efficacy of the proposed system.

$$Accuracy = \frac{Total\ correctly\ identified\ clusters}{Total\ no.\ of\ thermal\ breast\ images \times Total\ no.\ of\ cluster\ present\ per\ image} \times 100 \tag{11}$$

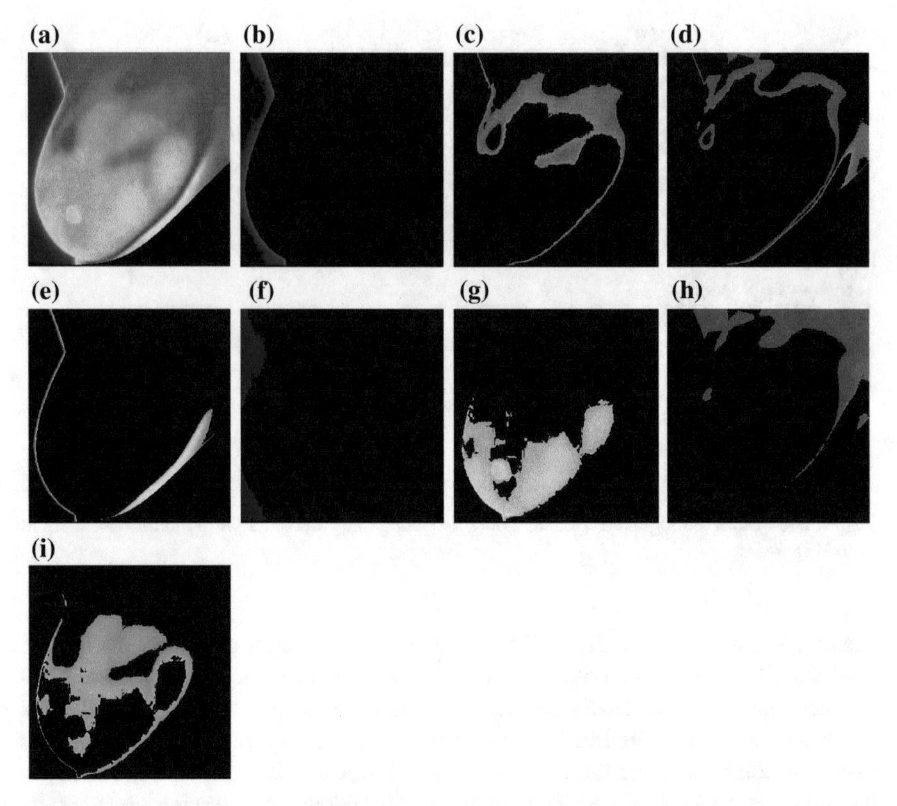

Fig. 9 Shows **a** manually segmented *right* view benign thermal breast image, **b–i** are the colors segmented by the method of EtehadTavakol et al. [7]

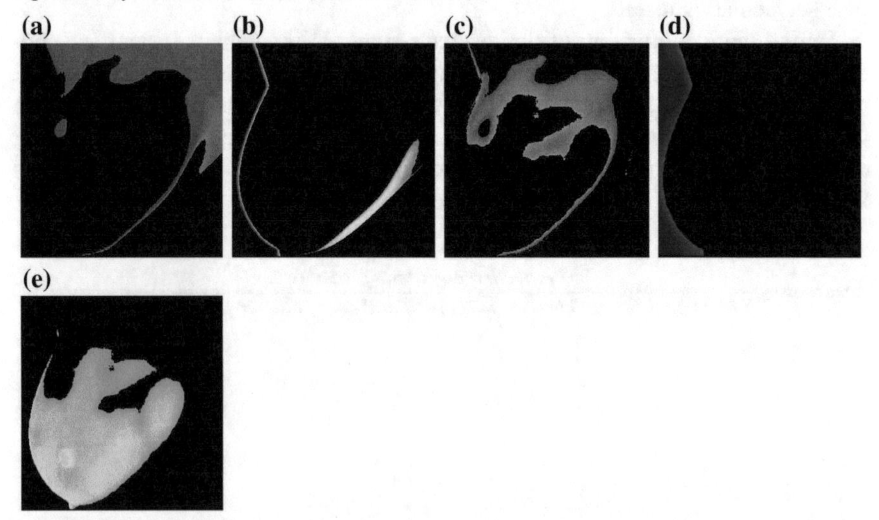

Fig. 10 **a–e** Shows the segmented clusters obtained using our proposed hybrid intelligence method of the Fig. 9a

6 Conclusion

Early breast cancer detection using breast thermal image have already got lots of attention by various researchers, since it has sensitivity of 90 %. Computer-aided analysis of breast thermal image involves multiple steps. Segmentation of the hottest region from the breast thermal image is a most significant step during analysis of the breast thermal image as it provides some useful information regarding the goodness of the breast. From the segmented hottest region, it is possible to identify the abnormal cases by comparing the hottest regions of the corresponding left and right breasts. In addition, it is also possible to determine the level of malignancy and the level of expansion of the tumor by extracting some characteristics from them. However, unfortunately very few amounts of research works have been delineated in the literature on the segmentation of this region. In this chapter, we try to highlight the effectiveness of hybrid intelligence system over the conventional methods for the segmentation of the hottest region. A hybrid computational intelligence technique is proposed to segment the different regions present in the breast thermal image and then identify the desired hottest region from them. The proposed method achieved 99 % accuracy for the identification of hottest region from a set of regions. Our method also provides better segmentation results compared to other methods. Although, we believe that the accurate segmentation of the hottest region may improve the accuracy and sensitivity of the breast thermal image in search of early breast abnormality detection.

Acknowledgments Authors are thankful to DBT, Govt. of India for funding a project with Grant no. BT/533/NE/TBP/2014. Sourav Pramanik is also thankful to Department of Electronics and Information Technology (DeitY), Govt. of India, for providing him PhD-Fellowship under Visvesvaraya PhD scheme.

References

1. International Agency for Research on Cancer (IARC) and World Health Organization (WHO). GLOBOCAN 2012: Estimated cancer incidence, mortality and prevalence worldwide in 2012 (2014). http://globocan.iarc.fr/Pages/fact-sheets-cancer.aspx
2. http://www.breastcancerindia.net/statistics/trends.html. Accessed 10 Jan 2016
3. Gautherie, M.: Thermopathology of breast cancer: measurement and analysis of in vivo temperature and blood flow. Ann. N. Y. Acad. Sci. **335**, 383–415 (1999)
4. Tan, T.Z., Quek, C., Ng, G.S., Ng, E.Y.K.: A novel cognitive interpretation of breast cancer thermography with complementary learning fuzzy neural memory structure. Expert Syst. Appl. **33**, 652–666 (2007)
5. Amula, W.C.: A review of breast thermography. Int. Acad. Clin. Thermol. (2003). http://www.iact-org.org/articles/articles-review-btherm.html
6. Borchartt, T.B., Conci, A., Lima, R.C.F., Resmini, R., Sanchez, A.: Breast thermography from an image processing viewpoint: a survey. Signal Process. **93**, 2785–2803 (2013)
7. EtehadTavakol, M., Sadri, S., Ng, E.Y.K.: Application of K- and fuzzy c-means for color segmentation of thermal infrared breast images. J. Med. Syst. **34**, 35–42 (2010)

8. Milosevic, M., Jankovic, D., Peulic, A.: Thermography based breast cancer detection using texture features and minimum variance quantization. EXCLI J. **13**, 1204–1215 (2014)
9. Pramanik, S., Bhattacharjee, D., Nasipuri, M.: Wavelet based thermogram analysis for breast cancer detection. In: Proceedings of IEEE International Symposium on Advanced Computing and Communication, Silchar, India, pp. 205–212 (2015)
10. de Oliveira, J.P.S., Conci, A., Perez, M.G., Andaluz, V.H.: Segmentation of infrared images: a new technology for early detection of breast diseases. In: Proceedings of IEEE International Conference on Industrial Technology, Seville, pp. 1765–1771 (2015)
11. https://ktlarkin.files.wordpress.com/2009/12/breast-anatomy2.gif?w=479. Accessed 21 Feb 2016
12. Sobin, L.H., Wittekind, C.H.: TNM classification of malignant tumours, 4th edn. Paris, Springer **100**, (1988)
13. http://www.cancer.org/cancer/breastcancer/detailedguide/breast-cancer-risk-factors. Accessed 10 Jan 2016
14. http://ww5.komen.org/BreastCancer/Statistics.html. Accessed 10th January 2016
15. http://ww5.komen.org/BreastCancer/FamilyHistoryofBreastOvarianorProstateCancer.html. Accessed 10 Jan 2016
16. http://ww5.komen.org/KomenPerspectives/Breast-Density-and-Breast-Cancer-Risk.html. Accessed 10 Jan 2016
17. http://www.breastcancer.org/risk/factors. Accessed 10th Jan 2016
18. Collaborative Group on Hormonal Factors in Breast Cancer: Breast cancer and breast feeding: collaborative reanalysis of individual data from 47 epidemiological studies in 30 countries, including 50,302 women with breast cancer and 96,973 women without the disease. Lancet **20**, 187–95 (2002)
19. Thermology. http://www.thermology.com/history.htm. Accessed 24 Jan 2016
20. Gautherine, M., Gros, C.: Contribution of infrared thermography to early diagnosis, pretherapeutic prognosis and post-irradiation follow-up of breast carcinomas. Med. Mundi **21**, 135–149 (1976)
21. Gautherie, M., Gros, C.M.: Breast thermography and cancer risk prediction. Cancer **45**(1), 51–56 (1980)
22. Kennedy, D., Lee, T., Seely, D.: A comparative review of thermography as a breast screening technique. Integr. Cancer Ther. **8**(1), 9–16 (2009)
23. Gogoi, U.R., Bhowmik, M.K., Bhattacharjee, D., Ghosh, A.K., Majumder, G.: A study and analysis of hybrid intelligent techniques for breast cancer detection using breast thermograms. In: Hybrid Soft Computing Approaches, pp. 329–359 (2015)
24. Amalu, W.C., Hobbins, W.B., Head, J.F., Elliot, R.L.: Infrared imaging of the breast: a review. In: Diakides Nicholas, A., Bronzino Joseph, D. (eds.) Medical Infrared Imaging, pp. 9-1–9-22. Taylor and Francis, Boca Raton (2007)
25. Gershen-Cohen, J., Haberman, J., Brueschke, E.E.: Medical thermography: a summary of current status. Radiol. Clin. North Am. **3**, 403–431 (1965)
26. Hoffman, R.: Thermography in the detection of breast malignancy. Am. J. Obstet. Gynecol. **98**, 681–686 (1967)
27. Amalric, R., Giraud, D., Altschule, C., Spitalier, J.M.: Value and interest of dynamic telethermography in detection of breast cance. Acta Thermogr. **1**, 89–96 (1976)
28. Hobbins, W.B.: Abnormal thermogram - significance in breast cancer. Interam. J. Rad. **12**, 337 (1987)
29. Spitalier, H., Giraud, D., Altschuler, C., Amalric, F., Spitalier, J.M., Brandone, H., Ayme, Y., Gardiol, A.: Does infrared thermography truly have a role in present-day breast cancer management? In: Biomedical Thermology (Proceedings of an International Symposium), pp. 269–278. Alan R. Liss, Inc., New York City (1982)
30. Nyirjesy, I., Ayme, T.: Clinical evaluation, mammography, and thermography in the diagnosis of breast carcinoma. Thermology **1**, 170–173 (1986)
31. Parisky, Y.R., Sardi, A., Hamm, R., Hughes, K., Esserman, L., Rust, S., Callahan, K.: Efficacy of computerized infrared imaging analysis to evaluate mammographically suspicious lesions. Am. J. Roentgenol. **180**, 263–269 (2003)

32. Borchartt, T.B., Resmini, R., Conci, A., Martins, A., Silva, A.C., Diniz, E.M., Paiva, A., Lima, R.C.F.: Thermal feature analysis to aid on breast disease diagnosis. In: Proceedings of 21st Brazilian Congress of Mechanical Engineering-COBEM2011, Natal, Brazil, pp. 1–8 (2011)
33. Gautherie, M., Kotewicz, A., Gueblez, P.: Accurate and objective evaluation of breast thermograms: Basic principles and new advances with special reference to an improved computer assisted scoring system. In: Thermal assessment of Breast Health, pp. 72–97. MTP Press Limited (1983)
34. Gautherie, M.: New protocol for the evaluation of breast thermograms. In: Thermological Methods, VCH mbH, pp. 227–235 (1985)
35. Amalu, W.C., Hobbins, W.B., Head, J.F., Elliot, R.L.: Infrared Imaging of the Breast - An Overview, Medical Devices and Systems, pp. 25.1–25.20. CRC Press, Boca Raton (2006)
36. http://www.adelphatherm.com/#!untitled/c1dyz. Accessed 1 Jan 2016
37. Ng, E.Y.K.: A review of thermography as promising non-invasive detection modality for breast tumor. Int. J. Thermal Sci. **48**(5), 849–859 (2009)
38. Ring, E.F.J.: Quantitative thermal imaging. Clin. Phys. Physiol. Meas. **11**, 87–95 (1990)
39. Thermography Guidelines (TG), Standards and Protocols in Clinical Thermographic Imaging, September 2002. http://www.iact-org.org/professionals/thermog-guidelines.html. Accessed Feb 2016
40. Qi, H., Head, J.F.: Asymmetry analysis using automatic segmentation and classification for breast cancer detection in thermograms. In: Proceedings of the 23rd Annual International Conference of the IEEE EMBS, 3, Turkey, pp. 2866-2869 (2001)
41. Arena, F., Barone, C., Di Cicco, T.: Use of digital infrared imaging in enhanced breast cancer detection and monitoring of the clinical response to treatment. In: Proceedings of the 25th Annual International Conference on Engineering in Medicine and Biology Society (EMBS), vol. 2, pp. 1129-1132 (2003)
42. Arora, N., Martins, D., Ruggerio, D., Tousimis, E., Swistel, A.J., Osborne, M.P., Simmons, R.M.: Effectiveness of a non-invasive digital infrared thermal imaging system in the detection of breast cancer. Am. J. Surg. **196**(4), 523–526 (2008)
43. Tang, X., Ding, H., Yuan, Y., Wang, Q.: Morphological measurement of localized temperature increase amplitudes in breast infrared thermograms and its clinical application. Biomed. Signal Process. Control **3**(1), 312–318 (2008)
44. Ng, E.Y.K., Kee, E.C.: Integrative computer-aided diagnostic with breast thermogram. J. Mech. Med. Biol. **7**(1), 1–10 (2007)
45. Agostini, V., Knaflitz, M., Molinari, F.: Motion artifact reduction in breast dynamic infrared imaging. IEEE Trans. Biomed. Eng. **56**(3), 903–906 (2009)
46. Wishart, G.C., Campisi, M., Boswell, M., Chapman, D., Shackleton, V., Iddles, S., Hallett, A., Britton, P.D.: The accuracy of digital infrared imaging for breast cancer detection in women undergoing breast biopsy. Eur. J Cancer Surg. **36**, 535–540 (2010)
47. Acharya, U.R., Ng, E.Y.K., Tan, J.H., Sree, S.V.: Thermography based breast cancer detection using texture features and support vector machine. J. Med. Syst. **36**(3), 1503–1510 (2010)
48. Nurhayati, O.D., Susanto, A., Widodo, T.S., Tjokronagoro, M.: Principal component analysis combined with first order statistical method for breast thermal images classification. Int. J. Comput. Sci. Technol. **2**(2), 12–18 (2011)
49. Kontos, M., Wilson, R., Fentiman, I.: Digital infrared thermal imaging (DITI) of breast lesions: sensitivity and specificity of detection of primary breast cancers. Clin. Radiol. **66**(2011), 536–539 (2011)
50. Zadeh, H.G., Haddadnia, J., Hashemian, M., Hassanpour, K.: Diagnosis of breast cancer using a combination of genetic algorithm and artificial neural network in medical infrared thermal imaging. Iran J. Med. Phys. **9**(4), 265–274 (2012)
51. PROENG. Image processing and image analyses applied to mastology (2012). http://visual.ic.uff.br/en/proeng/. Accessed 31 Jan 2016
52. Silva, L.F., Saade, D.C.M., Sequeiros-Olivera, G.O., Silva, A.C., Paiva, A.C., Bravo, R.S., Conci, A.: A new database for breast research with infrared image. J. Med. Imaging Health Inf. **4**(1), 92–100(9) (2014)

53. Tejerina, A.: Aula de Habilidadesy Simulacion em Patologia de la Mama' (in Spanish). ADEMAS Comunicacion Grafica, Madrid, Spain (2009)
54. Delgado, F.G., Luna, J.G.V.: Feasibility of new-generation infrared screening for breast cancer in rural communities. US Obstet. Gynecol. Touch Briefings **5**, 52–56 (2010)
55. Antonini, S., Kolaric, D., Nola, I.A., Herceg, Z., Ramljak, V., Kulis, T., Holjevac, J.K., Ferencic, Z.: Thermography surveillance after breast conserving surgery-three cases. In: 53rd International Symposium ELMAR, Croatia, pp. 317–319 (2011)
56. Kolaric, D., Herceg, Z., Nola, I.A., Ramljak, V., Kulis, T., Holjevac, J.K., Deutsch, J.A., Antonini, S.: Thermography-a feasible method for screening breast cancer. Coll. Antropol. **37**(2), 583–588 (2013)
57. Motta, L.S., Conci, A., Lima, R.C.F., Diniz, E.M.: Automatic segmentation on thermograms in order to aid diagnosis and 2D modeling. In: Proceedings of 10th Workshop em Informatica Medica, Belo Horizonte, MG, Brazil, vol. 1, pp. 1610-1619 (2010). http://www.visual.ic.uff.br/proeng Accessed 19 March 2015
58. Bharathi, G.B., Francis, S.V., Sasikala, M., Sandeep, J.D.: Feature analysis for abnormality detection in breast thermogram sequences subject to cold stress. In: Proceedings of the National Conference on Man Machine Interaction (NCMMI), 15-2 (2014)
59. Beware of Poor Breast Thermograms, ILSI Thermography Service. http://www.doctormedesign.com/HTMLcontent/Beware/of/Poor/Thermograms.htm
60. Etehadtavakol, M., Ng, E.Y.K.: Breast thermography as a potential non-contact method in the early detection of cancer: a review. J. Mech. Med. Biol. **2**(13), 1330001.1–1330001.20 (2013)
61. Ville Marie Medical Center. http://www.villemariemed.com. Accessed 31 Jan 2016
62. Koay, J., Herry, C., Frize, M.: Analysis of breast thermography with an artificial neural network. Eng. Med. Biol. Soc. IEMBS **1**(1), 1159–1162 (2004)
63. BioBD. Bancode Dadosde Pesquisa Biomedica (2012). http://150.161.110.168/termo. Accessed 23 Jan 2016
64. Araujo, M.C., Lima, R.C.F., Santos, F.: Desenvolvimento de um banco de dados como ferramenta auxiliary na deteccao precoce de cancer de mama'30 Iberian-Latin-American Congress on Computational Methods in Engineering, vol. 1, pp. 1–15. Armacao dos Buzios, RJ, Brazil (2009)
65. Conci, A., Lima, R.C.F., Fontes, C.A.P., Motta, L.S., Resmini, R.: A new method for automatic segmentation of the region of interest of thermographic breast image. Thermol. Int. **20**(4), 134–135 (2010)
66. Conci, A., Lima, R.C.F., Fontes, C.A.P., Borchartt, T.B., Resmini, R.: A new method to aid to the breast diagnosis using fractal geometry. Thermol. Int. **20**(4), 135–136 (2010)
67. Borchartt, T.B., Resmini, R., Motta, L.S., Clua, E.W.G., Conci, A., Viana, M.J.A., Santos, L.C., Lima, R.C.F., Sanchez, A.: Combining approaches for early diagnosis of breast diseases using thermal imaging'. Int. J. Innov. Comput. Appl. **3**(4), 163–183 (2012)
68. Gerasimova, E., Audit, B., Roux, S.G., Khalil, A., Gileva, O., Argoul, F., Naimark, O., Arneodo, A.: Wavelet-based multifractal analysis of dynamic infrared thermograms to assist in early breast cancer diagnosis. Front. Physiol. **5**(176), 1–11 (2014)
69. Gileva, O.S., Freynd, G.G., Orlov, O.A., Libik, T.V., Gerasimova, E.I., Plekhov, O.A.: Interdisciplinary approaches to early diagnosis and screening of tumors and precancerous diseases (for example, breast cancer). RFBR J. **74–75**, 93–99 (2012)
70. Qi, H., Kuruganti, P.T., Snyder, W.E.: Detecting breast cancer from thermal infrared images by asymmetry analysis. Biomedical Engineering Handbook **27**(1–27), 14 (2006)
71. Jin-Yu, Z., Yan, C., Xian-Xiang, H.: IR thermal image segmentation based on enhanced genetic algorithms and two- dimensional classes square error. In: Second International Conference on Information and Computation Science, vol. 2(1), pp. 309–312 (2009)
72. Kapoor, P., Prasad, S.V.A.V.: Image processing for early diagnosis of breast cancer using infrared images. In: 2nd International Conference on Computer and Automation Engineering, vol. 3(1), pp. 564–566 (2010)
73. Zadeh, H.G., Kazerouni, I.A., Haddadnia, J.: Distinguish breast cancer based on thermal features in infrared images. Canadian J. Image Process. Comput. Vis. **2**(6), 54–58 (2011)

74. Golestani, N., EtehadTavakol, M., Ng, E.Y.K.: Level set method for segmentation of infrared breast thermograms. EXCLI J. **13**, 241–251 (2014)
75. Ji, Z., Liu, J., Cao, G., Sun, Q., Chen, Q.: Robust spatially constrained fuzzy c-means algorithm for brain MR image segmentation. Pattern Recogn. **47**(7), 2454–2466 (2014)
76. Torbati, N., Ayatollahi, A., Kermani, A.: An efficient neural network based method for medical image segmentation. Comput. Biol. Med. **44**, 76–87 (2014)
77. Krawczyk, B., Schaefer, G., Woźniak, M.: A hybrid cost-sensitive ensemble for imbalanced breast thermogram classification. Artif. Intell. Med. **65**(3), 219–227 (2015)
78. Li, H., Burgess, A.E.: Evaluation of signal detection performance with pseudo-color display and lumpy backgrounds. In: Kundel, H.L. (ed.) SPIE, Medical Imaging: Image Perception, Newport Beach, vol. 3036, pp. 143–149 (1997)
79. Bezdek, J.C., Keller, J., Krisnapuram, R., Pal, N.R.: Fuzzy models and algorithms for pattern recognition and image processing. Kluwer, Norwell (1999)
80. Davies, D.L., Bouldin, D.W.: A cluster separation measure. IEEE Trans. Pattern Anal. Mach. Intell. **PAMI–1**(2), 224–227 (1979)

Modeling of High-Dimensional Data for Applications of Image Segmentation in Image Retrieval and Recognition Tasks

Dariusz Jakóbczak

Abstract Probabilistic Features Combination method (PFC), is the approach of multidimensional data modeling, extrapolation and interpolation using the set of high-dimensional feature vectors. This method is a hybridization of numerical methods and probabilistic methods with N-dimensional data interpolation for feature vectors. Each feature is treated as a random variable.

Keywords Probabilistic Features Combination method · Multidimensional data modeling · Image Segmentation · Image Retrieval

1 Introduction

Multidimensional data modeling appears in science and industry, for example in image retrieval and data reconstruction. The chapter is dealing with these questions via modeling of high-dimensional data for applications of image segmentation. This chapter is concerned with two parts. Image retrieval is based on probabilistic modeling of unknown features via combination of N-dimensional probability distribution function for each feature treated as random variable. The sketch of proposed Probabilistic Features Combination (PFC) is built from some points, which are described in this chapter using feature vectors and modeling functions. So high-dimensional data interpolation in handwriting identification [1] consists for example of personalized handwriting [2–4] and many others approaches [5–17], also based on Hidden Markov Model [18, 19]. So hybrid soft computing is essential in N-dimensional data interpolation and multidimensional reconstruction [20].

Current approaches are dealing with polynomials, which is not always appropriate road to high-dimensional modeling and data reconstruction [21–25]. The chapter consists of proposed method—Probabilistic Features Combination (PFC) in high-dimensional reconstruction and multidimensional data retrieval. PFC requires infor-

D. Jakóbczak (✉)
Department of Electronics and Computer Science, Koszalin University
of Technology, Sniadeckich 2, 75-453 Koszalin, Poland
e-mail: dariusz.jakobczak@tu.koszalin.pl

S. Bhattacharyya et al. (eds.), *Hybrid Soft Computing for Image Segmentation*, DOI 10.1007/978-3-319-47223-2_12

mation about data (image, object, curve) as the set of N-dimensional feature vectors. Proposed PFC method is applied in image retrieval and recognition tasks via different coefficients for each feature as random variable. So the author tries to retrieve the image using N-dimensional feature vectors and high-dimensional nodes via hybrid soft computing.

2 Hybrid Multidimensional Modeling of Feature Vectors

The method of PFC is computing (interpolating) unknown (unclear, noised, or destroyed) values of features between two successive nodes (N-dimensional vectors of features) using hybridization of probabilistic methods and numerical methods. Calculated values (unknown or noised features such as coordinates, colors, textures, or any coefficients of pixels, voxels, and doxels or image parameters) are calculated for real number $\alpha_i \in [0; 1]$ $(i = 1, 2, \ldots N - 1)$ between two successive values of feature. PFC is dealing with N-dimensional feature vectors $p_1 = (x_1, y_1, \ldots, z_1)$, $p_2 = (x_2, y_2, \ldots, z_2), \ldots, p_n = (x_n, y_n, \ldots z_n)$ as $h(p_1, p_2, \ldots, p_m)$ and $m = 1, 2, \ldots n$ to interpolate unknown value of feature (for example y) for the rest of coordinates

$c_1 = \alpha_1 \cdot x_k + (1 - \alpha_1) \cdot x_{k+1}, \ldots\ldots, c_{N-1} = \alpha_{N-1} \cdot z_k + (1 - \alpha_{N-1}) \cdot z_{k+1}$, $k = 1, 2, \ldots, n - 1$

$c = (c_1, \ldots, c_{N-1}), \alpha = (\alpha_1, \ldots, \alpha_{N-1}), \gamma_i = F_i(\alpha_i) \in [0; 1], i = 1, 2, \ldots, N - 1$

$$y(c) = \gamma \cdot y_k + (1 - \gamma)y_{k+1} + \gamma(1 - \gamma) \cdot h(p_1, p_2, \ldots, p_m), \tag{1}$$

$\alpha_i \in [0; 1], \gamma = F(\alpha) = F(\alpha_1, \ldots, \alpha_{N-1}) \in [0; 1]$.

The basic structure of Eq. (1) is built on modeling function $\gamma = F(\alpha)$ which is used for points' interpolation between the nodes. Then N-1 features $c_1, \ldots, c_{N-1}$ are parameterized by $\alpha_1, \ldots, \alpha_{N-1}$ between two nodes and the last feature (for example y) is interpolated via formula (1). Of course there can be calculated $x(c)$ or $z(c)$ using (1). Additionally for better reconstruction and modeling there is a factor with nodes combination h and function γ, for example [26] that origins from some calculations with orthogonal matrices

$$h(p_1, p_2) = \frac{y_1}{x_1}x_2 + \frac{y_2}{x_2}x_1 \tag{2}$$

or

$$h(p_1, p_2, p_3, p_4) = \frac{1}{x_1^2 + x_3^2}(x_1x_2y_1 + x_2x_3y_3 + x_3x_4y_1 - x_1x_4y_3) + \frac{1}{x_2^2 + x_4^2}(x_1x_2y_2 + x_1x_4y_4 + x_3x_4y_2 - x_2x_3y_4).$$

The simplest nodes combination is

$$h(p_1, p_2, \ldots, p_m) = 0 \tag{3}$$

and then there is a formula of interpolation:

$$y(c) = \gamma \cdot y_i + (1 - \gamma)y_{i+1}.$$

Formula (1) gives the infinite number of calculations for unknown feature and PFC data-object modeling.

2.1 N-Dimensional Probability Distribution Functions in PFC Modeling

Unknown values of features are calculated via PFC approach (1). Figure 1 is a linear interpolation ($\gamma = \alpha, h = 0$) of function $y = 2^x$.

MHR method [26] is the example of PFC modeling for feature vector of dimension $N = 2$. Distributions of random variables α_i and γ in (1)

$\gamma = F(\alpha), F : [0; 1]^{N-1}[0; 1], F(0, \ldots, 0) = 0, F(1, \ldots, 1) = 1$
$\gamma_i = \alpha_i^s, \gamma_i = sin(\alpha_i^s \pi/2), \gamma_i = sin^s(\alpha_i \pi/2), \gamma_i = 1 - cos(\alpha_i^s \pi/2),$
$\gamma_i = 1 - cos^s(\alpha_i \pi/2), \gamma_i = tan(\alpha_i^s \pi/4),$
$\gamma_i = tan^s(\alpha_i \pi/4), \gamma_i = log_2(\alpha_i^s + 1), \gamma_i = log_2^s(\alpha_i + 1), \gamma_i = (2^\alpha - 1)^s,$

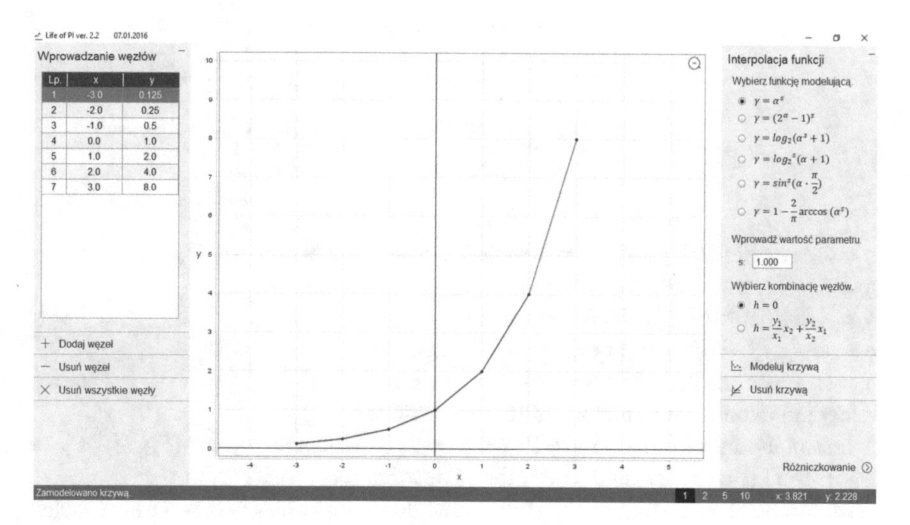

Fig. 1 Example of PFC modeling for function $y = 2^x$ with linear version and seven nodes

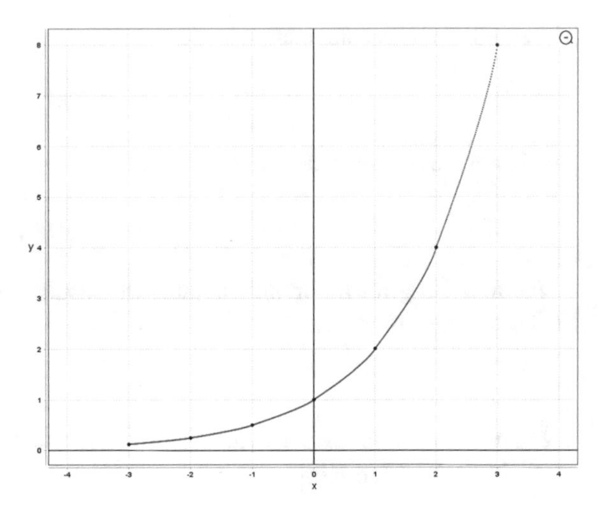

Fig. 2 PFC two-dimensional modeling of function $y = 2^x$ with seven nodes as Fig. 1 and $h = 0$, $\gamma = \alpha^{0.8}$

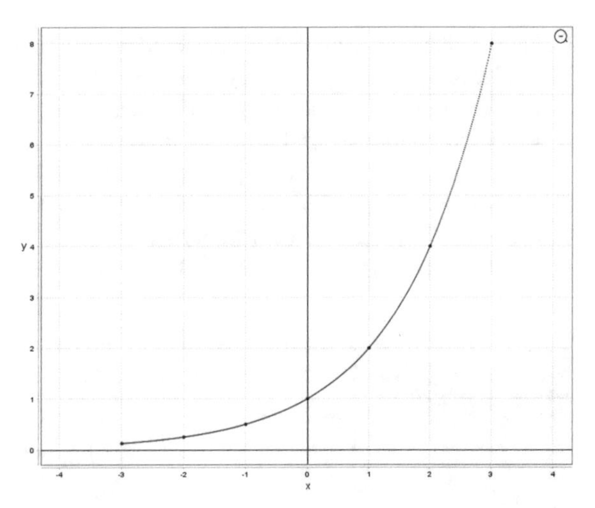

Fig. 3 PFC two-dimensional reconstruction of function $y = 2^x$ with seven nodes as Fig. 1 and $h = 0$, $\gamma = log_2(\alpha + 1)$

$\gamma_i = 2/\pi arcsin(\alpha_i^s)$, $\gamma_i = (2/\pi arcsin\alpha_i)^s$, $\gamma_i = 1 - 2/\pi arccos(\alpha_i^s)$,
$\gamma_i = 1 - (2/\pi arccos\alpha_i)^s$,
$\gamma_i = 4/\pi arctan(\alpha_i^s)$, $\gamma_i = (4/\pi arctan\alpha_i)^s$,
$\gamma_i = ctg(\pi/2 - \alpha_i^s \pi/4)$, $\gamma_i = ctg^s(\pi/2 - \alpha_i \pi/4)$,
$\gamma_i = 2 - 4/\pi arcctg(\alpha_i^s)$, $\gamma_i = (2 - 4/\pi arcctg\alpha_i)^s$
or any monotonic combinations of these functions.

Interpolations of function $y = 2^x$ for $N = 2$, $h = 0$ and $\gamma = \alpha^{0.8}$ (Fig. 2) or $\gamma = log_2(\alpha + 1)$ (Fig. 3) are quite better then linear interpolation (Fig. 1).

Main advantage and superiority of PFC method comparing with known approaches are that there is no method connecting all these ten points below together (see Sect. 3)

1. Interpolation of some complicated functions using combinations of a simple function;
2. Only local changes of the curve if one node is exchanged;
3. No matter if the curve is opened or closed;
4. Data extrapolation is computed via the same formulas as interpolation;
5. Object modeling in any dimension N;
6. Curve parameterization;
7. Modeling of specific and nontypical curves: signatures, fonts, symbols, characters or handwriting (for example Figs. 4, 5, 6, 12, 13, 14, 15, 16, 17 and 18 or 22, 23 and 24);
8. Reconstruction of irregular shapes (Figs. 25, 26, 27, 28, 29 and 30);
9. Applications in numerical analysis because of very precise interpolation of unknown values;
10. Even for only two nodes a curve can be modeled.

N-dimensional probability distribution is for example:

$$\gamma = \frac{1}{N-1}\sum_{i-1}^{N-1}\gamma_i, \gamma = \prod_{i=1}^{N-1}\gamma_i$$

and every monotonic combination of γ_i such as
$\gamma = F(\alpha), F : [0; 1]^{N-1}[0; 1], F(0, \ldots, 0) = 0, F(1, \ldots, 1) = 1.$
For example when $N = 3$ there is a bilinear interpolation:

$$\gamma_1 = \alpha_1, \gamma_2 = \alpha_2, \gamma = (\alpha_1 + \alpha_2) \tag{4}$$

or a bi-quadratic interpolation

$$\gamma_1 = {\alpha_1}^2, \gamma_2 = {\alpha_2}^2, \gamma = ({\alpha_1}^2 + {\alpha_2}^2) \tag{5}$$

or a bi-cubic interpolation

$$\gamma_1 = {\alpha_1}^3, \gamma_2 = {\alpha_2}^3, \gamma = ({\alpha_1}^3 + {\alpha_2}^3) \tag{6}$$

or others modeling functions γ.

Three examples of PFC reconstruction (Figs. 4, 5 and 6) for $N = 2$ and four nodes: $(-1.5; -1)$, $(1.25; 3.15)$, $(4.4; 6.8)$ and $(8; 7)$. Formula of the curve is not given.

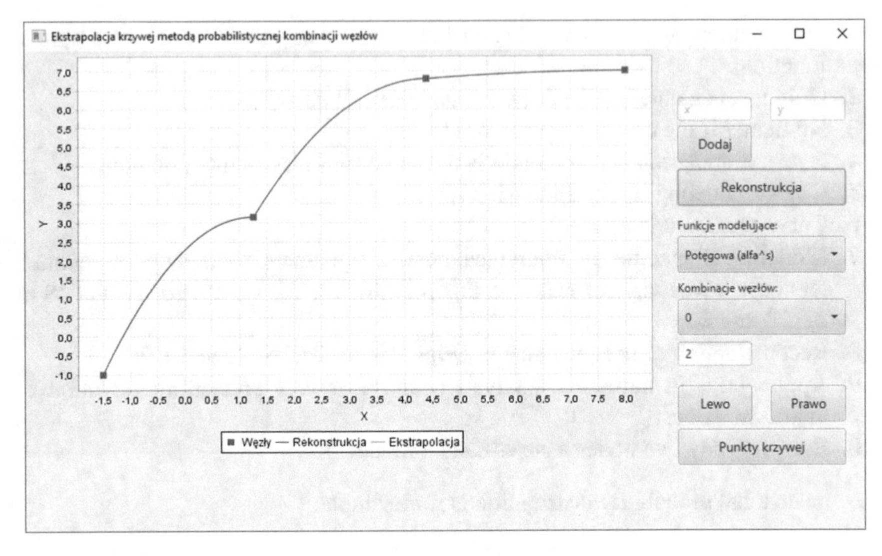

Fig. 4 A curve in PFC modeling for $\gamma = \alpha^2$ and $h = 0$

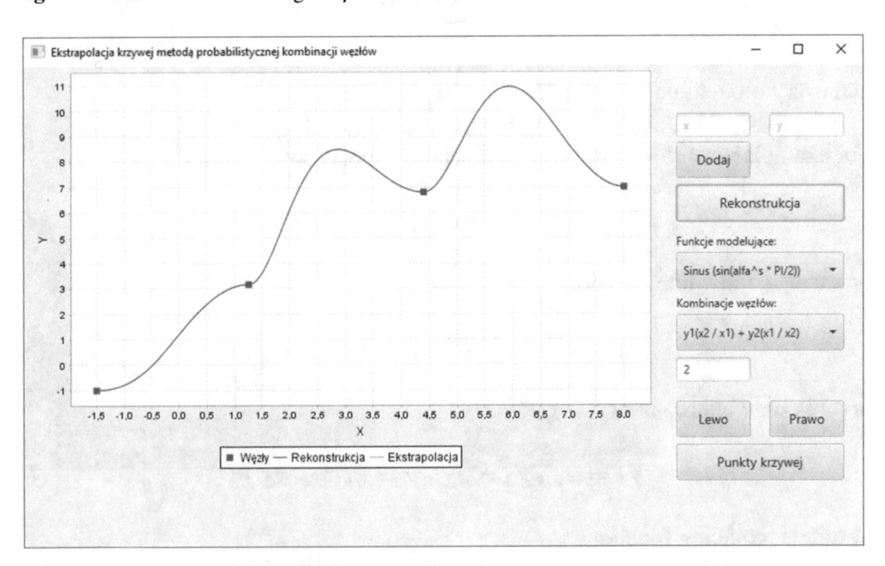

Fig. 5 The example of PFC reconstruction for $\gamma = sin(\alpha^2\pi/2)$ and h in (2)

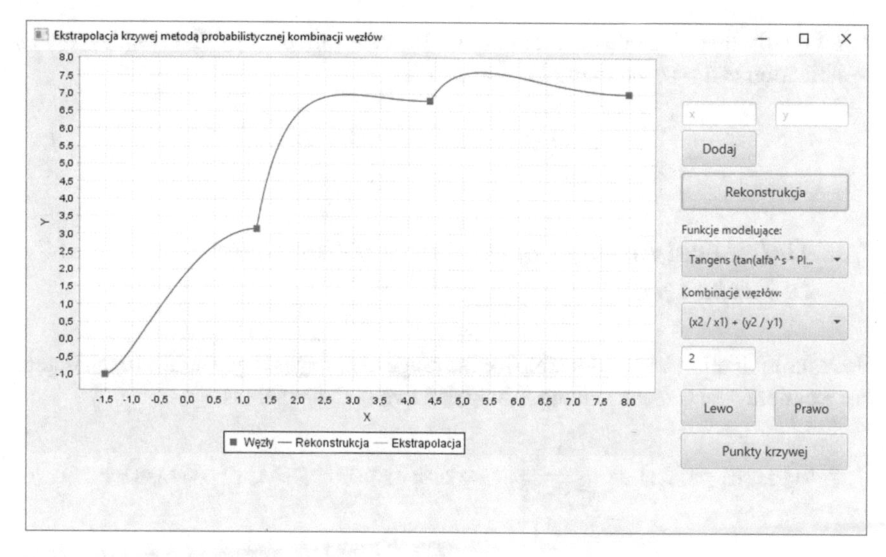

Fig. 6 Data PFC interpolation for $\gamma = tan(\alpha^2\pi/4)$ and $h = (x_2/x_1) + (y_2/y_1)$

3 Discussion in Details Over PFC Approach

What are the unique features of PFC method comparing with other methods of function interpolation, curve modeling and data extrapolation? This paragraph is answer this question.

3.1 Interpolation of Some Complicated Functions Using Combinations of a Simple Function

Some mathematical formulas of functions are very complicated and have very high complexity of calculations. Then there is necessity of modeling via any simple function. Of course one can take a linear function between two nodes but this is noneffective approach. The idea of PFC formula:

$$y(c) = \gamma \cdot y_k + (1-\gamma)y_{k+1} + \gamma(1-\gamma) \cdot h(p_1, p_2, \ldots, p_m)$$

is to calculate unknown value or coordinate as follows—take another modeling function (not linear) between two nodes

$$c = \alpha \cdot x_k + (1-\alpha) \cdot x_{k+1}.$$

Additionally there is nodes combination h for better modeling. The simplest nodes combination is $h = 0$ and then PFC formula is

$$y(c) = \gamma \cdot y_i + (1 - \gamma) y_{i+1}.$$

3.2 Only Local Changes of the Curve if One Node Is Exchanged

Nodes combination h is responsible for the range of changes if one node is exchanged. For example $h = 0$ means changes between two nodes whereas

$$h(p_1, p_2, p_3, p_4) = \frac{1}{x_1^2 + x_3^2}(x_1x_2y_1 + x_2x_3y_3 + x_3x_4y_1 - x_1x_4y_3) + \frac{1}{x_2^2 + x_4^2}(x_1x_2y_2 + x_1x_4y_4 + x_3x_4y_2 - x_2x_3y_4).$$

means changes between four nodes.

3.3 No Matter if the Curve Is Opened or Closed

PFC formulas require the order and numbering of nodes exactly like on the curve, for example a graph of function. The only assumption for closed curve is that first node and last node are the same.

3.4 Data Extrapolation Is Computed via the Same Formulas As Interpolation

Extrapolation is computed for $\alpha \notin [0; 1]$. Then modeling function $\gamma = F(\alpha)$ has to be chosen for the situation when $\alpha < 0$ or $\alpha > 1$. Sometimes one can take a parallel version of PFC formulas

$$y(c) = \gamma \cdot y_{k+1} + (1 - \gamma) y_k + \gamma(1 - \gamma) \cdot h(p_1, p_2, \ldots, p_m),$$

$c = \alpha \cdot x_{k+1} + (1 - \alpha) \cdot x_k$.
when for example calculations for $\alpha < 0$ are impossible.

3.5 Object Modeling in any Dimension N

Modeling functions $\gamma_i = F_i(\alpha_i)$ are valid for any feature i = 1, 2, …, N − 1.

3.6 Curve Parameterization

Parameterization of the curve between each pair of nodes is connected with random variable α.

3.7 Modeling of Specific and Nontypical Curves: Signatures, Fonts, Symbols, Characters, or Handwriting

Figures 4, 5, 6, 12, 13, 14, 15, 16, 17, 18, 22, 23 and 24 show the examples of PFC modeling. In the individual cases one can take for each pair of nodes different $\gamma = F(\alpha)$ and h.

3.8 Reconstruction of Irregular Shapes

Nontypical and irregular shapes are modeled via the choice of modeling functions $\gamma_i = F_i(\alpha_i)$.

3.9 Applications in Numerical Analysis Because of Very Precise Interpolation of Unknown Values

All numerical methods for numerical analysis (quadratures, derivatives, nonlinear equations, etc.) are based on the values of function given in the table. PFC method enables precise interpolation of the function (for example Fig. 2).

3.10 Even for only Two Nodes a Curve Can Be Modeled

Thankfully that PFC is modeling the curve between each pair of nodes, even two nodes are enough in some cases for interpolation and reconstruction.

This chapter was concerned on some aspects and features of PFC approach from mathematical and computational points of view.

4 Image Retrieval via PFC High-Dimensional Feature Reconstruction

After the process of image segmentation and during the next steps of retrieval, recognition or identification, there is a huge number of features included in N-dimensional feature vector. These vectors can be treated as "points" in N-dimensional feature space. The paper is dealing with multidimensional feature spaces that are used in computer vision, image processing and machine learning.

Having monochromatic (binary) image which consists of some objects, there is only two-dimensional feature space (x_i, y_i)—coordinates of black pixels or coordinates of white pixels. No other parameters are needed. Thus any object can be described by a contour (closed binary curve). Binary images are attractive in processing (fast and easy) but don't include important information. If the image has grey shades, there is 3-dimensional feature space (x_i, y_i, z_i) with grey shade z_i. For example most of medical images are written in grey shades to get quite fast processing. But when there are color images (three parameters for RGB or other color systems) with textures or medical data or some parameters, then it is N-dimensional feature space. Dealing with the problem of classification learning for high-dimensional feature spaces in artificial intelligence and machine learning (for example text classification and recognition), there are some methods: decision trees, k-nearest neighbors, perceptrons, nave Bayes, or neural networks methods. All of these methods are struggling with the curse of dimensionality: the problem of having too many features. And there are many approaches to get less number of features and to reduce the dimension of feature space for faster and less expensive calculations.

This paper aims at inverse problem to the curse of dimensionality: dimension N of feature space (i.e., number of features) is unchanged, but number of feature vectors (i.e., "points" in N-dimensional feature space) is reduced into the set of nodes. **So the main problem is as follows: how to fix the set of feature vectors for the image and how to retrieve the features between the "nodes"?** This paper aims in giving the answer of this question.

4.1 Grey Scale Image Retrieval Using PFC 3D Method

Binary images are just the case of 2D points (x,y): 0 or 1, black or white, so retrieval of monochromatic images is done for the closed curves (first and last node are the same) as the contours of the objects for $N = 2$ and examples as Figs. 1, 2, 3, 4, 5 and 6. Grey scale images are the case of 3D points (x, y, s) with s as the shade of grey. So the grey scale between the nodes $p_1 = (x_1, y_1, s_1)$ and $p_2 = (x_2, y_2, s_2)$ is computed with

$\gamma = F(\alpha) = F(\alpha_1, \alpha_2)$ as (1) and for example (4)–(6)) or others modeling functions γ_i. As the simple example two successive nodes of the image are: left upper corner with coordinates $p_1 = (x_1, y_1, 2)$ and right down corner $p_2 = (x_2, y_2, 10)$. The image retrieval with the grey scale 2–10 between p_1 and p_2 looks as follows for a bilinear interpolation (4) (Fig. 7); or for other modeling functions γ_i (Fig. 8).

The feature vector of dimension $N = 3$ is called a voxel.

4.2 Color Image Retrieval via PFC Method

Color images in for example RGB color system (r, g, b) are the set of points (x, y, r, g, b) in a feature space of dimension $N = 5$. There can be more features, for example texture t, and then one pixel (x, y, r, g, b, t) exists in a feature space of dimension $N = 6$. But there are the subspaces of a feature space of dimension $N_1 < N$, for example, (x, y, r), (x, y, g), (x, y, b) or (x, y, t) are points in a feature subspace of dimension $N_1 = 3$. Reconstruction and interpolation of color coordinates or texture parameters is done like in Sect. 3.1 for dimension $N = 3$. Appropriate combination of α_1 and α_2 leads to modeling of color r, g, b or texture t or another feature between the nodes. For example, (x, y, r, t), (x, y, g, t), (x, y, b, t) are points in a feature subspace of dimension $N_1 = 4$ called doxels. Appropriate combination of α_1, α_2 and α_3 leads to modeling of texture t or another feature between the nodes. For example, color image, given as the set of doxels (x, y, r, t), is described for coordinates (x, y) via pairs (r, t) interpolated between nodes $(x_1, y_1, 2, 1)$, and $(x_2, y_2, 10, 9)$ as follows (Fig. 9).

So dealing with feature space of dimension N and using PFC method there is no problem called "the curse of dimensionality" and no problem called "feature selection" because each feature is important. There is no need to reduce the dimension N and no need to establish which feature is "more important" or "less important". Every feature that depends from $N_1 - 1$ other features can be interpolated (reconstructed) in the feature subspace of dimension $N_1 < N$ via PFC method. But having a feature space of dimension N and using PFC method there is another problem: how to reduce the number of feature vectors and how to interpolate (retrieve) the features between the known vectors (called nodes).

Difference between two given approaches (the curse of dimensionality with feature selection and PFC interpolation) can be illustrated as follows. There is a feature matrix of dimension $N \times M$: N means the number of features (dimension of feature space) and M is the number of feature vectors (interpolation nodes)—columns are feature vectors of dimension N. One approach (Fig. 10): the curse of dimensionality with feature selection wants to eliminate some rows from the feature matrix and to reduce dimension N to $N_1 < N$. Second approach (Fig. 11) for PFC method wants to eliminate some columns from the feature matrix and to reduce dimension M to $M_1 < M$.

So after feature selection (Fig. 10) there are nine feature vectors (columns): $M = 9$ in a feature subspace of dimension $N_1 = 6 < N$ (three features are fixed as less important and reduced). But feature elimination is a very unclear matter. And what to do if every feature is denoted as meaningful and then no feature is to be reduced? For PFC method (Fig. 11) there are seven feature vectors (columns): $M_1 = 7 < M$ in a feature space of dimension $N = 9$. Then no feature is eliminated and the main problem is dealing with interpolation or extrapolation of feature values, like for example image retrieval (Figs. 7, 8 and 9).

Fig. 7 Reconstructed grey scale numbered at each pixel

2	3	4	5	6	7	8	9	10
2	3	4	5	6	7	8	9	10
2	3	4	5	6	7	8	9	10
2	3	4	5	6	7	8	9	10
2	3	4	5	6	7	8	9	10
2	3	4	5	6	7	8	9	10
2	3	4	5	6	7	8	9	10
2	3	4	5	6	7	8	9	10
2	3	4	5	6	7	8	9	10

Fig. 8 Grey scale image with shades of grey retrieved at each pixel

2	2	2	2	2	2	2	2	2
2	3	3	3	3	3	3	3	3
2	3	4	4	4	4	4	4	4
2	3	4	5	5	5	5	5	5
2	3	4	5	6	6	6	6	6
2	3	4	5	6	7	7	7	7
2	3	4	5	6	7	8	8	8
2	3	4	5	6	7	8	9	9
2	3	4	5	6	7	8	9	10

2,1	3,1	4,1	5,1	6,1	7,1	8,1	9,1	10,1
2,2	3,2	4,2	5,2	6,2	7,2	8,2	9,2	10,2
2,3	3,3	4,3	5,3	6,3	7,3	8,3	9,3	10,3
2,4	3,4	4,4	5,4	6,4	7,4	8,4	9,4	10,4
2,5	3,5	4,5	5,5	6,5	7,5	8,5	9,5	10,5
2,6	3,6	4,6	5,6	6,6	7,6	8,6	9,6	10,6
2,7	3,7	4,7	5,7	6,7	7,7	8,7	9,7	10,7
2,8	3,8	4,8	5,8	6,8	7,8	8,8	9,8	10,8
2,9	3,9	4,9	5,9	6,9	7,9	8,9	9,9	10,9

Fig. 9 Color image with color and texture parameters (r, t) interpolated at each pixel

2	2	2	2	2	2	2	2	2
2	3	3	3	3	3	3	3	3
2	3	4	4	4	4	4	4	4
2	3	4	5	5	5	5	5	5
2	3	4	5	6	6	6	6	6
2	3	4	5	6	7	7	7	7
2	3	4	5	6	7	8	8	8
2	3	4	5	6	7	8	9	9
2	3	4	5	6	7	8	9	10

→

2	2	2	2	2	2	2	2	2
2	3	3	3	3	3	3	3	3
2	3	4	4	4	4	4	4	4
2	3	4	5	5	5	5	5	5
2	3	4	5	6	6	6	6	6
2	3	4	5	6	7	7	7	7

Fig. 10 The curse of dimensionality with feature selection wants to eliminate some rows from the feature matrix and to reduce dimension N

2	2	2	2	2	2	2	2	2
2	3	3	3	3	3	3	3	3
2	3	4	4	4	4	4	4	4
2	3	4	5	5	5	5	5	5
2	3	4	5	6	6	6	6	6
2	3	4	5	6	7	7	7	7
2	3	4	5	6	7	8	8	8
2	3	4	5	6	7	8	9	9
2	3	4	5	6	7	8	9	10

→

2	2	2	2	2	2	2
2	3	3	3	3	3	3
2	3	4	4	4	4	4
2	3	4	5	5	5	5
2	3	4	5	6	6	6
2	3	4	5	6	7	7
2	3	4	5	6	7	8
2	3	4	5	6	7	8
2	3	4	5	6	7	8

Fig. 11 PFC method wants to eliminate some columns from the feature matrix and to reduce dimension M

5 Recognition Tasks via High-Dimensional Feature Vectors' Interpolation

Human signature or handwriting consists of nontypical curves and irregular shapes (for example Figs. 4, 5 and 6). Therefore it is very important to calculate values of features (pen pressure, speed, pen angle, etc.) appearing in high-dimensional feature vectors fixing the values in feature vectors for unknown signature or handwritten words: N-dimensional feature vectors (x, y, p, s, a, t) with x, y-points' coordinates, p-pen pressure, s-speed of writing, a-pen angle, or any other features t.

5.1 Signature Modeling and Multidimensional Recognition

Many features in multidimensional feature space are not visible but used in recognition process (for example p-pen pressure, s-speed of writing, a-pen angle). Here are some examples of nontypical curves and irregular shapes as the whole signature or a part of signature, reconstructed via PFC method for seven nodes (x, y) see Figs. 1, 2 and 3.

Figures 12, 13, 14, 15, 16, 17 and 18 are two-dimensional subspace of N-dimensional feature space, for example (x, y, p, s, a, t) when $N = 6$. If the recognition process is working "offline" and features p-pen pressure, s-speed of writing, a-pen angle, or another feature t are not given, the only information before recognition is situated in x, y-points' coordinates.

If the recognition process is "online" and features p-pen pressure, s-speed of writing, a- pen angle, or some feature t are given, then there is more information in the process of author recognition, identification or verification in a feature space (x, y, p, s, a, t) of dimension $N = 6$ or others. Some person may know how the signature of another man looks like (for example Figs. 4, 5 and 6 or Figs. 12, 13, 14, 15, 16, 17 and 18), but other extremely important features p, s, a, t are not visible. Dimension N of a feature space may be very high, but this is no problem. As it is illustrated (Figs. 10 and 11) the problem connected with the curse of dimensionality with feature selection does not matter. There is no need to fix which feature is less important and can be eliminated. Every feature is very important and each of them can be interpolated between the nodes using PFC high-dimensional interpolation. For example pressure of the pen p differs during the signature writing and p is changing for particular letters or fragments of the signature. Then feature vector (x, y, p) of dimension $N_1 = 3$ is dealing with p interpolation at the point (x, y) via modeling functions (4)–(6) or others. If angle of the pen a differs during the signature writing and a is changing for particular letters or fragments of the signature, then feature vector (x, y, a) of dimension $N_1 = 3$ is dealing with a interpolation at the point (x, y)

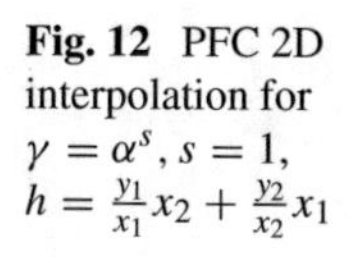

Fig. 12 PFC 2D interpolation for $\gamma = \alpha^s, s = 1$, $h = \frac{y_1}{x_1}x_2 + \frac{y_2}{x_2}x_1$

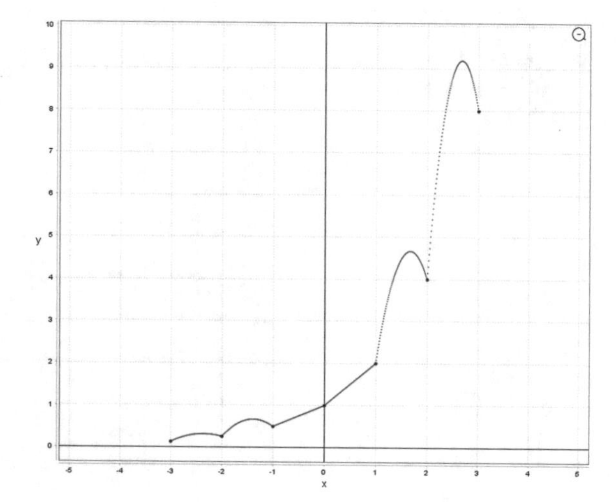

Fig. 13 PFC 2D interpolation for $\gamma = \alpha^s, s = 0.8$, $h = \frac{y_1}{x_1}x_2 + \frac{y_2}{x_2}x_1$

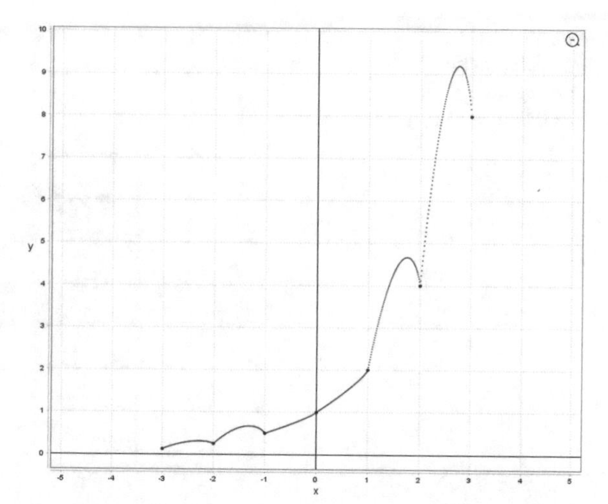

via modeling functions (4)–(6) or others. If speed of the writing s differs during the signature writing and s is changing for particular letters or fragments of the signature, then feature vector (x, y, s) of dimension $N_1 = 3$ is dealing with s interpolation at the point (x, y) via modeling functions (4)–(6) or others. This PFC 3D interpolation is the same like in Sect. 3.1 grey scale image retrieval but for selected pairs (α_1, α_2)—only for the points of signature between $(x_1, y_1, 2)$ and $(x_2, y_2, 10)$.

If a feature subspace is dimension $N_1 = 4$ and feature vector is for example (x, y, p, s), then PFC 4D interpolation is the same like in Sect. 3.2 color image retrieval but for selected pairs (α_1, α_2)—only for the points of signature between

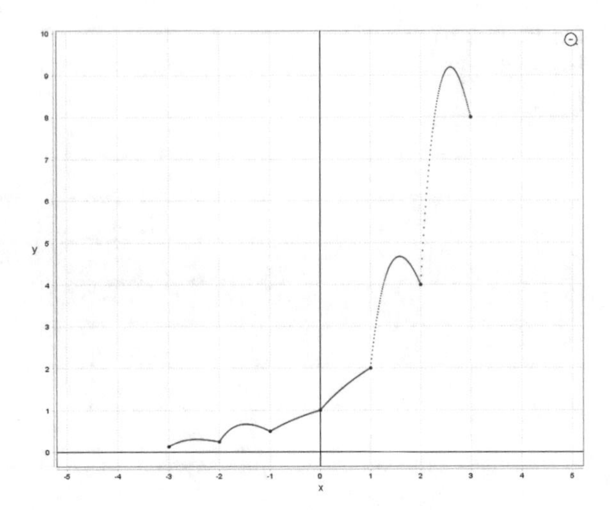

Fig. 14 PFC 2D interpolation for $\gamma = (2^{\alpha} - 1)^{s}, s = 1, h = \frac{y_1}{x_1}x_2 + \frac{y_2}{x_2}x_1$

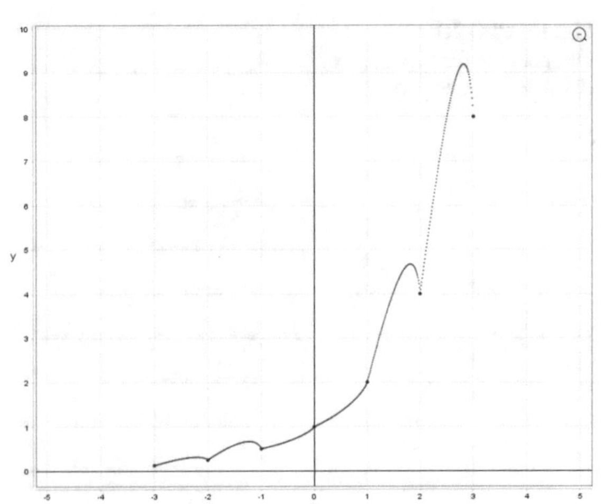

Fig. 15 PFC 2D interpolation for $\gamma = log_2(\alpha^{s} + 1)$, $s = 0.8, h = \frac{y_1}{x_1}x_2 + \frac{y_2}{x_2}x_1$

$(x_1, y_1, 2, 1)$ and $(x_2, y_2, 10, 9)$—Fig. 20. If a feature subspace is dimension $N_1 = 5$ and feature vector is for example (x, y, p, s, a), then PFC 5D interpolation is the same like in Sect. 3.2 color image retrieval but for selected pairs (α_1, α_2)—only for the points of signature between $(x_1, y_1, 2, 1, 30)$ and $(x_2, y_2, 10, 9, 60)$—Fig. 21.

Figures 19, 20 and 21 are the examples of denotation for the features that are not visible during the signing or handwriting but very important in the process of "online" recognition, identification or verification.

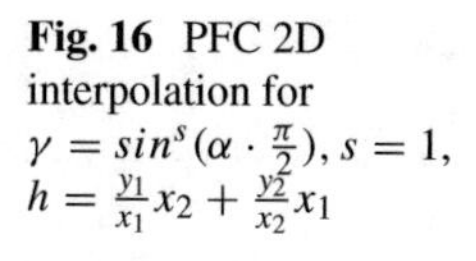

Fig. 16 PFC 2D interpolation for $\gamma = sin^s(\alpha \cdot \frac{\pi}{2}), s = 1, h = \frac{y_1}{x_1}x_2 + \frac{y_2}{x_2}x_1$

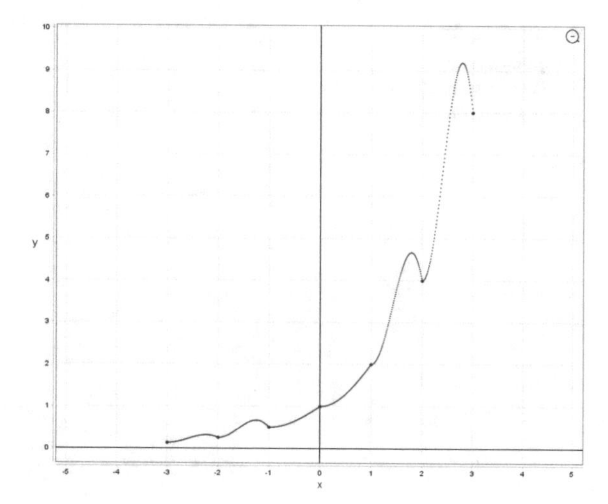

Fig. 17 PFC 2D interpolation for $\gamma = sin^s(\alpha \cdot \frac{\pi}{2}), s = 0.8, h = 0$

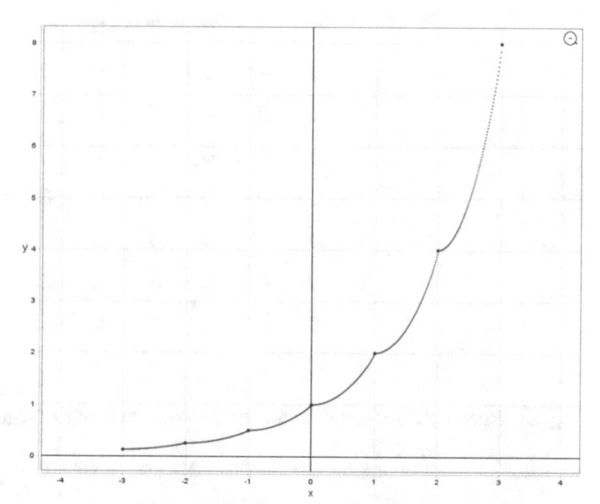

Even if from technical reason or other reasons only some points of signature or handwriting (feature nodes) are given in the process of "online" recognition, identification, or verification, the values of features between nodes are computed via multidimensional PFC interpolation like for example between $(x_1, y_1, 2)$ and $(x_2, y_2, 10)$ on Fig. 19, between $(x_1, y_1, 2, 1)$ and $(x_2, y_2, 10, 9)$ on Fig. 20 or between $(x_1, y_1, 2, 1, 30)$ and $(x_2, y_2, 10, 9, 60)$ on Fig. 21. Reconstructed features are compared with the features in the basis of patterns and appropriate criterion gives the result.

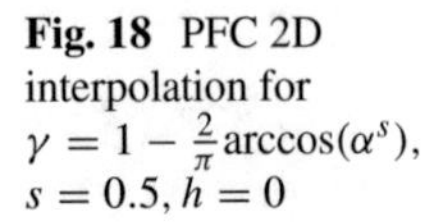

Fig. 18 PFC 2D interpolation for $\gamma = 1 - \frac{2}{\pi}\arccos(\alpha^s)$, $s = 0.5, h = 0$

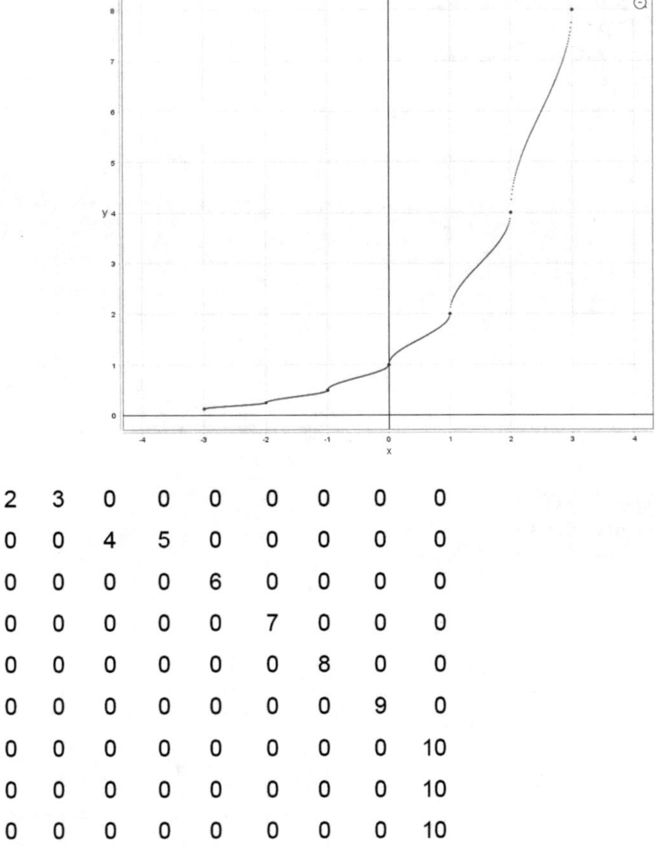

2	3	0	0	0	0	0	0	0
0	0	4	5	0	0	0	0	0
0	0	0	0	6	0	0	0	0
0	0	0	0	0	7	0	0	0
0	0	0	0	0	0	8	0	0
0	0	0	0	0	0	0	9	0
0	0	0	0	0	0	0	0	10
0	0	0	0	0	0	0	0	10
0	0	0	0	0	0	0	0	10

Fig. 19 Reconstructed speed of the writing s at the pixels of signature

2,1	3,1	0	0	0	0	0	0	0
0	0	4,2	5,2	0	0	0	0	0
0	0	0	0	6,3	0	0	0	0
0	0	0	0	0	7,4	0	0	0
0	0	0	0	0	0	8,5	0	0
0	0	0	0	0	0	0	9,6	0
0	0	0	0	0	0	0	0	10,7
0	0	0	0	0	0	0	0	10,8
0	0	0	0	0	0	0	0	10,9

Fig. 20 Reconstructed pen pressure p and speed of the writing s as (p, s) at the pixels of signature

2,1,30	3,1,30	0	0	0	0	0	0	0
0	0	4,2,32	5,2,34	0	0	0	0	0
0	0	0	0	6,3,37	0	0	0	0
0	0	0	0	0	7,4,43	0	0	0
0	0	0	0	0	0	8,5,45	0	0
0	0	0	0	0	0	0	9,6,46	0
0	0	0	0	0	0	0	0	10,7,53
0	0	0	0	0	0	0	0	10,8,56
0	0	0	0	0	0	0	0	10,9,60

Fig. 21 Reconstructed pen pressure p, speed of the writing s and angle a as (p, s, a) at the pixels of signature

5.2 *Modeling and Interpolation of Nontypical Curves and Irregular Shapes*

PFC two dimensional interpolation and modeling enables to solve classic problem in numerical methods (for example Fig. 2): how to parameterize and to model known function. But having the set of nodes there is another problem connected with handwriting and human signing: how to model or to reconstruct the curve which is the part

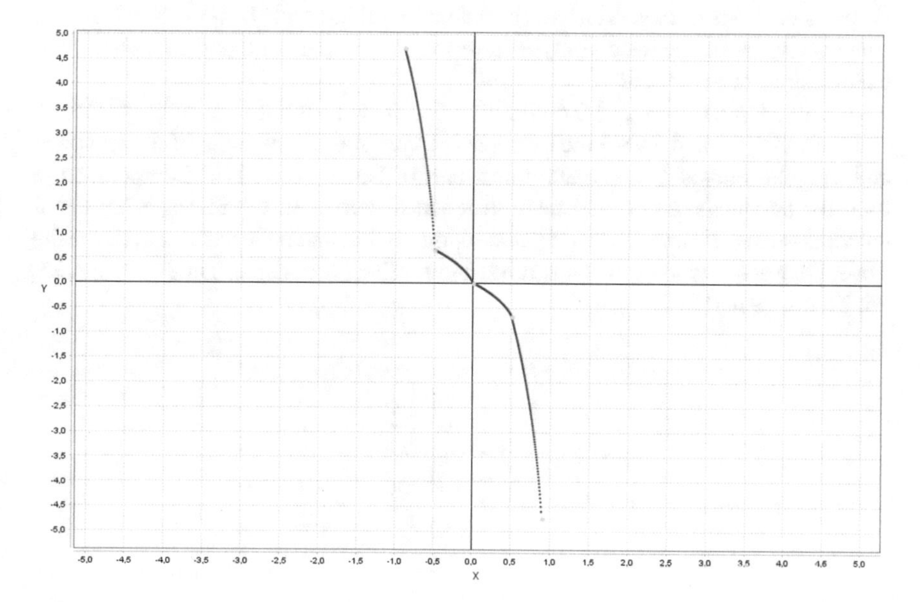

Fig. 22 PFC 2D interpolation for $\gamma = log_2{}^s\,(\alpha + 1)\,, s = 0.8, h = 0$

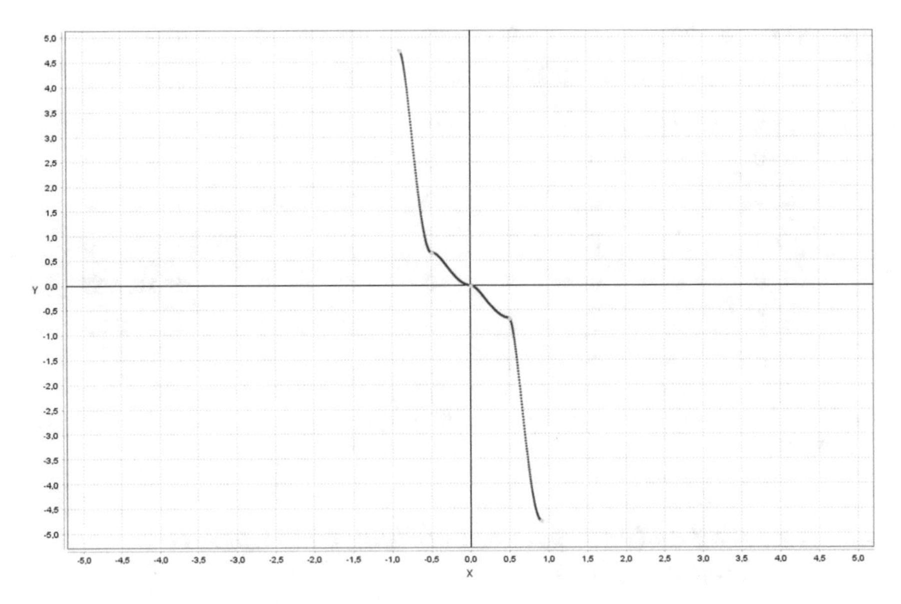

Fig. 23 PFC 2D interpolation for $\gamma = \sin\left(\alpha^s * \frac{\pi}{2}\right)$, $s = 1.8$, $h = 0$

of signature or handwriting but which is nontypical or irregular. Human signature and handwriting consists of nontypical curves and irregular shapes. PFC method (1) is the way of modeling and interpolation for nontypical curves and irregular shapes—contours as closed curves (if first node and last node is the same). Here are some examples of modeled nontypical or irregular curves as a part of signature or handwriting for five nodes

$(-0.9; ; 4.736)$, $(-0.5; ; 0.666)$, $(0; ; 0)$, $(0.5; ; -0.666)$, $(0.9; ; -4.736)$

Figures 22, 23 and 24 are the examples of very specific modeling for nontypical and irregular curves as a signature or a part of handwriting. PFC interpolation is used for parameterization and reconstruction of curves in the plane. What about contours—closed curves? The only assumption is that first node and last node are the same. Here are examples of shape modeling for five (or rather six) nodes—Figs. 25, 26, 27, 28, 29 and 30.

Lp.	x	y
1	1	2.5
2	1.7	4
3	2.6	3
4	2.3	0.5
5	1.5	1

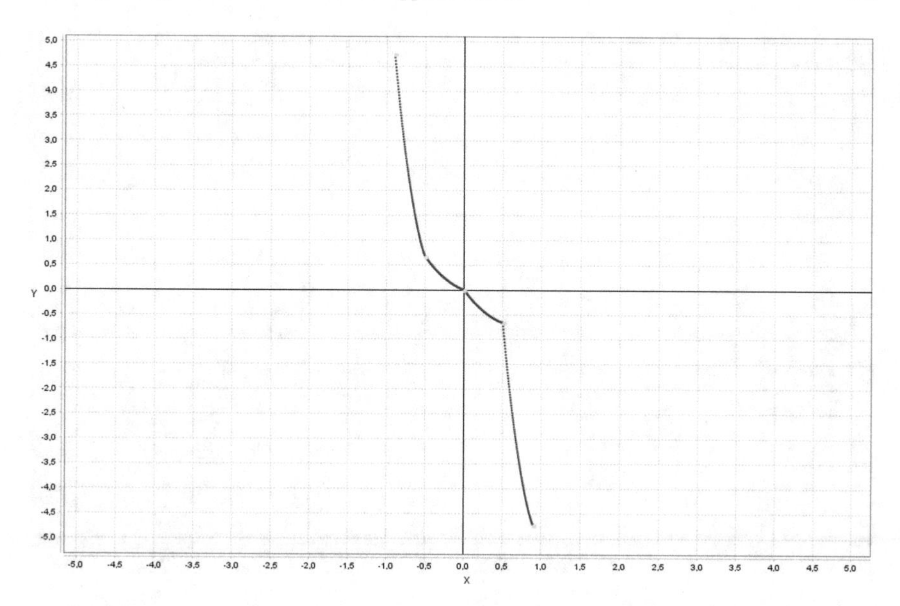

Fig. 24 PFC 2D interpolation for $\gamma = (2^{\alpha} - 1)^{s}$, $s = 1.2$, $h = 0$

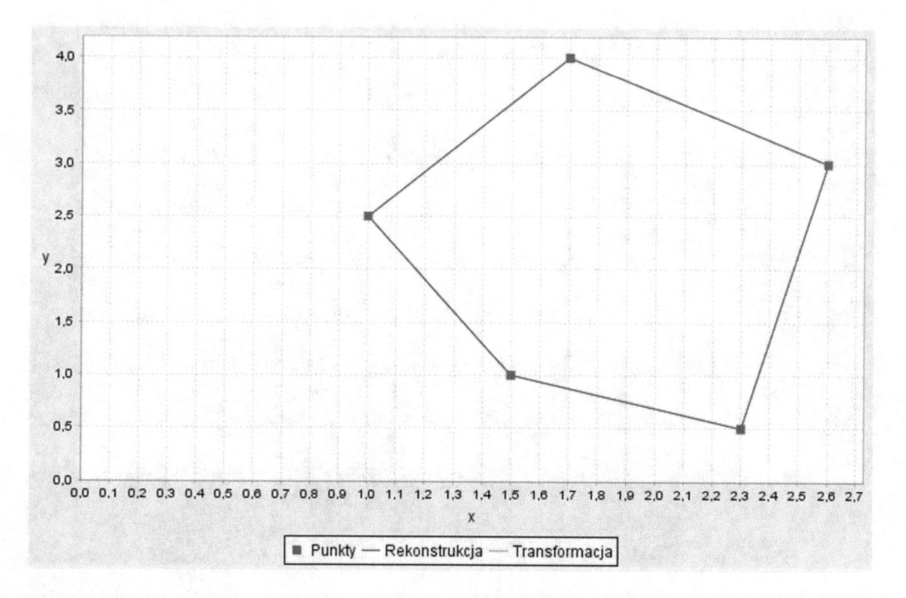

Fig. 25 PFC 2D modeling for $s = 1$, $h = 0$, $\gamma = \alpha^{s}$ (linear modeling)

For the same set of nodes contour is reconstructed via modeling function γ and parameters s and h.

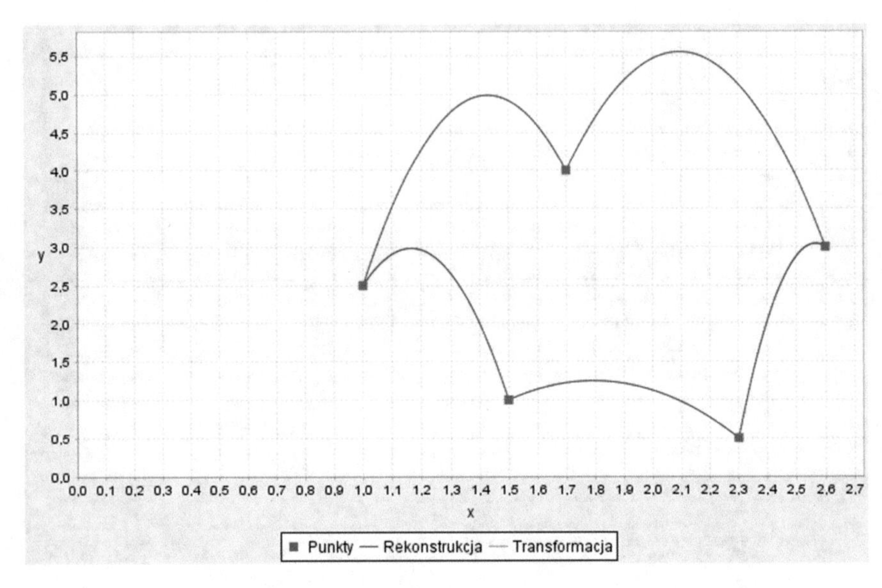

Fig. 26 PFC 2D modeling for $s = 1$, $h = x_1x_2 + y_1y_2$, $\gamma = \alpha^s$

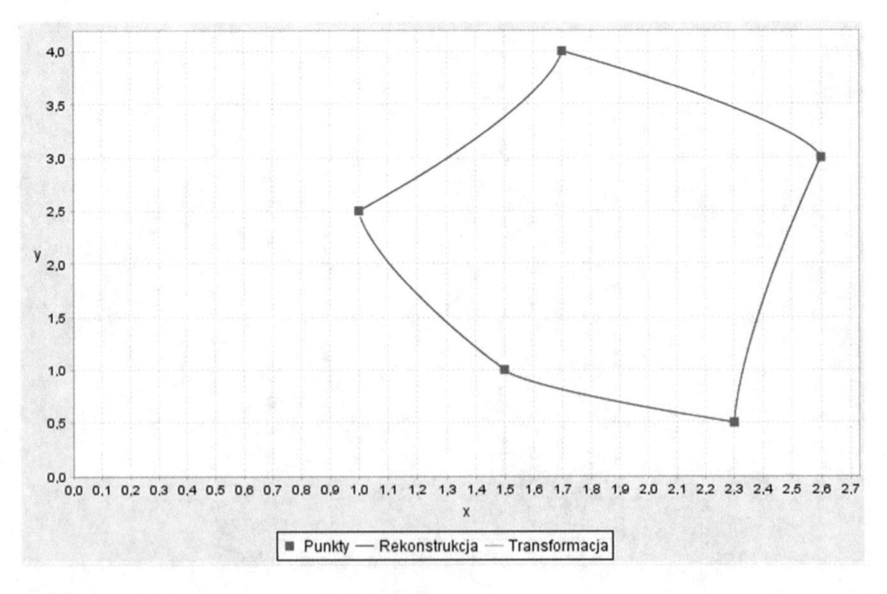

Fig. 27 PFC 2D modeling for $s = 0.7$, $h = 0$, $\gamma = \alpha^s$

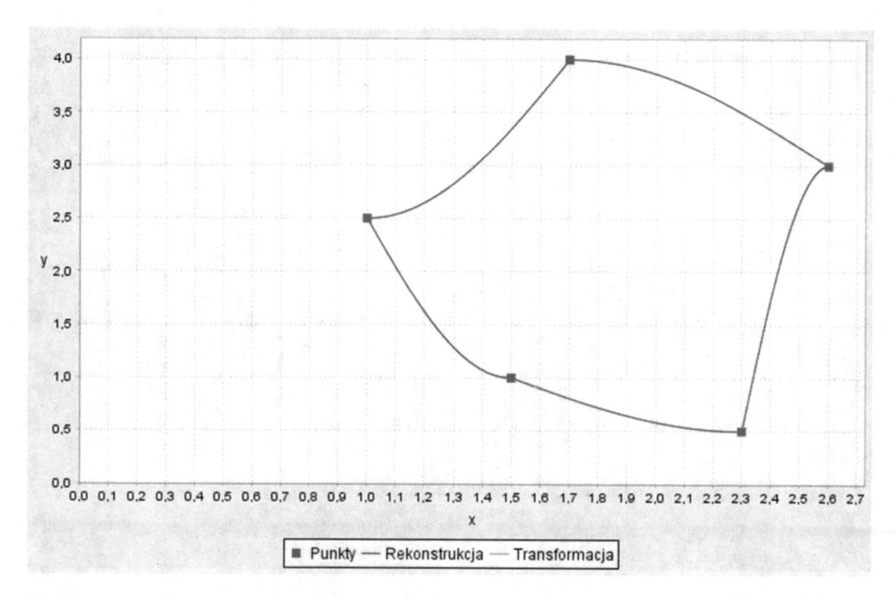

Fig. 28 PFC 2D modeling for $s = 1, h = 0, \gamma = \sin(\alpha^s \cdot \frac{\pi}{2})$

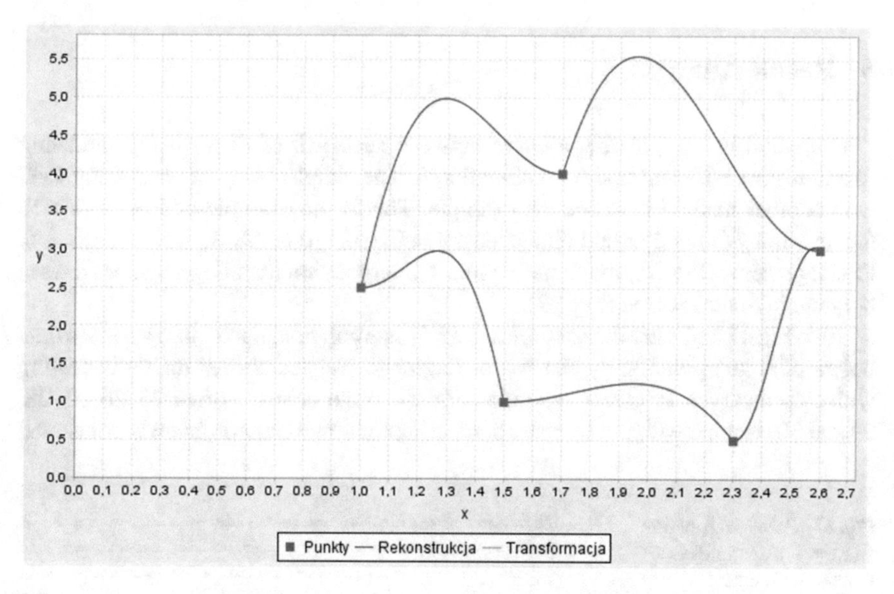

Fig. 29 PFC 2D modeling for $s = 1, h = x_1x_2 + y_1y_2, \gamma = 1 - \cos(\alpha^s \cdot \frac{\pi}{2})$

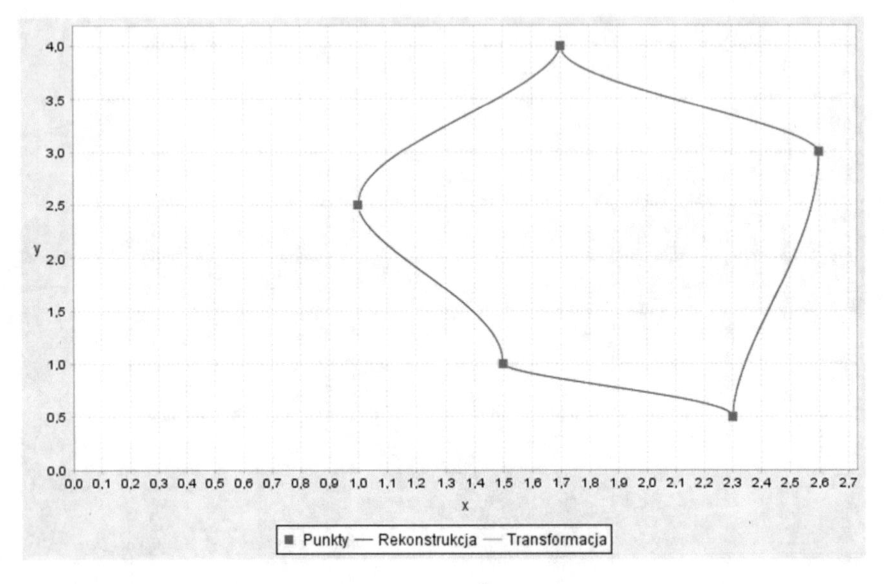

Fig. 30 PFC 2D modeling for $s = 0.6, h = 0, \gamma = \frac{2}{\pi} \cdot \arcsin(\alpha^s)$

6 Result Analysis

PFC method is interpolating a curve between each pair of nodes using modeling function $\gamma = \mathrm{F}(\alpha)$ and nodes combination h. The simplest way of comparing PFC with another method is to see the example. This is the example of PFC approach for function $f(x) = 1/x$ and nine nodes: $y = 0.2, 0.4, 0.6, 0.8, 1, 1.2, 1.4, 1.6, 1.8$. PFC represents (Fig. 31) much more precise interpolation than Lagrange or Newton polynomial interpolation (Fig. 32).

Also Figs. 2 and 3 are the examples of PFC interpolation much more accurate than polynomial interpolation by Newton or Lagrange. Very important matter is dealing with closed curves. PFC reconstruction of the contour or shape (Figs. 25, 26, 27, 28, 29 and 30) is done with the same formulas. Another important problem is connected

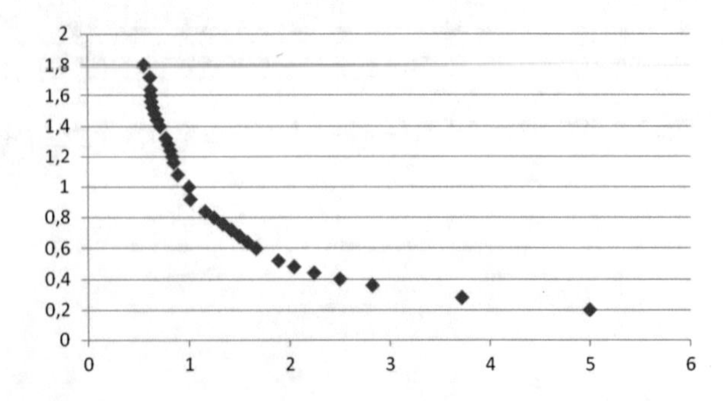

Fig. 31 Points of function $f(x) = 1/x$ using PFC method with nine nodes—better than polynomial interpolation

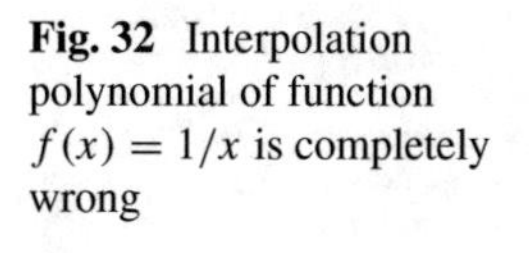

Fig. 32 Interpolation polynomial of function $f(x) = 1/x$ is completely wrong

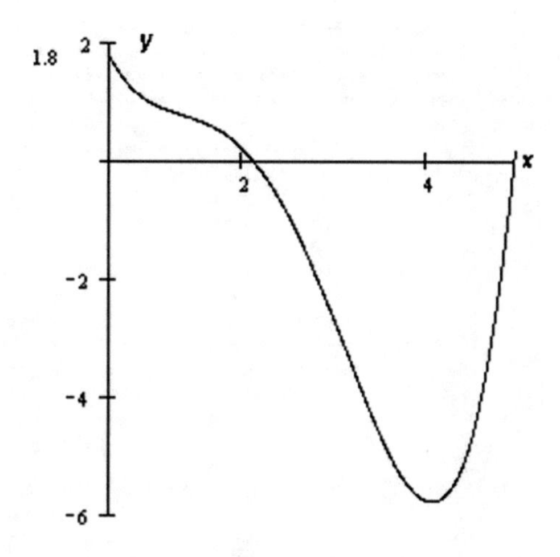

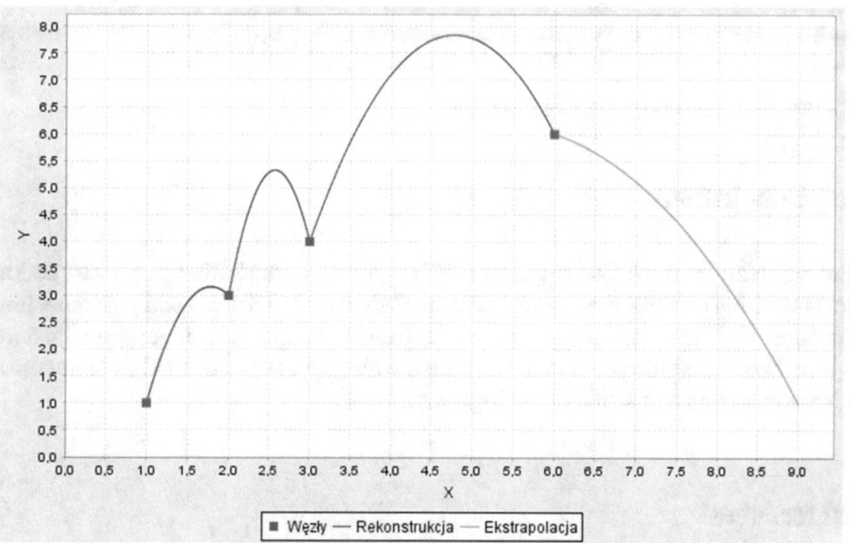

Fig. 33 Extrapolation of data right of the last node

with extrapolation. PFC method gives the tool of data anticipation or prediction. PFC is new proposition in data interpolation, reconstruction, and extrapolation (Figs. 33 and 34).

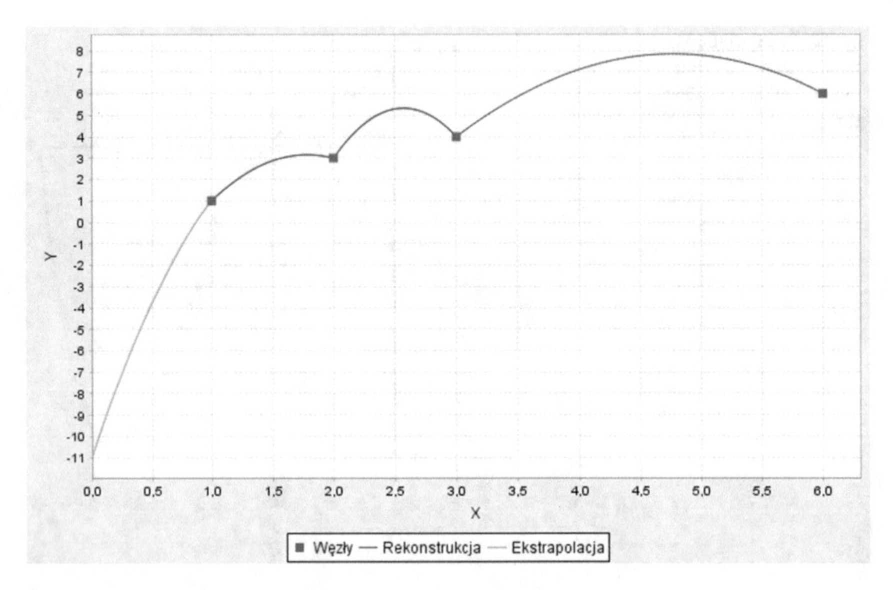

Fig. 34 Prediction of values left of first node

7 Conclusions

Probabilistic Features Combination (PFC) is a novel proposition in reconstruction and modeling of high-dimensional data via features coefficients h and γ as modeling function. PFC leads to image retrieval via feature vectors in N-dimensional feature space. Next scientific papers are connected with applications of PFC in biometrics, computer vision and artificial intelligence.

References

1. Marti, U.-V., Bunke, H.: The IAM-database: an English sentence database for offline handwriting recognition. Int. J. Doc. Anal. Recognit. **5**, 39–46 (2002)
2. Djeddi, C., Souici-Meslati, L.: A texture based approach for Arabic writer identification and verification. In: International Conference on Machine and Web Intelligence, pp. 115–120 (2010)
3. Djeddi, C., Souici-Meslati, L.: Artificial immune recognition system for Arabic writer identification. In: International Symposium on Innovation in Information and Communication Technology, pp. 159–165 (2011)
4. Nosary, A., Heutte, L., Paquet, T.: Unsupervised writer adaption applied to handwritten text recognition. Pattern Recogn. Lett. **37**(2), 385–388 (2004)
5. Bulacu, M., Schomaker, L., Brink, A.: Text-independent writer identification and verification on off-line Arabic handwriting. In: International Conference on Document Analysis and Recognition, pp. 769–773 (2007)
6. Chen, J., Cheng, W., Lopresti, D.: Using perturbed handwriting to support writer identification in the presence of severe data constraints. In: Document Recognition and Retrieval, pp. 1–10 (2011)

7. Chen, J., Lopresti, D., Kavallieratou, E.: The impact of ruling lines on writer identification. In: International Conference on Frontiers in Handwriting Recognition, pp. 439–444 (2010)
8. Galloway, M.M.: Texture analysis using gray level run lengths. Comput. Graphics Image Process. **4**(2), 172–179 (1975)
9. Garain, U., Paquet, T.: Off-line multi-script writer identification using AR coefficients. In: International Conference on Document Analysis and Recognition, pp. 991–995 (2009)
10. Ghiasi, G., Safabakhsh, R.: Offline text-independent writer identification using codebook and efficient code extraction methods. Image Vis. Comput. **31**, 379–391 (2013)
11. Ozaki, M., Adachi, Y., Ishii, N.: Examination of effects of character size on accuracy of writer recognition by new local arc method. In: International Conference on Knowledge-Based Intelligent Information and Engineering Systems, pp. 1170–1175 (2006)
12. Schlapbach, A., Bunke, H.: A writer identification and verification system using HMM based recognizers. Pattern Anal. Appl. **10**, 33–43 (2007)
13. Schomaker, L., Franke, K., Bulacu, M.: Using codebooks of fragmented connected- component contours in forensic and historic writer identification. Pattern Recogn. Lett. **28**(6), 719–727 (2007)
14. Shahabinejad, F., Rahmati, M.: A new method for writer identification and verification based on Farsi/Arabic handwritten texts, Ninth International Conference on Document Analysis and Recognition (ICDAR 2007), pp. 829–833 (2007)
15. Siddiqi, I., Cloppet, F., Vincent, N.: Contour based features for the classification of ancient manuscripts. In: Conference of the International Graphonomics Society, pp. 226–229 (2009)
16. Siddiqi, I., Vincent, N.: Text independent writer recognition using redundant writing patterns with contour-based orientation and curvature features. Pattern Recogn. Lett. **43**(11), 3853–3865 (2010)
17. Van, E.M., Vuurpijl, L., Franke, K., Schomaker, L.: The WANDA measurement tool for forensic document examination. J. Forensic Doc. Exam. **16**, 103–118 (2005)
18. Schlapbach, A., Bunke, H.: Off-line writer identification using Gaussian mixture models. In: International Conference on Pattern Recognition, pp. 992–995 (2006)
19. Schlapbach, A., Bunke, H.: Using HMM based recognizers for writer identification and verification. In: 9th International Workshop on Frontiers in Handwriting Recognition, pp. 167–172 (2004)
20. Bulacu, M., Schomaker, L.: Text-independent writer identification and verification using textural and allographic features. IEEE Trans. Pattern Anal. Mach. Intell. **29**(4), 701–717 (2007)
21. Chapra, S.C.: Applied Numerical Methods. McGraw-Hill, Boston (2012)
22. Collins II, G.W.: Fundamental Numerical Methods and Data Analysis. Case Western Reserve University (2003)
23. Ralston, A., Rabinowitz, P.: A First Course in Numerical Analysis, 2nd edn. Dover Publications, New York (2001)
24. Schumaker, L.L.: Spline Functions: Basic Theory. Cambridge Mathematical Library (2007)
25. Zhang, D., Lu, G.: Review of shape representation and description techniques. Pattern Recogn. **1**(37), 1–19 (2004)
26. Jakobczak, D.J.: 2D Curve Modeling via the Method of Probabilistic Nodes Combination - Shape Representation, Object Modeling and Curve Interpolation-Extrapolation with the Applications. LAP Lambert Academic Publishing, Saarbrucken (2014)

Index

S. Bhattacharyya et al. (eds.), *Hybrid Soft Computing for Image Segmentation*, DOI 10.1007/978-3-319-47223-2